UNDERSTANDING PLANT ANATOMY

UNDERSTANDING PLANT ANATOMY

By

Dr. Shubhrata R. Mishra

Department of Botany

Vikram University

Ujjain (M.P.)

(India)

DISCOVERY PUBLISHING HOUSE PVT. LTD.

NEW DELHI-110 002

First Published - 2009

Reprinted - 2018

ISBN: 978-81-8356-457-1

Understanding Plant Anatomy

Published by:

DISCOVERY PUBLISHING HOUSE PVT. LTD.
4383/4B, Ansari Road, Darya Ganj
New Delhi-110 002 (India)
Phone: +91-11-23279245, 43596064-65
Fax: +91-11-23253475
E-mail: discoverypublishinghouse@gmail.com
sales@discoverypublishinggroup.com
web: www.discoverypublishinggroup.com

Printed at:
Infinity Imaging Systems
Delhi

Preface

The present title "Understanding Plant Anatomy" has been written for those students interested in careers in diverse fields of biological sciences. It provides a structured approach to learning by covering all the important topics in a uniform, systematic format. The book has been comprehensively designed incorporating recent advances in this fast moving field. It also provides accessible information on plant anatomy in compact form for undergraduate students in biology and related life sciences. It is intelligible to the educated layman, though it deals with some complex ideas. It is an adequate text for all the requirements of students in this area. In addition, busy lecturers who require a quick reference compendium will find it useful, particularly for tuitional planning. Simple, yet hopefully clear figures and tables are provided throughout the book.

The over-riding goal of this book, and indeed of the whole *Understanding series,* is to present the essential information concerning plant anatomy in a compact, readily accessible form which leads itself to student learning and revision. The convergence of various approaches has generated a rich panorama of detail, the significance of which we are still attempting to unraval. The present text has been written as an introduction to this rapidly growing field.

To make the work more comprehensive and informative, the author has consulted many authoritative books, research journals, abstracts, monographs etc., so there can be no claim to originality except in the manner of treatment.

The author expresses his thanks to his friends and colleagues whose continue inspirations have initiated him to bring out this book.

The author expresses his gratitude to Mr. Wasan and staff of M/s Discovery Publishing House Pvt. Ltd. for their whole hearted cooperation in the publication of this book.

In the mean time, the author will remain sincerely responsible for any shortcomings of the book and be grateful to the readers for their suggestions and constructive criticism for the continuous betterment of the book. He takes this opportunity to appeal to the readers to send their suggestions straightaway to his Publisher.

Author

CONTENTS

1

INTRODUCTION

The study of plant tissues and the cells which compose them is termed *plant histology*. One important point must be made at the outset, namely that the study of plant histology is more concerned with the architecture of the cell in relation to the function which is performs than with its internal composition, and therefore it is the study of cell walls rather than the study of the whole cell. Moreover, as we shall see later, many of the types of cells which form tissues are really dead structures, containing no protoplasm, although they still serve important functions. For this reason the amount of material which is actually living and growing in a higher plant is a varying and uncertain quantity, depending on the proportion of living to non-living cells. This proportion is greatest in early life and it diminishes progressively with age.

In the embryonic stage all the body cells are living and actively growing, but in the course of development the growth activity becomes very soon localized, particularly at the tips of shoots and roots, where the tissues retain the embryonic character throughout the whole life of the individual. Plants thus present an opposite condition to that in animals, in which growth is general in all organs until a specifically limited size is attained, after which it ceases. In plants, on the contrary, growth is localized from an early stage, but is maintained at these points more or less continuously. The size reached by a plant is therefore, in favourable circumstances, simply a function of its age.

Among many lower plants there is little differentiation in kinds of cells; only simple vegetative and reproductive cells are present. Groups or masses of these cells that are alike in origin, structure and function form *tissues*. The plant body consists of "vegetative tissue" and "reproductive tissue." In the higher plants the body is complex in cellular structure, made up of many kinds of cells, of most varied form and function and of different origin. For these plants a definition of "tissue" that is less rigid than one that requires similarity of origin, structure, and function is necessary.

Morphologically, a *tissue* is a continuous, organized mass of cells, alike in origin and in principal function. Within a tissue there may be great diversity of cellular form and function, but the cells that make up a tissue must be contiguous (not dispersed among other cells) and must form a structural part of the plant.

Classification of Tissues

Tissues are classified on several bases—position in the plant body kind of constituent cells function, method or place of origin, stage of development. Each classification is made primarily or wholly on one of these bases. The classification can be divided into those of descriptive anatomy and morphology and those of physiological anatomy. Continuity of cells in the tissue is essential from the morphological viewpoint, whereas it is not important from the physiological view point where function alone binds cells together as a tissue. This major difference in basis can be made clear by examples. Of the morphological tissue wood, or *xylem*, only part—the actual conducting cells–forms the "wood" of physiological anatomy, the supporting cells belong to a different tissue, the storage cells to still another. Morphologically the epidermis forms a single tissue, but in physiological treatments the majority of the cells of the epidermis are grouped with the periderm to form "dermal tissues," and the guard and accessory cells of the stomata form a part of the "aerating tissues." Scattered and isolated cells and groups of cells form many of the tissue of physiological anatomy.

Tissue Types Based on Stage of Development

Meristematic and Permanent tissues. Tissues are distinguished as "meristematic" and "permanent" on the same basis as are cells. Meristematic tissue are immature tissues in which growth is taking place; permanent tissues are those in which growth has ceased, at least comporarily, permanent tissues—as a whole or in part—may again become meristematic.

Tissue Types Based on Kind of Constituent Cells

Simple and Complex tissues. From the standpoint of number of kinds of cells making up a tissue, tissues are *simple or complex*: they are simple when homogenous, consisting of one kind of cell; complex when heterogenous, consisting of more than one kind of cell. This classification aids in detailed descriptions of tissues. There are very few simple tissues in plants. The common ones are *parenchyma*, *collenchyma* and *sclerenchyma*. These terms are applied also to cells; for example, a cell may be called "a parenchyma cell." Such a cell is either a unit of the simple tissue, parenchyma or a cell with the characters of the cells of parenchyma but a member of a complex tissue. Parenchyma and sclerenchyma cells are common constituents of complex tissues; collenchyma cells do not occur with other kinds of cells.

The adjectives "parenchymatous" and "sclerenchymatous" indicate cells which possess come of the characters of the cells of parenchyma and of sclerenchyma, respectively, but which do not belong definitely in those tissues. There are parenchymatous fibres, sclerenchymatous cork cells, etc. The conducting tissues, *xylem* and *phloem*, are the prominent complex tissue types. They need separate discussion here because of their great complexity and variety and because of their structural and functional conveninence. Other complex tissues can be interpreted as combinations in parenchyma and sclerenchyma or modifications of these tissues. Xylem and phloem are considered in this chapter as to general structure and function; they are treated further in other chapters.

Parenchyma

Simple vegetative tissue, that is, tissue which is usually not complex or elaborate in structure or form, like that which takes up the body mass of lower plants and the nonspecialized portions the more complex plants, is called *parenchyma*. Parenchyma is a rather, closely used term in that it is applied to all generally unspecialized and firstly simple tissues which are concerned largely with the ordinary vegetative activities of a plant. Parenchyma is obviously, phylogenetically the primitive tissue since the lower plants have undoubtedly given are to the higher plants through specialization and since the single type in the few types of cells found in the lower plants have become by specialization the many and elaborate types of higher plants.

Further, all *cristematic* tissue is unspecialized. Hence, it is parenchyma-like—in face is often called parenchyma—and it can be said that, ontogenetically also, parenchyma is the primitive tissue. The general characters of parenchyma cells are diameters essentially equal, walls thin, protoplast present, and a capability for cell division even when the cells are permanent cells. Exceptions occur in all these characters. Parenchyma makes up large parts of various organs in many plants. Pith, the mesophyll of leaves, and the pulp of fruits consist chiefly of parenchyma; the cortex and pericycle are often wholly or in large part parenchyma and parenchyma cells occur freely in xylem and phloem.

In early studies of plant structure all tissues were divided on the basis of general form and function into *parenchyma* and *prosenchyma*. The latter was distinguished from parenchyma chiefly by its edongate, pointed, thick-walled cells and its specialized functions of support, protection, and conduction. But entirely different types of tissue are present in "prosenchyma" and the term became obsolete. For conveninence, cells and tissues are still frequently described as "prosenchymatous" in contrast with "parenchymatous."

Collenchyma

Collenchyma is a living tissue and is composed of more or less elongated cells with thick primary *non-lignified walls*. Just as in parenchyma, even in collenchyma the protoplast as is living. The word collenchyma is derived from the Greek word *colla*, meaning glue, which refers to the thick glistening wall of collenchyma.

ORIGIN

According to Ambronn (1881) Haberlandt (1914) and Majumdar (1941), collenchyma originates jointly with the vascular tissues from the procambium. According to Esau (1936) however, collenchyma independently originates from the procambium.

Distribution

Collenchyma characteristically occurs in the peripheral position of stems, leaves, floral parts etc. It may also occur in root contex particularly if it is exposed to light. In the stems and leaves of dicotyledons, collenchyma is hypodermal, situated immediately below the epidermis or separated from it by one or two layers of parenchyma. Collenchyma is absent in the stems and leaves of many monocotyledons.

Cells Shapes

Collenchyma cells are considerably elongated. Sometimes, cells are 2 mm., long resembling fibres. The shorter collenchyma cells are prismatic and appear polygonal in a transection. In a long section the cells appear rectangular.

Cell Structure and Function

The structure of the cell wall is the most characteristic feature of the collenchyma cells. The thickenings are deposited unevenly, mainly deposited at the corners where several cells join together. In long sections collenchyma cells show thin and thick wall portions depending on the direction of the cut. The walls contain mainly of cellulosic and pectic compounds and consist much water. Ultrastructurally the thickenings of collenchyma show an alternation of layers of longitudinally oriented microfibrils and non-cellulosic material.

Collenchyma cells have living protoplasts and may have chloroplasts. Chloraplasts are less in number in long and narrow cells and more in cells which approach parenchyma in form. Tannins may be present in collenchyma cells. Collenchyma is primarily a mechanical tissue adapted for support of growing organs. The thick walls and close packing, make collenchyma a strong tissue. At the same time the peculiarities of growth and the structure of its walls impart a considerable tensile strength with flexibility and plasticity.

In many stems, support in early stages is given in large part by a soft strong tissue known as *collenchyma*. Important characteristics of this tissue are its early development and its adaptability to elongate in the rapidly growing organs, especially these of increase in length. When it becomes functional, no other strongly supporting tissues have appeared. It consists of elongate cells, various in shape, with unevenly thickened walls, rectangular, oblique, or tapering ends, and persistent protoplasts. The cells overlap and interlock in varying degrees, forming strands similar to those of fibres.

The walls consist of cellulose and pectin and have a high water content. (The presence of pectin probably accounts for the high water-absorbing capacity). They are plastic, extensible, and readily adapted to rapid growth. The strands are at first of small diameter but are added to as growth continues, from surrounding meristematic tissue. Cells on the borders of the strands may be transitional in structure, passing into the parenchyma type. The areas of greater wall thickness are in the form of longitudinal strips which occupy the corners of the cells cover the tangential walls or are confined to those parts of the walls that about on intercellular spaces. Where spaces are surrounded by thickened wall strips, hollow rod-like structures are formed, and the collenchyma appears to consist of thick-walled calls among thin-walled cells.

Sclerenchyma

Another type of supporting tissue, which is also in large measure protecting is *Sclerenchyma*. The cells of this tissue, in contrast with those of collenchyma have hard, usually lignified walls with a low percentage of water. At maturity they usually have no protoplasts. The walls are uniformly and strongly thickened. In shape and size, sclerenchyma cells are most diverse, but two general types are recognized—*fibres* and *sclereids*.The distinction of these forms of sclerenchyma is one of convenience in description; it has no morphological significance. Intermediate forms are many and both types may occur in the same tissue, even intermingled, and serve the same general function.

Fibres

Fibres are elongated sclerenchyma cells, usually with pointed ends. Chemically, the walls are usually lignified although there are fibres with walls largely of cellulose and others with gelatinous walls. The pits of fibres are always small, round or slit-like in outline and, in mature cells, unless a protoplast is present, are doubtless functionless. Pits may be numerous but are common rather few in number, and in fibre types with excessively thick walls they may be few or present only as vestigial structures. The lumen of fibres is small; it often is a more channel through the center of the fibre, and this opening may be blocked in spots so that at certain levels in the fibre no lumen exists, only a line or spot representing, in cross section, its former position. In the development of fibers the protoplast often becomes multinucleate. In most kinds of fibres, however, the protoplast disappears as the cell reaches maturity, and the permanent cell is dead and empty. Fibres that retain their protoplasts and other types of fibres are discussed in more detail under the tissues in which they occur.

Classification of fibres

If the usual and loose use of the term "fibre" such as is covered by the above definition—is accepted, fibres may occur in nearly all parts of the plant. They are most generally found and are most abundant in the cortex, pericycle, phloem, and xylem. Morphologically, there are two distinct types. The fibres of the cortex, pericycle, and phloem possess simple pits and are thus different from those of the xylem which have bordered pits (although these pits may be so reduced as to be essentially simple) since xylem fibres are morphologically reduced tracheids.

Fibres are sometimes divided into classes: "bast fibres" and wood fibres. These groups are essentially the same as the two just discussed, but the terms are poor since the word "bast" is involved and has, unfortunately, several uses. In its most common use, the term "bast" is synonymous with phloem or refers to fibres of the secondary phloem.

"Bast fibres" in the above classification include fibres of the cortex and pericycle as well as of the phloem. Fibres can best be designated by means of the tissue or region in which they occur, as cortical fibres, pericyclic fibres, phloem fibres, wood fibres etc. Fibres occur singly or in small groups scattered among other cells. Usually they form strands or sheets of tissue extending longitudinally for considerable distances. Their value as strenghth-ening tissue is largely due to their arrangement in these long masses and to the overlapping and interlocking of the cells. They serve also to give general firmness to tissues.

Fibres develop in two ways. In those only a few millimeters long—for example, those of the Manila hemp, *Agave*, *Sansevieria*—all parts of the cell are always at the same stage of development at the same time. As in most cells, the secondary wall is deposited simultaneously throughout the cell when full size is attained. In much longer fibres, for example, those of flax and hemp, the cell elongates apically, keeping pace with the growth of surrounding cells, and the secondary wall develops in part of the cell while the apex is soil growing. The term "fibre" is popularly applied to various plant structures that are not morphologically fibres: hairs (cotton); strands of cortical or phloem fibres (flax, hemp): foliar vascular bundles with their sclerenchyma sheaths or caps, or the caps alone (Manila hemp, New Zealand flax, sisal); strands of collenchyma (celery); wood cells generally (paper pulp); fragments of leaves and woody tissues (various drugs); bits of "seed" coasts (wheat flour).

Sclerids

These are short isodiametric cells but sometimes may be elongated also. The cell walls are very thick and lignified but the cells do not have tapering ends. The cells are dead and have a very narrow lumen. The secondary walls show prominent pits which are simple. Sometimes there is light overarching of the secondary wall over the pit chamber. The lignin deposition is uniform, but may be uneven as in the seed coats of some *Leguminosae*. Sclereids arise through a belated sclerosis of ordinary parenchyma cells or directly from cells called sclerid primordia.

Stone cells are also called *brachysclereids*. Sclereids more extreme in form have been distinguished as: *macrosclereids* (rod cells) more or less columnar in shape, constituting the "palisade layer" of many seeds and fruits and occurring in some xerophytic leaves and stem cortices; *osteosclereids* (bone cells) of brone-like or barrel shape contituting hypodermal layers in many seeds and fruits and frequent in xerophytic leaves; *astrosclereids* with extreme lobes or arms, in leaves and stems of xerophytic organs; *trichosclereids* ("internal hairs"), branched sclereids with lobes projecting like hairs, into intercellular spaces, in leaves and stems of hydrophytes. The last two types are structurally closely alike. A classification of sclereids on shape and structure is of little value. The character of "branching" or "non-branching pit cavities," which has been used, in part, in distinguishing sclereid types, is largely dependent upon the thickness of the cell wall. Sclereids may occur almost anywhere in the plant body but are most abundant in the cortex, in the phloem, and in fruits and seeds.

Like fibres, they occur singly, in groups of a few cells, or in large masses. The hard parts of seeds, nuts, and hard fruits generally are made up largely of stone cells of various types. Where stone cells are scattered, they merely give firmness, as in leaves and the flesh of fruits. In masses, they give hardness and mechanical protection, as in many kinds of bark and in the shells of nuts. The wall of sclereids is very thick and strongly lignified. Occasionally, it is suberized or cutinized.

The pits are very small with round apertures, and their cavities have the form of more or less branching canels because of the fact that, as the area of the cell wall is reduced on the inside by the increased thickening, the pits are brought together. Two or even several pits thus fuse to form one structure which has only one aperture in each cell but has as many arms as there were original pits. Sclereids are usually dead cells. The shriveled remains of protoplasm and inclusions of the protoplast, such as tannin and mucilage, may be present.

The Important Complex Tissues

The structural and functional prominence of the vascular system renders its tissues of great importance as tissue types. These vascular tissues are, in fact, the only complex tissues which need separate discussion; all others may be interpreted as combinations of parenchyma and sclerenchyma and, of modifications of these tissues. The following treatment of vascular tissues is of general nature only since a description of xylem and phloem as primary and secondary tissues and as specially modified in various organs.

XYLEM

The Tracheid

The fundamental cell type in xylem is the *tracheid*. The tracheid is an elongate cell with tapering ends which when matures is nonliving, that is, without a protoplast. The walls are hard but not thick and are usually lignified. In cross section the tracheid is typically angular, though more or less rounded forms occur. The tracheids of secondary xylem, owing to their method of arrangement, have fewer sides than those of primary xylem and are often more sharply angular. The ends of the tracheids do not taper uniformly to the tip in all planes, but the tapering is confined largely to the radial plane and usually to one side of the cell only. The end of a tracheid of the secondary wood is more or less chisel-like. The tapering is seen in tangential sections of the trachied; radial sections do not show tapering, the end of the cell in such sections being rectangular or somewhat rounded.

The pits are abundant and of the bordered type, though they vary in size, in outline, and in distribution over the walls. The lumen of a tracheid in large and free of contents of any kind. The tracheid is apparently well adapted structurally to its functions which are water conduction, primarily, and support, secondarily. It is a long, empty, firm-walled tube extending parallel with the long axis of the organ. It is in communication with contiguous tracheids–as well as with other types of cells, living and nonliving–by means of numerous, well-developed pits. These thin areas permit ready diffusion into adjacent cells. The arrangement of tracheids is always such that contiguous cell overlap at least over the tapering portions, and over these areas of the wall, the pits are often abundant. Channels for longitudinal conduction are thus provided through a series of *lumina* which form a more or less direct line or an anastomosing system.

The continuity of the walls of the tracheid

The individual lumina are shut off from one another by thick walls and, in the pits, by the closing membranes. The closing membranes, however, except in the central region, the torus, are very delicate and, at least in some of the conifers, are perforated. In *Larix* and *Sequoai*, for example, the perforations, though microscopically minute, are so numerous that the torus is suspended by a meshwork of strands. The opening in perforated pit membranes are not readily seen unless the membrane is well stained, but their presence is demonstrable by the passing of solid particles such as those of carbon in India ink under pressure from one tracheid to another through the pits.

Pits of Tracheid walls

The position of pits in the wall of the tracheid and the size and shape of pits depend upon the position and nature of the contiguous cells. The various larger plant groups have more or less constant types of bordered pits which have characteristic shape, torus, and extent of border. The ferns and the clubmosses have transversely elongated pits with narrow borders and little or no torus. The pits lie close together, covering the wall, giving it a ladder-like appearance and the cells are called *scalariform tracheids*, or better, *scalariformpitted tracheids*. The former term is undesirable in this connection since it is also used with other meanings and where definite pits do not occur (*protoxylem*).

In other gymnosperms and most angiosperms the bordered pits are chiefly rounded with wide borders those of the angiosperms are mostly much the smaller. The best development of the torus is in the gymnosperms where the bordered pit probably reaches its highest development. The closing membrane of these pits is so constructed that its position in the pit cavity can be readily changed from a median position to a lateral one with the torus closely appressed to the aperture of the pit. A valve-like action is thus secured, the pit being freely open when the torus is in the median position, diffusion– or in perforated closing membranes, direct passage–taking place around the torus through the peripheral part of the closing membrane. When one of the pit moths is closed by the placing of the torus against it, direct passage is largely or wholly shut off, and diffusion must take place through the thicker and denser torus. Thus, apparently by changes in the position of the closing membranes of bordered pits, there may be some control over the passage of fluids in xylem. It is significant that the type of pit in which this control of the passage of fluids is present is characteristic of water-conducting cells and does not occur elsewhere; also that in the structurally reduced pits of fibre-tracheids and fibres the pit membranes have lost the capability for movement. In simple pits no structural complexity occurs.

Function of the Tracheid

The tracheid is structurally adapted, both in lumen and in wall, to the function of conduction. The thick and firm walls of tracheids also aid in support, and, where there are no fibres or other supporting cells, the tracheids play a prominent part in the support of an organ. For this the overlapping and interlocking of the cells and their union into strands and cylinders are as important as are thick walls.

Tracheids alone perhaps made up the xylem of very ancient plants, but in living forms wood is a complex tissue, with parenchy-matous wood rays even in the simplest forms of secondary xylem; and primary xylem always contains parenchyma cells. The more complex types of xylem may contain several kinds of cells–tracheids, *fibres* of one or more kinds, vessels of one or two types, parenchyma cells (known as xylem parenchyma or wood parenchyma) and wood-ray parenchyma.

The trăcheid clearly serves both as a conducting and as a supporting cell. But evolutionary development in xylem has resulted in a specialization of this once simple tissue in such a way that the original functions of the tissue have become segregated in distinct cell types—support to fibres of various kinds and conduction (of an apparently more efficient type) to vessels. The new function of food storage has been acquired, and this is carried on by wood parenchyma. The wood-ray parenchyma is also concerned with food storage; the ray is, however, probably primarily concerned with lateral conduction. In complex xylem in which the difinitely dissociated functions of support and conduction are given over to fibres and vessels respectively, tracheids are usually not found. But in the wood of *Quercus* and some other general all three kinds of cells are present.

The Term "Vessel"

Since the perforations of the cell occur usually in the end walls, the development of end walls transverse to the long axis of the cell brings a series of cells into a definite tube-like system, which provides for transmission in a more or less nearly straight line. Such a condition is in contrast with the indirect lines of conduction in a group of tracheids. A tube-like series of cells thus formed has long been known as a *vessel*, or *trachea*.

Both these terms have unfortunately been applied in two ways: to the series or system of cells, and to individual cells with perforate walls that serve in direct water conduction. The former has the support of priority and long usage and should be continued. The unit cells making up such a series are called *vessel elements*, *vessel members*, or *vessel units*. (The term "vessel segment," which has been in frequent use is inappropriate and should be discarded because it implies that the series is a unit which has been divided to form the cells.) Some difficulties arise in the application of these terms under all conditions. In plants in which the water-conducting cells are tracheid-like in shape, union of the cells forms a mesh-system, and vessels of the tube type are not present.

The terms "vessel element" and "vessel member" can be applied here to the individual cells. Vessels are present in the sense of continuous series of fused, perforate cells although individual tubes cannot be distinguished.

Types of vessel perforations

The openings in vessel-element walls, known as *perforations*, are restricted to the end walls except in certain slender, tapering types where, definite end walls cannot be distinguished, and they are said to be present on the side walls. Most vessel elements are perforated in two areas, one at each end, but three and even four such areas occur in some vessel elements. Where there are three or four, connections are made with other vessels or with branches of a meshwork system of vessel elements. The more or less definitely delimited or outlined area of the wall in which perforation occurs is the *perforations plate*. Commonly this is an end wall. The portion of the plate remaining after perforation has occurred is the *perforation rim*; the strips of cell wall between scalariform perforations are the *perforation bars*. Perforation plates are described as having a *simple perforation* if there is but one opening; *multiple perforations* if there are two or more openings.

Multiple perforations are grouped as *scalariform*, where the openings are more or less elongate and parallel; and *reticulate*, where the arrangement of the openings makes a mesh-like structure. The common types are the simple and scalariform. The reticulate type is infrequent and not clearly separable from the scalariform. Small vessel elements occasionally have one type of perforation at one end, another at the other. Simple perforations are usually round but range in form in slender elements to narrowly elliptical. In annular and spiral vessels the perforation is of either type, but most frequently is simple. Commonly, end walls that are transverse have simple perforations, and those that are oblique have scalariform openings, but there are many exceptions. Phylogenetically the scalariform type is primitive and forms transitional to the simple are frequent.

Vessels are characteristic of the angiosperms; in only a few are they lacking—the Winteraceae, Trochodendraceae, and Tetracentraceae and in some xerophytic, parasitic, and aquatic general where they have been lost in reduction. In many monocotyledons they are absent from the stems and leaves; in others from either the stems or the leaves. They are present in some species of *Selaginella*; among the ferns, in two species of *Pteridium*; among the gymnosperms, in the Gnetales. In each of these groups the vessel has clearly arisen independently, in the angiosperms probably more than once.

Vessel length and width

Vessels extend for distances that vary with the kind of plant, the type of xylem, the type of vessel element, the location in the organ, and apparently with the rate of growth of the organ. The limits of an individual vessel are difficult to determine. In climbing plants and in trees with ring-porous wood and simple-perforate elements they may be several, perhaps many meters long. But they are often less than one meter and frequently only a few centimeters long. In a single tree they are apparently progressively longer from near the pith outward. It is highly doubtful that they extend "from root tip to stem tip" in any plant. Where branching and interlocking of elements occur, no question of length arises.

Even the usually direct-series type may branch, and determination of unit length becomes uncertain. The terminal elements have but one perforation and taper at the blind end. Although vessels characteristically are wide, the more slender forms may be even narrower than are typical tracheids. From this diameter they range to a maximum of somewhat over one milimeter. The wider vessels are characteristic of certain herbs such as Zea, many woody vines and lianas, and of some trees, such as, *Castanea*, *Quercus*, and *Fraxinus*.

Ontogeny of the vessel

Vessels are formed from series of xylem mother cells—procambium cells or cambium derivatives—by the fusion of the cells end to end during the last stages of development. This fusion involves the loss of the end walls, or parts of those walls, so that the lumina of the series of cells are freely open into one another, and, with the walls, form a long tube. From the meristematic stage the vessel elements enlarge rapidly, increasing greatly in diameter. Those with scalariform perforations and the more elongate, simply perforate types may increase in length somewhat, the tips forming "talls" which penetrate between surrounding cells. Those that become of great diameter, and especially those developing from stratified cambium initials, do not elongate and may even become somewhat shorter.

During the rapid growth in cell size, the primary cell wall, although greatly and rapidly increased in extent, remains constant in thickness except in the areas which later disintegrate in the formation of the perforations. These areas become thicker and marginally limited. In section they are lens-shaped (in many herbs) or plate-like (in many woody plants) and can often be seen to be three-layered, consisting of the primary walls of the two adjacent cells and the middle lamella. The primary walls are chiefly of cellulose, with doubtless some pectic substances; the middle lamella is largely pectic. Multinucleate stages have been reported in the development of primary vessel elements in the Euphorbiaceae, but the uninucleate condition is apparently characteristic of all types of vessel elements. Throughout the enlargement of the cell the cytoplasm remains abundant and active. As maturity is reached, it begins a slow disintegration.

In some woody plants the nucleus becomes small and greatly flattened and lies in scant cytoplasm against the wall where performation is about to occur in other plants it maintains a more or less nearly central position in the cell. After the primary wall is mature, and in some wood plants the secondary wall is partly, perhaps even fully formed the perforation of the end wall and loss of the protoplast begin. The wall in the perforation area becomes thinner, and, as the cytoplasm gradually goes to pieces, it also disintegrates–in same plants simultaneously throughout, in others beginning in the centre. The disappearance of this piece of wall is in some plants at first by a thinning, suggesting a dissolving, followed by a breaking up into several delicate layers. Statements that these walls are ruptured early in vessel development under tension as the cells rapidly increase in diameter and that the broken walls retract to form the rim are based on inaccurate observation.

Maturation does not proceed simultaneously in the members of a vessel series but progresses from one end to the other. The statement that a vessel matures simultaneously from the base of a tree trunk to its tip is not borne out by careful study. The cambium, which forms the rows of vessel-element initials, itself progresses in activity from one region to another in various directions. The progress of performation of vessel elements seems to be closely similar in the protoxylem and metaxylem of herbs and the secondary xylem of trees. It apparently has not been studied in detail elsewhere.

Sieve Cell and Sieve Tube Element

Sieve cells and sieve tube elements are morphologically equivalent and are alike in fundamental structure and in function. They differ in that the perforations of the walls of the sieve cell and their cytoplasmic strands are all alike whereas those of the sieve-tube element are of two degrees of specialization, and in that sieve cells are not arranged in series as are sieve tube elements. These sieve cells and sieve tube elements are elongate living cells with a thin cellulose wall. The protoplast has a large central vacuole and a thin peripheral layer of cytoplasm. No nucleus is present when the cell is mature. The cytoplasm contains leucoplasts which in some plants accumulate starch or similar substances. In some dicotyle-dons the vacuole contains slimy materials (of proteinaceous nature) which may be distributed throughout the cell sap or accumulate in masses in various places. The slime plugs seen in sections in some plants are doubtless artifacts, aggregations of slime caused by injury to the tissues. The wall apparently is composed of primary layers only. In some genera, for example, *Liridendron*, it may become thick at some stages, perhaps serving then as a place of food storage.

Sieve areas and sieve plates

Sieve cells and sieve-tube elements are unique as living, functioning cells in the absence of a nucleus and the presence in the walls of fine pores through which extend strands of cytoplasm that resemble plasmodesmata but are commonly of much greater diametre and, unlike plasmodesmata, are sheathed with callus. The connecting strands of cytoplasm of sieve elements are clustered in areas of the wall known as *sieve areas*. These areas, which are unspecialized in gymnosperms and pteridophytes, become highly specialized in angiosperms by enlargement of the strands and by sharper limitation. Such elaborated areas occupy more or less definitely restricted parts of the cell wall–usually of the end walls–which are known as *sieve plates*. Two types of sieve plates are distinguished–*simple*, with one sieve area; *compound*, with several sieve areas.

In simple sieve plates the pores and strands tend to be very large, and the plate occupies all or nearly all of the end wall, which is commonly transverse. In compound sieve plates the pores tend to be smaller than those of simple sieve plates, and the plates occupy only part of the usually oblique end wall. The sieve areas are scattered over the side and end walls or are in part or wholly restricted to certain walls, as to the radial and end walls in sieve tubes with long-tapering ends and to the end walls where the cell tapers more abruptly or has a transverse end wall. The number of sieve areas on the side walls varies greatly; there are usually few or none where the end wall is transverse or nearly so where the end wall is long-tapering, the side wall–as well as the end wall from which it is hardly to be distinguished–may be completely covered by set sieve areas. The number and position of sieve areas is to a large extent controlled by the position and arrangement of the surrounding sieve tubes. Sieve tube elements may be somewhat lobed or forked and may even have three definite ends. Such cells stand at points of branching of a sieve tube.

In phloem with tubes of the more primitive type such branching may be frequent, the tubes forming a loose mesh system rather than linear unbranched or rarely branched series, as in the highest type. Where well-marked sieve areas are lacking on the side walls, vestigial sieve plates, known as *lattices*, are often present. These resemble typical sieve areas and plates of various types but are indefinite in outline and often ghost-like; the perforations are exceedingly minute, being usually no larger than normal plasmo-desmata and often apparently lacking. Differences in the size of the pores in sieve plates and lattices of woody plants is shown by the following measurements: in *Juglans nigra* size of pores in the sieve plate is 1.8 to 3.5 micra, that in the lattices, 0.5 to 0.6 micra; in *Populus deltoides*, the sizes are respectively, 3.5 to 5.5 micra and 0.5 to 0.6 micra; in Salix nigra 2.0 to 3.0 micra and 0.4 to 0.5 micra. Stages intermediate between typical sieve areas and lattices may be found. Lattices are common in sieve tubes of intermediate type; they are usually absent in most herbaceous plants. The term "*sieve field*," sometimes applied to lattices, has become confused in use and is probably unnecessary.

Sieve tubes

Union of the sieve-tube elements to form a sieve tube is secured structurally by modification in shape and arrangement so as to form a linear tube and functionally by the greater specialization of the connecting strands in the end walls of the elements. In evolutionary development the sieve tube elements have become increased in diameter and shortened. They have a range in shape similar to that of the tracheid vessel series: in the most primitive type the ends are long-tapering, and an end wall as distinct from a side wall is hardly apparent; commonly the well-defined end wall is oblique or, in the most-specialized forms, transverse. Also in evolutionary development of sieve tubes their sieve area decrease in number. The primitive type of sieve tube has many sieve areas on a plate occupying a long, oblique end wall, and these areas closely resemble the many areas of the side walls; the most advanced type has one area in a plate occupying nearly all of a transverse end wall, and areas on the side walls are scarce or absent. The diameter of the connecting cytoplasmic strands in this highest type is much greater than that of other types. The sieve tube type is not constant in families or sometimes even in genera.

The primitive type is found in some families considered advanced, such as the Caprifoliaceae, and the most advanced type occurs in such primitive families as the Moraceae and Ulmaceae. Evolutionary advance in sieve-tube type has apparently taken place within families (Fagaceae, Rosaceae, Leguminosae) and within genera (*Fraxinus Prunus*). Although the sieve tubes of vines and herbs are usually of the highest, type all types occur in woody and in herbaceous forms.

Ontogeny of the sieve elements

The mother cells of sieve elements vary in shape from short cylindric to elongate, slender, and tapering. Primary pit-fields are numerous on the walls, especially in secondary phloem, as on young tracheid walls. As these cells differentiate, they elongate; the cytoplasm becomes highly vacuolate and streams actively; the wall thickens; sieve areas develop from the pit-fields, and the cytoplasmic strands become prominent and increase in size; callus develops around the connecting strands. The relation of pit-fields to sieve areas in ontogeny is not altogether clear. Apparently on pit-field may from one or more sieve areas, or several pit-fields may form one area where the sieve plate is simple. In plates of this simple type when the connecting strands are very large, one or more pit-fields perhaps take part in the formation of the pore, each strand representing one greatly enlarged plasmadesmata thread or the fused threads of one pit-field. In the development of callus, rings are first formed about the strands at the apertures of the pores; increasing deposition forms cylinders that sheath the strands. As the sieve tube elements reaches maturity in size, the wall becomes thinner, the nucleus disintegrates the connecting strands continue to increase in diameter, the streaming of the cytoplasm ceases, the peripheral layer of cytoplasm becomes extremely thin, the boundary between the cytoplasm and the vacuole becomes indistinct, and the semipermeable properties of the cell appear to be lost.

At this stage the functioning period of the cell as a conducting structure apparently begins. In all plants conduction by sieve tubes probably lasts only a short time—from a few days in the early-formed primary phloem to a year or perhaps more in the secondary phloem of woody plants. During the functioning life of the sieve tube, the callus increases in amount, lengthening the cylinders about the strands. Callus is deposited also on the wall around and between the strands forming with the cylinder a cushion-like mass over the sieve area. In the late stages of thickening in this cushion, the strands become attenuate, and some or all of them may be broken off. This condition seems to accompany the death of the protoplast. The callus cushion is for this reason known as the *definitive callus*. In some woody plants, where sieve tubes seem to function during a second growing season—perhaps even longer—the protoplast does not die, and the definitive callus is dissolved as activity in conduction is renewed.

Under these conditions the sieve strands must remain unbroken. Little is known in detail concerning the structure of sieve tubes that appear to function through more than one growing season. Callus cushions forms only weakly or not all over the sieve areas of lattices. This with the presence of few weak connecting strands, or none, is evidence of the vestigial nature of these structures. Lattices form a transition stage in lateral wall structure between the numerous sieve areas of the primitive sieve tube element and the absence or great reduction in number of the highest type.

Obliteration of sieve tubes

After the loss of the protoplast of the sieve tube element the empty cell wall, at this stage as thin as in early stages of development, is crushed or collapses under the pressure and tension of the surrounding tissues which are brought about by increase in diameter of length of the organ of which they are a part. In many herbaceous plants the callus cushion and cylinders are still present when crushing takes place; in secondary phloem of woody plants they have disappeared. The crushing of the deal sieve tubes and their companion cells commonly becomes so complete that these cells are soon represented only by strands or sheets of structureless material, and this material may be soon absorbed. In most plants, monocotyledons probably excepted, the surrounding living cells crowd into the space formerly occupied by the crushed cells, and evidence of the earlier structure of this tissue is difficult to find. The crushing and absorption of sieve and companion cell is known as *obliteration*.

Sieve cells of gymnosperms

Sieve cells have not been studied so intensively as sieve tube elements, and the structure of their sieve areas is more difficult to ascertain. The connecting strands are said to be arranged in small groups, and the callus cylinders appear to surround these groups rather than individual strands as in angiosperms. The history of development, functioning period, and obliteration is much the same as that of sieve tube elements. It is possible that the sieve cells of vascular cryptogams have a much longer functioning period —even several years—than those of seed plants.

Companion cells

The *companion cell* is a specialized type of parenchyma cell which is closely associated in origin, position, and function with sieve-tube elements. These cells occur only in the angiosperms but in these plants accompany most sieve-tube elements. Perhaps protophloem sometimes lacks them, and they appear to be scarce in the primary and early secondary phloem of some woody plants. In highly specialized phloem, such as that of many monocotyledons, they are abundant, together with sieve tubes making up the entire tissue. Companion cells are formed by longitudinal, or oblique-longitudinal, division of the mother cell of the sieve tube element before specialization of this cell begins. One daughter cell may become a companion cell and the other a sieve-tube element; or the latter may divide further, forming more companion cells. Transverse divisions in the companion-cell initial may form a row of companion cells so that one to several may accompany each element.

A companion cell or a row of a few such cell formed by transverse division of a single companion cell initial may extend the full length of the sieve tube element. In a species the number of companion cells accompanying a sieve-tube element is fairly constant. Solitary, long companion cells are common is primary phloem and herbaceous plants; short and numerous companion cells appear to be characteristic of the secondary phloem of woody plants. Companion cells have abundant grandular cytoplasm and a prominent nucleus which is retained through the life of the cell. Thery do not contain starch at any time. They live only so long as the sieve tube element with which they are associated and they are crushed with those cells. When seen in transverse section the companion cell is usually a small, triangular, rounded, or rectangular cell beside a sieve-tube element. Often it seems to lie as if within the limits of the sieve tube elements. If it does not, it may be impossible to determine with which particular sieve tube it is associated for it may be in contact with more than one. In secondary phloem of many gymnosperms the marginal cells of the rays, called *aluminous cells*, differ markedly from the other ray cells.

Phloem Parenchyma

Phloem typically contains parenchyma cells of other types than companion cells. Various terms have been given to these on a basis of shape and probable function, but these terms are confused in usuage and the differences hardly merit distinction. In shape these cells range from elongate and tapering to broadly cylindrical, sub-spherical, and polyhedral. The elongate cell, while young, may divide forming a row of cells that retain in their mature form and position evidence of this original. In content also these cells show great variety: crystals, tanniferous substances, mucilage, latex, etc. Most of the parenchyma cells are filled with starch or oil during dormant periods. They remain alive until cut off from the inner living tissues by periderm formation. Those of secondary phloem in age may be transformed into sclereids, as in Quercus. Parenchyma may be lacking in Phloem—the tissue consisting of sieve tubes and companion cells only, as in vascular bundles of many monocotyledons.

Phloem Fibres and Sclereids

Sclerenchymatous cells are rare or absent in phloem of living pteridophytes, and they are absent from this tissue in some gymnosperms and angiosperms. But in many seed plants, fibres form a prominent part of both the primary and the secondary phloem. The fibres of the primary phloem–or part of these cells–have in some plants been mistaken for pericyclic fibres. Phloem fibres differ from xylem fibres in that the pits, which have smaller, linear, or rounded apertures, are always simple. The walls are lignified. In development the long-tapering ends become interlocked and strong strands are formed. The fibres often form tangential sheets or cylinders enclosing the inner tissues. These are obviously of structural importance as layers protective to the soft cambium region within, and also to some extend as longitudinal strengthening tissues. In some herbs and occasional woody plants, such as *Dirca palustris*, they apparently are of more importance in supporting the stem than is the xylem cylinder. Fibres of the protophloem, forming the outermost cells of the primary phloem, are prominent features of many stems, both woody and herbaceous.

In the early stages of stem development these fibres may contribute largely to the support of the stem. They are arranged in various ways: as continuous, uniform, or irregular bands; as scattered, isolated strands; as clusters "capping" the primary phloem strands. Such fibres are usually lignified, as are those of *Cannabis* (hemp), but may be of cellulose, as in *Linum* (flax). These primary phloem fibres are often similar to fibres of the cortex and to those of the secondary phloem. All or any of them with other fibres and sometimes vascular bundles constitute commercial "bast."

The term "bast". Because of the strength of strands of phloem fibres, these have long been used in the manufacture of cords and ropes and in the weaving of matting and cloth. Fibrous tissue used in this way has been known since early days as *bast* or *bass*. The term was originally applied to any fibres obtained from the outer part of a plant, though a large part of such material came from the secondary phloem, as in the bassword, *Tilia*. When the secondary phloem was recognized as distinct from the cortex, the term "bast" was applied to this phloem, which was the common source of fibres. And in this sense, that is, as synonymous with phloem—as wood is synonymous with xylem— "bast" is still in frequent use. With the term used in this morphological senset the fibres of the phloem become bast fibres. But the term "*bast fibres*," or simply "bast," is also applied to any fibres from the outer parts of a plant. This is topographical not histological or morphological use; such fibres may be a part of the cortex or pericycle. Secondary phloem is also frequently "divided into hard phloem, or bast, and soft phloem"; and "bark" is divided into "bast and living phloem."

The term "bast" is used to such an extent without accurate botanical meaning that it should be discontinued as a technical term. Furthermore, it is superfluous, since the term "phloem," "phloem fibres," "pericycle fibres," and "cortical fibres" cover accurately all its uses. Primary phloem occasionally contain sclereids, and the older secondary phloem of many trees contains abundant cells of this type. These cells develop from parenchyma cells as the tissue ages and the sieve tubes cease to function.

Pitting in Cells of the Phloem

In phloem the complex cellular relationship, the delicate walls of many of its cells, and the similarly of pits with plasmodesmata to sieve areas with cytoplasmic strands make the determination of pitting difficult. Statements concerning wall structure in some positions vary greatly, and much additional information is needed. The pitting of the thicker walled cells is clear: the pit-pairs are the simple pit-pairs of parenchyma and sclerenchyma, except that in some types of parenchyma the pits resemble sieve areas. Between sieve tube and parenchyma cells there is a sieve area on the side of the former and a pit on the side of the latter. Between the sieve tube and companion cell the wall is usually very thin and specialized areas are not seen; rarely, this wall is thick, and "pits" have been reported. It has been generally believed that companion cells are not pitted with parenchyma, but pit-pairs have been described in some plants.

Function of Phloem

The chief function of phloem is the conduction of elaborated foodstuffs, both proteins and carbohydrates. The sieve elements are believed to be the cells concerned in this conduction with the companion cells, or the albuminous cells, related in some way to their activities. The sclerenchymatous tissues—fibres and sclereids—serve or some measure in support or organs and protection of soft, underlying tissues. Many parenchyma cells are starch storage cells at certain periods; other perhaps serve in conduction of some substanes; some are crystal storage regions.

The term "phloem" is sometimes used–as in the term "xylem"—to indicate only the conducting cells of a complex tissue whose chief function is conduction. In this usuage only the sieve elements of the tissue usually called phloem are indicated as "phloem". Occasionally all the "soft cells" are labelled "phloem", the sclerenchyma being excluded as "bast". The restriction of "phloem" to sieve tubes and companion cells is especially applied to protophloem. But histologically and morphologically the entire, continuous tissue is the phloem, and the term is so used in this book. For example: protophloem does not consist only of the sieve-tube elements that constitute its first maturing cells but includes the much later maturing fibres that surround them.

Transfusion tissue

A peculiar type of conducting tissue, which consists largely of short tracheids with thin, cellulose walls and boradered pits or reticulate or scalariform thickenings, often accompanies typical vascular tissue in the leaves of the gymnosperms. These cells are trached-like in the character of their pitting and in the absence of a protoplast but otherwise suggest elongate parenchyma cells. They lie adjacent to typical xylem at the sides of the bundle and may partly or even completely surround it. Since these cells seem to serve as conducting tissue, connecting the veins and the mesophyll of the leaves, taking the place of the usual very small branches of the veins, they form together what is called *transfusion tissue*. Though the function of this tissue is still uncertain, it undoubtedly represents modified vascular tissue.

Arrangements of Tissue

All the tissues of a plant which perform the same general function, regardless of position or continuity in the body, may be considered to form, together, a *tissue system*. In this sense the term is wholly a physiological one. There are found in physiologically treatments of anatomy such tissue systems as the "mechanical system", the "absorbing system," and the "storage system." The various parts of most of such systems, however, are bound together only by function; they have little or no structural or morphological unity. Continuity and similarity of nature or of origin may be lacking. From a morphological viewpoint the grouping of tissues into tissue systems is sometimes convenient.

In the morphological sense, a tissue system must be a complex of cells extending continuously throughout the plant body or over a considerable portion thereof. It may be so simple as to consist of only one type of cell or one type of tissue or may consist of two or more types of tissues. Very few systems in any was structurally distinct can be distinguished. The older students of anatomy distinguished an *epidermal*, or *tegumentary system*; a *fundamental*, or *ground system*; and a vascular system. The distinction of these systems is not often made in present-day anatomy, doubtless because the systems have, in use, been morphologically too inclusive or indefinite or have, been non morphological. The epidermal system has sometimes included a hypodermis and other cortical tissues—even periderm; the fundamental system, the cortex, pericycle, and pith. Recognition of these systems has value in a strictly topographical sense and is especially useful in the interpretation of meristematic tissues.

In young organs, the epidermal system consists of the dermatogen (or protoderm) alone; the vascular system consists of the procambium and the earliest formed xylem and phloem elements; the fundamental system consists of ground meristem—all the remaining tissues, which at this stage show little or no differentiation. In the mature primary body these form, respectively, the epidermis, the vascular tissues, and the cortex, pericycle, pith, and mesophyll. The epidermis and vascular system are tissue system of such uniformity and continuity of structure and of such constancy of function that they constitute important gross structural features of the plant body. The terms "epidermal system"—if used to cover epidermis only—and "vascular system" are convenient and valuable. "Ground tissue system" covers the heterogeneous remaining parts.

Secretory Tissue

All cells directly concerned with the secretion of gums, resins, essential oils, nectar, and similar substances are together frequently referred to as "secretory tissue." Such a classification is purely a physiological one, as secretory cells and tissues often do not have common origin or morphological continuity. Secretory cells frequently are isolated from other similar cells, embedded in pith, xylem, phloem, cortex, or any region. On the other hand, secretory cells may be aggregated forming a tissue in the strict morphological sense. Not infrequently such cells constitute a definite, organized secretory structure or gland.

Secretory Cells

Secretory cells are of two general types: those in which the secretion formed is exuded from the secreting cell, as in glandular hairs and secretory surfaces, such as nectaries and the epithelium of resin and oil canals; and those in which the secretion formed is stored within the secretory cell. The term *excretory cells* is often applied to the first type. This type is cell is usually characterized by a protoplast with richly granular cytoplasm and a conspicuous nucleus; the secretory cell is usually large with inconspicuous cytoplasm and large lumen filled with the secretion. This type of cell contains in various plants many different substances, such as essential oils in *Liriodendron*, *Sassafras*, and *Zingiber*, and mucilage in many ferns. Glandular hairs often show elaborate specialization related to various often show elaborate specialization related to various functions, as, for example, the stinging hairs of the nettles, *Urtica*. Such specialized hairs are often multicellular.

Glands

Glands are classified physiologically as either secondary as excretory, but it is difficult to draw a clear line between the two functions. Broadly speaking a secretion is directly valuable to the plant and an execution is a superfluous a deterians substance. Plants have no regular method of evacyating their excretion as animals have. They dispose of them introductly, building thems into their structure usually in superannuated tissues. Thus they may sometimes become indirectly useful to the plant. For example, mineral materials may serve to strength tissues..

Again, the aromatic terpenes, to which plant odours are due, are superfluous substances from the point of view of metabolism, but they serve an essential biological purpose in attracting insects to flowers, whereby pollination is effected. From the anatomical standpoint it is simpler to classify glands as either superficial or internal. The superficial glands are usually modified hairs in which one or more cells contain the secreted material. A gland of this type is the stinging hair of the Nettle (Urtica dioica).

The secretory cells is flask-shaped the bottom of which is seated on a pedestal of small cells. The neck is drawn out into a long, tapering cone, the wall of which is impregnated with Calcium carbonate, and the tip is bent sideways and forms a small silicified bulb. When touched this bulb breaks off, leaving a sharp point, with a lateral opening like a hypodermal needle, which pierces the skin. Pressure forces the poison upwards from the flask-like base of the cell and into the subcutaneous tissues. The gland contains a mixture of histamine and acetyl choline. The former causes itching and the combination produces the prolonged burning sensation. Acetyl choline alone is without effect. *Urtica stimulans*, the Java Nettle, causes serious illness, comparable in severity to a snake bite.

The glands produced on the leaves of the Insectivorous Plants are generally glandular hairs and they function both for the capture and the digestion of the insect prey. Superficial glands, if not formed from hairs, may be simply modified patches of epidermis, as, for instance, in the *nectaries*, found in most flowers and sometimes also on leaves or stems, where they are called *extra-floral nectaries*. The epidermis of these gland areas has no cuticle and consists of a layer of narrow, highly protoplasmic cells called *epithem*.

Digestive glands

In the majority of plants, enzyme secretion is not confined to specialized cells or tissues but is a characteristic of most living cells. In certain so-called "insectivorous" and "carnivorous" plants there are special glands which secrete protein-digesting enzymes; these enzymes act upon insects or other organisms so that the products of digestion can be absorbed by the plant. In *Droscra*, the secretory tissue is at the tips of the leaf "hairs" or "tentacles", structures which also serve to imprison the insects. Here, in addition to the digestive enzymes, there are secreted *viscid* substances which hold the insects. In such plants as *Nepenthes* and *Sarracenia*, which normally have pitcher-like *transpartly* filled with liquid, the glands are sessile and secrete the enzymes into the liquid from which the products of digestion are absorbed. In some other general, for example, Dionaea and Pinguicula, the glands are inactive except when stimulated by contact with animal matter. Less specialized glandular tissue of this type is found in the embryos of certain seeds, but such tissue is usually not clearly differentiated.

Nectaries

Many entomophilous plants produce nectar which attracts insects. This substance is secreted by specialized cells either on the floral parts themselves or, more rarely, on bracts or other structures outside the flower. Usually, the secretion of nectar is from specialized epidermal cells which cover certain regions of the flower, rather him from number the organs adapted to secretion alone. Definite and elavorate structures do, however, occur in certain families, for example, the Euphorbiaceae. In the less specialized nectaries the secreting cells are superficial upon the floral parts and in most plants closely resemble the other epidermal cells of the region but lack a cuticle. Sometimes they are set off from the surrounding epidermal cells by a somewhat more columnar or papillose shape and by denser cytoplasm. The nectar is exuded through the wall and exposed upon the outer or nectariferous surface. The secreting cells of stigmatic surfaces are of the same nature as those of nectaries but are often not clearly set off from normal epidermal cells.

The *septal nectaries* or *glands* of many monocotyledonous flowers are pockets in the septal walls of syncarpous ovaries where the fusion of the carpel walls is incomplete and the epidermal cells are glandular. These glands may be simple, slit-like cavities or deep pockets with canal-like passage ways to the surface of the ovary.

Hydathodes

Many plants possess structurally modified regions where water is exuded under conditions of low transpiration and abundant soil moisture. These are known as *hydathodes* or sometimes as water pores or water *stomata*. Morphologically, they are considered to be enlarged stomata which serve for water secretion. Structurally, they may resemble stomata closely but often they show elaborate structural specialization. Such structures do not "secrete" the fluid but merely provide and control the openings through which it escapes Hydathodes occur commonly at the tips of leaves as in grasses, at the apices of serrations on the margins of leaves, and in other positions. They are found mostly on plants of humid climates.

Resin, Oil and Gum Ducts

In the gymnosperms generally and in many angiosperm families, resins, oils, gums, and other substances are secreted and conducted in ducts. In some plants, as in *Pinus*, these ducts or canals may form extensive systems extending both vertically and horizontally. In other plants the ducts may be local in occurrence and limited in extent, as in the fruits of the Umbelliferae. In *Pinus* and closely related genera, the resin ducts are schizogenous in nature, and when mature, have the structure of a tube with an epithelial lining. The oil ducts of the Umbelliferae are of the same general nature. The secretory cells lining these cavities are thin-walled parenchyma with dense protoplasm.

In general, these cells are elongate with the long dimension extending parallel with the long dimension of the duct. The substances secreted are various in nature and in some plants—for example the resin of *Pinus* and of *Agathis* (Kaurigum), and some essential oils–are of much economic importance. Another type of gland is that found in the aim of citrus fruits. Here there is a lysigenous cavity filled with essential oil and other substances that have been formed by the disintegration of the cells and as definite sometimes before the breaking down of the tissues. The origin of this secretion is not well understood. These glands are the source of the essential oils of lemon and orange.

Laticiferous Ducts

Latex is found in a considerable number of angiosperm families. This substance appears as a white, yellow or reddish, sometimes slightly viscous fluid which has been shown to be an emulsion of proteins, sugars, gums, alkaloids, enzymes, rubber and other substances suspended in a matrix of watery fluid. Starch grains may be abundantly present. The function of latex is not well understood. It is apparently secreted by the cells in which it contained, and is conducted by them throughout the plant body. The latex of some plants is of great importance, especially as a source of rubber (Heeca, Ficus, etc.) chicle (Achras), and papain (Carica), as well as for other substances. Laticiferous ducts are of two types–one known as nonarticulate latex ducts or latex cells and the other as articulate latex ducts or *latex vessels*. The function and the contents of the two kinds of ducts are essentially the same, but the morphological nature and development are different.

Nonarticulate latex ducts

These ducts are individual cells that extend as ramifying structures for long distances through the plant body. The walls are soft and often thick, and the cytoplasm contains great numbers of nuclei. These ducts arise as typical small meristematic cells among other promeristem cells. They quickly become elongate, and their tips keep pace in growth with the surrounding meristem, penetrating among the new cells. They branch and extend through all the tissues of the plant, even in some general through secondary xylem.

Although the branches of a cell come in contact with those of other cells, no anastomoses occur. Two kinds of these nonarticulate ducts are described but the differences lie in the number per plant, the position of origan and extend in the plant body rather than in structure or function. In one kind the cells arise only once, in small number, in the embryo. These few cells, extending with the meristems, ramify through the entire plant. The ducts of the *Asclepiadeceae*, most of the Apocynaceae and Euphorbiaceae, and those of some other families are of this kind. Ducts of the other kind arise repeatedly in the meristems and ramify into adjacent tissue but are restricted in their growth to one internode and attached leaf and branch. The ducts of the *Urticaceae* and those of some of the Apocynaceae (vinca) are of this type.

Articulate latex ducts

These ducts originate in the meristems from rows of cells by absorption, complete or partial of the separating walls early in the ontogeny of the cells. By branching and frequent anastomoses, a highly complex system is built up. A duct of this type resembles a xylem vessel in that it is made up of a series of cells united to form a tube by the dissolution of walls, but the latex tube is living and coenocytic. The Papaveraceae (poppy), Caricaceae (papaya), Compositate (dandilion), Musaceae (banana) and the genus Hevea (Braziliar rubber tree) have this type of laticiferous ducts.

Meristem

The embryonic tissue which forms the growing regions of the shoots and roots is called meristem (mristos =divisible), because these are the sites of rapid cell-multiplication. A certain proportion of the cells produced, subsequent enlarge, cease to divide, and acquire thickened walls or undergo other changes and are then added to the *permanent tissues* of the plant. This change from meristematic to permanent form is called *cell differentiation*. A certain number of residual cells, however, always retain the meristematic character, so that meristem, in addition to forming permanent tissues, is also self-perpetuating.

In most Pteridophyta the function of producing new cells is localized in a single *apical cell* instead of being distributed through a meristematic tissue. In this resemble certain Thallophyta. The tip of a stem is usually occupied by a bud, which is composed of immature leaves produced by the growing point itself. The latter lies in the centre of the bud, and a median longitudinal section will show that it is in fact the actual apex of the stem, which tapers rapidly within the bud, to end at this point. The meristem of the growing point is extremely soft, delicate and juicy, as seen with the naked eye. A similar tissue occupies the tip of the root, but no bud surrounds it in this case, because roots have on leaves.

Microscopically the cells of the growing point are small, fairly regular in size and shape and have thin, delicate cell walls formed of a mixture of cellulose and protein, with an intercellular layer of mucilaginous pectin. They have prominent spherical nucleus, which are very large relative to the size of the cells, and the cytoplasm is dense and completely fills the cells. When a meristem cell divides, each half quickly grows to the size of the original cell, thus doubling the amount of protoplasm. It follows that it is here that the chief synthesis of protoplasm goes on. The cells fit closely together, without intercellular spaces and their shapes are those of plastic spheres in natural compression, due to their active growth. They are either dodecahedra or tetridecahedra. The youngest cells have no vacuoles and they never possess chloroplasts, though transitory starch grains may sometimes appear. Opinion varies as to the exact limits of the term meristem. Some confine it entirely to the non-vacuolated cells, but there is a considerable interval between the first vacuolization of the cell and its completed differentiation, and many prefer to extend the term generally to all those cells which continue to divide, even when somewhat altered from their youngest form. The meristem of the growing point is called the apical meristem, and the term primary meristem is applied both to it and to all meristem tissues derived directly from it. Thus the cambium in Dicotyledons and the meristems of lateral shoots are also actually primary meristems.

In many Monocotyledons there are intercalary meristems at the bases of levels, and similar embryonic regions occur in older parts of Dicotyledons. All these are primary meristems. True secondary meristems, i.e. those formed from tissues previously differentiated, are rare, the chief example being the cork cambium. When we consider that the development of the meristem cells is responsible for determining the specific form of the plant and the structural anatomy of the older portions, it is obvious that it is the centre of many activities of great interest, about which we know very little. The evidence seems to show, however, that almost without exceptions the divisions of meristem cells are equational, and therefore that the potentialities of the cells produced are equal. There is little sign in plants of the primary segregation of germ cells and soma cells which Weismann suggested in animals.

Theoretically every cell of the plant is totipotential, that is, it retains it itself the possibility of regenerating the whole plant, is suitable circumstances. Any actual limitations of this power are imposed by special conditions. The differentiation of cells is therefore explicable, not by inherent differences between them but by the influence of their environment, that is to say by differences of water and food supply, oxidation etc., due to the position of the cell in the tissue. Even within the meristem some differentiation is apparent. In 1868 Hanstein named three zones, which he considered to be typical and which he called *histogens* or "tissue producers". The outermost was a single layer of cells named the dermatogen, which gave rise to the epidermis. Beneath this came an intermediate zone, the periblem, giving rise to the cortex and the inner tissues of the leaf. Centrally was the core or plerome, producing pith and vascular tissues. In cases where such layers can be seen camės still have a descriptive value and the frequently used, but the developmental significance assigned to them by their author is not consistent with fact.

Frequently, indeed, the three zones cannot be distinguished, especially at the apex of the stem. Modern studies have interpreted the meristem of stems as composed of two zones only. These are *tunica*, which consists of one or several peripheral layers of cells, enclosing the *corpus* or central core of tissue. The distinction between them reflects two different modes of growth. In the tunica, surface growth predominates and cell divisions are uniformly anticlinal, that is, perpendicular to the surface, with the result that the cells are regularly oblong in longitudinal section and each tunica layer remains distinct. In the corpus, by contrast, volume growth is predominant and the planes of cell division and cell arrangement tend to the highly irregular.

The maintenance of balance between surface and volume growth leads to constant adjustments which find their expression in the rhythmical alternation of minimal and maximal areas of growing point, the latter being associated with the formation of external folds or leaf primordia. These zones are not "histogens" in Hanstein's sense but represent the distribution of growth patterns, which may vary considerably in different species. Immediately below the corpus may often be seen in tissue which is called the *rib meristem*, distinguished by short, vertical rows of cells in process of vacuolization. These are formed from cells of the corpus which have repeatedly divided horizontally. As each row is produced from one original cell it has a common wall surrounding it. This tissue continues downwards into the pith

Classification of Meristems

Meristems are classified on several bases: stage of development, structure, position in plant, origin, function, topography. The bases of three classifications are not mutually exclusive, and the definitions cannot be rigid. A discussion of the more important types of meristems follows.

Meristems Based on Stage or Method of Development

Promeristem. The region of new growth in a plant body where the *foundation* of new organs or parts of organs is initiated constitutes promeristem. (The terms "primordial meristem" "Urmeristem," and "embryonic meristem" have also been applied to this initiating region.) Structurally this region consists of the initials and their immediate derivatives, all young cells with diameters much alike, walls thin, with early stages of pits; cytoplasm active, vacuolate or nonvaculoate; nuclei large; and intercellular spaces absent or minute. The promeristem of a given organ or region is of rather limited extent, varying in amount in different plants, in different organs and under differing growth conditions. It is of course not definitely set off from the somewhat older meristematic tissue into which it merges. As soon as cells of the promeristem begin to change in size, shape, and character of wall and cytoplasm, setting off the beginnings of tissue differentiation, they are no longer a part of typical promeristem; they have passed beyond that earliest stage. For example, there is at the tip of an organ a meristem of some length, only the youngest part of this, a small apical portion, is promeristem. The remainder of the meristem represents the early stages of the tissues formed by the promeristem. No term exists for this partly developed region in which segregation of tissues is beginning, but cell division continues freely.

Mass, Plate, and Rib Meristems

On the basis of plane of division, *mass*, *plate*, and *rib meristems* have been distinguished as "growth forms" of meristem. In mass meristem, growth is by three-plane or all-plane division and produces increase in mass; plate meristem is by divisions–chiefly in two planes–so that there is plate-like increase in area; rib meristem, by continuing divisions (anticunal) in only one plane, produces rows or columns of cells, functioning chiefly in the increase of organs in length. Example of mass meristem are early stages of many embryos; developing sporangia; the endosperm of many plants; young pith and cortex of some plants. Plate meristem forms epidermis, and is prominent is leaf development where in early stages divisions in the plane of the leaf blade and at right angles to that plane build up great increase in blade area with little increase in thickness. Rib meristem plays a prominent part in the development of young roots and of the pith and cortex or young stems.

The term "rib" meristem is in frequent use at present but is unfortunately often applied not to regions of cell initiation but to those of cell differentiation where the cells stand in long rows. In general, these "meristem" are merely parts of meristematic tissue-masses distinguished by planes of cell division. Recently the term file meristem has to some extent replaced rib meristem; this is a more accurately descriptive term. It is obvious that these types can be distinguished only when the cells are chiefly of typical parenchyma-cell form; meristems will cell of prosenchymatous form cannot be thus classified. Cells that are beginning to elongate constitute the first stages of procambium, of collenchyma, or of strands of fibres.

MERISTEMS BASED ON HISTORY OF INITIATING CELLS

Primary and Secondary Meristems

On the basis of type of tissue in which origin occurs, meristems are classified as *primary* and *secondary*. Primary meristems are those that build up the fundamental part of the plant and consist in part of promeristem. (In some uses, promeristems are considered distinct from primary meristems). In primary meristems, promeristem is always the earliest stage, and transition stages to mature tissue constitute the remainder of the meristems.

The possession of promeristem continuously from an early embryonic origin is characteristic of primary meristems; no stage wherein all or some of the cells have become permanent, or modified from the meristematic, has entered their history. The chief primary meristems are the apices of stems and roots and the primordia of leaves and similar appendages. Exceptions to the promeristem continuity of primary meristems from early embryonic tissue are found in adventitious bud and roots and in some types of wound tissue. (Many so-called "adventious buds" arise from buried dormant bud initials, not *de novo* in permanent tissue.) The originating meristems of true adventitious organs arise secondarily in more or less nearly permanent tissues, but they are, because of their structure and behaviour, primary meristems. Once established they may persist indefinitely.

Secondary meristems are set apart from primary meristems in that they always arise in permanent tissues; in their history there is interposed a stage of permanent tissue or at least of partial development toward permanency. They have no typical promeristem, though their initiating layers may to some extent resemble this tissue. *Secondary meristems* are so called because they arise as new meristem in tissue which is not meristematic. The cork cambium is an example of secondary meristem; it is formed from mature cells–cortical, epidermal, or phloem cells or from any mature cells that are not too highly-specialized.

The primary meristems build up the early and, for a time, structurally and functionally completed plant body. The secondary meristems later add to that body, forming supplementary tissues that functionally replace or reinforce the early formed tissues or serve in protection and repair of wounded regions. The cambium, one of the most important meristems, does not fall definitely in either group. It arise from apical meristem of which it is a late and specialized stage. The tissues formed by the cambium are secondary, as are all those formed by secondary meristems, whereas other primary meristems form only primary tissues.

The classification of meristems as primary and secondary has sometimes been considered of little value but it is especially helpful in an understanding of the manner in which the complex mature plant body is attained and modified with continuing growth.

Meristems Based on Position in Plant Body

On the basis of position in the plant body meristems are usually classified as *apical*, "*intercalary*," and *lateral*. Apical meristems are those which lie at the apices of the axis, and of the appendage and are commonly called growing points. "Intercalary meristems" are those which lie between regions of permanent tissue, as for example, at the base of the leaves of many monocotyledons. Lateral meristems, as the name implies, are situated laterally in an organ. The cambium and the cork cambium are lateral meristems.

Apical Meristems

Apical meristems, or growing points, occur universally at the tip of the roots and stems, and often of the leaves, of vascular plants. The activity of these meristems brings about increases in the length of these organs, laying down the primary body of the plant. Initiation of growth is by one or more cells situated at the tip of the organ which maintain their individuality and position and are known as apical initials or apical cells. These cells may be strictly terminal or terminal and subterminal.

Apical cells

Among vascular plants, solitary apical cells (except those of leaves) occur in the horsetails, most of the ferns, and a few other pteridophytes. In other vascular plants a group or groups of apical, or apical and subapical cells constitute the initiating body. The behaviour of these cells in tissue formation has not been extensively and critically studied until recent years and is even yet only partly understood. Since the initials apparently differ little, if at all, from their recently formed daughter cells and since all of these cells may continue to divide freely it is most different to determine the number and limits of the initials. The number of initials is probably fairly constant for a given organ and a given species, but it is probable that at least in some plants, the number of initials and, to some extent, the form of the group vary from time to time even in the same organ.

Meristems are highly plastic. Variations are related to vigour of growth, to seasonal conditions, and to morphology of the organ developing. It is now known that in seed plants there are many types of apical-meristem differentiation. For each major plant group ther eis apparently a fairly characteristic plan of development, but this plan may not be consistently followed in detail even within an individual plant. Initials, where solitary, appear to persist indefinitely; where there are several or many, they may function for a time and then be replaced by new cells. During periods of rapid growth, additional cells may serve more or less temporarily as initials.

Types of apical cells

Solitary apical cells of several types are distinguished by shape and by the number of sides from which new cells are cut off. The two most common are the lenticular, or two-sided cell (not found in vascular plants), and the pyramidal, or three-sided cell. The former is strictly a three sided cell in shape and the latter a four-sided cell, but new cells are formed only from two and three sides respectively, hence terms have been given which apply to the activity of the cell rather than to its shape. The side from which no new cells are formed lies in the direction of growth. In roots, cells are cut off on this face also. The size of the apical cell is reduced only temporarity by the formation of daughter cells, and its position is maintained indefinitely.

The behaviour of solitary apical cells in segmentation, and of the cells which are cut off by the apical cells, is of importance in the study of the morphology, both vegetative and reproductive,of the bryophytes. In these plants, groups of cells in certain positions and of certain origin always form the same tissues or organs. That this must also be true in general of vascular plants was formerly believed, but such is clearly not the case. The behaviour of a given segment of an apical meristem may have little morphological significance in the groups of plants above the bryophytes.

Intercalary Meristems

So-called "intercalary meristems" are merely portions of apical meristems that have become separated from the apex during development by layers of more mature or permanent tissues and left behind as the apical meristems moves on in growth. The more mature layers are the nodal regions and the "intercalary meristem" are therefore internodal. At early stages the internode is uniformly meristematic; later some, part of it matures more rapidly than the rest, and a definite sequence in development within the internode is set up and maintained. The region of youngest cells is in most plants at the base of the internode, but it maybe in the middle or at the top. Where the youngest region is basal, the order of maturing in the tissues of the internodes is of course acropetal, where it is at the top, the order is basipetal; and where it is median, the sequence is both upward and downward. It has been generally believed that "intercalary meristems" are separated from the mother meristem at so early a stage that they retain a layer of the promeristem of the apical region and that this promeristem continues, after its segregation, to function as promeristem, initiating complete new segments of the organ. But it has been shown that this is not true in some grasses, and it is perhaps not so in any plants. Although these regions consist partly of promeristem cells, vascular strands, with protoxylem and protophloem cells in various stages of maturation run through them. With such structure they are not typical meristems but meristematic regions and should be so called. The best known intercalary meristematic regions are those of the stems of grasses and other monocots and of horsetails, where they are basal, and those of some mints, where they lie just below the node. In some peduncles, such as those of *Taraxacum* and *Plantago*, they are said to occur at the top. Leaves of many monocotyledons (grasses, Iris) and some other plants, such as *Pinus*, have basal meristematic regions. Intercalary meristematic regions ultimately disappear; they become wholly transformed into permanent tissues.

Lateral Meristems

Lateral meristems are composed of initials that divide chiefly in one plane (periclinally) and increase the diameter of an organ. They add to existing tissues or build new tissues. The cambium and the cork cambium are meristems of this type. The former has been called "the lateral meristem", but phellogen layers also belong here. The briefly functioning "marginal meristems" of some leaves have been called lateral but fall in none of the above groups.

Meristems Based on Function: Theories of Structural Development and Differentiation

The progress of differentiation in meristems brings about in cellular organization a structural segregation, or zonation, with regions distinguishable by:

(a) the number and position of the initiating cells,

(b) the planes of division and consequent arrangemen of cells,

(c) the size, shape and content of the cells, and

(d) the ratio of maturation of the cells.

Several theories dealing with the methods of origin of the patterns formed by this zonation and their histological and morphological significance have been proposed.

The Apical Cell Theory

Solitary apical cells occurs in many of the algae, in the bryophytes, and, among vascular plants, in the Psilotaceae, the horsetails, most of the ferns, and some species of *Selaginella*. Early studies of the apical meristems of bryophytes and of the pteridophytes with a single apical cell demonstrated morphological significance in the segmentation of the apex. It was believed that the same conditions, although difficult to demonstrate in seed plants, held for all higher plants, and the *apical cell theory* was proposed as the basis for an understanding of the method of growth and morphology in all groups. But controversy arose over the interpretation of the complex apices of by gymnosperms, and it became evident that the theory was not applicable to seed plants.

The Histogen Theory

As a basis for the interpretation of the growing points of seed plants, the *histogen theory* replaced the older theory. Under this theory the more or less distinct major regions of the stem and root apex were called *histogens*, that is, tissue builders, in the sense of builders of specific parts of an organ. The histogens were the *plerome*, a central core; the *dermatogen*, a uniseriate, external layer; the periblem, the region between the plerome and the dermatogen. The histogens, or their individual layers, were believed developed by separate initials. This theory, in contrast to the apical cell theory, placed the origin of axis apices in a group of initials. It differed from the earlier theory also in morphological interpretations or implications—the plerome formed the pith and primary vascular tissues; the dermatogen, the epidermis; the periblem, the cotex.

The viewpoint and the terminology of the histogen theory have long dominated anatomical interpretation of tissues, regions, and even organs in vascular plants. But the distinction of these histogens in an apex cannot be made in some plants, and in others the regions have no morphological significance. The plerome may form only the pith or the entire central cylinder and part of the cortex, the periblem may form the cortex and the outer stelar tissues or part of the cortex. Even in homologous axes on the same plant the histogens may form different parts. The terms clearly have no morphological value, and, because, their prominence has rested largely on such value, they are becoming less used. They are found occasionally in histological descriptions of stems, commonly in these of roots, but serve chiefly to indicate regions in a rather loose or topographical sense.

The Tunica-Corpus Theory

Interest in the development of seed-plant axes reawakened about twenty years ago with the formulation of the *tunica-carpus theory* and increasing attention has been given to the ontogeny of the stem tips of vascular plants. According to the tunica-corpus thoery (which has been applied only to the leafy stem, or shoot) different rates and method of growth in the apex set apart two regions of unlike structure and appearance a central core, the corpus, and an outer enveloping layer, around and above the corpus, the tunica. In the corpus the cells are large, with arrangement and planes of cell division irregular, and increase in volume; in the tunica the cells are usually smaller than those of the corpus and lie in layers or sheets, with divisions strictly, or chiefly, anticlinal, and growth is primarily in area. These regions vary greatly in the distinctness of their limits and in their form and relative size; their boundaries are often not clear-cut; the corpus may be massive or slender; the tunica, many-, few-, or one-layered.

The imoportant elements in such an elaboraton of regional development are the position and number of initials and the planes of division in these cells and their derivatives. The number of initials is few to several or many. Rarely, in small, very slender apices, such as those of grass seedlings, there may be only one or two in tunica and about two in the corpus. The determination of initials as self-maintaining and persistent cells is often impossible because they differ little or not at all from their recently formed daughter cells. Certainly it is unlikely that the initials remain constant in number, form, and sequence of divisions. Where all the initials are on the surface, their divisions are both anticlinal and periclinal, and no clearly defined tunica and corpus are formed. Where the initials are in a cluster or mound, the outer ones, if their divisions are strictly anticlinal, form a definite tunica, and the inner ones, with divisions in more than one plane, form a corpus.

Under these conditions tunica and corpus have completely independent origins and are fairly distinct. Between these extremes lie, many intermediate conditions in which distinction of tunica and corpus is weal or doubtful. Below the initials or below the direct derivatives of these cells (the two groups constituting promeristem), elaboration of cells in size, form, and arrangement initiates tissue specialization and the building of the framework of the primary body. The various regions, or stages, merge on their margins and overlap longitudinally. Major interruptions in the progressively more advanced stages back from the apex are brought in by the appearance of nodal areas and lateral structures.

Types of Stem Apices

In vascular plants, zonation in stem-apex differentiation follows more or less definite patterns that appear to be characteristic of the major groups. These patterns show increasing complexity from the lower to the higher groups and seem to represent a series in specialization from simplicity to complexity. In lower groups the initials form a uniseriate surface layer, and there is no evidence of differentiation into tunica and corpus; in the highest groups the initials form a two-storied cluster of which the superficial members build a tunica–which, in its greatest specialization, is probably uniseriate–and the inner members from their corpus. The majority of seed plants are intermediate in apex structure between these types.

The primitive type of stem apex

Among the pteridophytes, Lycopodium, Isoetes, and some species of *Selaginella*, and among primitive gymnosperms, the cycads, have simple apices with surface initials and no distinction of tunica and corpus.

Lycopodium

A serve as an example of this primitive type. The initiating layer in a weakly defined, uniseriate surface area which divides freely both anticlinally and periclinally. No definable initials can be distinguished; all the cells of the layer are morphologically alike. The anticlinal divisions increase the area of the surface layer; the periclinal division forms on inner core.

The stem apex with weak tunica-corpus segregation

Beginnings in distinction of tunica and corpus are perhaps present in some of the lower conifers. In the *Pinaceae* (Abies, Pinus), the initials form a terminal uniseriate group. From these, by both periclinal and anticlinal divisions, a central core and an enveloping uniseriate layer are formed. The latter suggests a tunica in appearance; but it has frequent periclinal divisions even on the flanks of the apex, and there is no clear line of separation between the tissues formed by the two regions.

Some higher conifers have a somewhat more specialized apex. As in the Pinaceae, the initials are a small group of surface cells in one tier, with divisions both anticlinal and periclinal. These divisions form, respectively, a dermatogen-like layer and a central mass, but in the outer layer the periclinal divisions are rare or absent, except close to the apex. The part played by the outer layer in building the body of the stem is restricted to the cells at the very apex. The outer layer early become tunica-like in its restriction of planes of division.

Of the conifer genera thus far critically studied, only *Crytomeria* and *Taxodium* have a "dermatogen" in which there are no, or alomost no, periclinal divisions. The apices of conifers have the structural appearance of tunical-corpus segregation but in those so far studied there is only one tier of initials and therefore no independent meristematic regions. The conifers show much variety in apex structure. Even within a species (*Taxodium distichum*, *Sequria sempervirens*) there are marked differences, correlated in part with morphology of the stem concerned, vigour of growth, and other conditions.

The stem apex with definite tunica and corpus

In the angiosperms the segregation of apical-meristem zones is usually more definite than in lower groups; there are two sets of initials, one above the other, which give rise to tunica and corpus that are in large measure more wholly independent. The tunica has no only rare periclinal divisions and ranges in thickness from several layers to one, with two or three layers probably most frequent. The large numbers of tunica layers occur more frequently in the dicotyledons. A single-layered tunica such as that of grasses (Avena, Trificum) probably represents the most specialized condition, but even in this, occasional periclinal divisions may occurs, as in Zea. The corpus varies from a large complex type to a slender, simple type. The number of layer in the tunica may vary even in an individual plant. The limits of the two groups of initials and, of the tunica and corpus is often definite but may be difficult or impossible to determine and in some genera the apex shows no distinction of tunica and corpus, resembling the apex of the primitive gymnosperms. *Sinocalamus Beecheyana* and *Vinca rosea* serve as examples. In the former of tunica is two-layered in the latter it is four-layered.

Discussion, Tunica-corpus Theory

The tunica-corpus theory has served well in the establishment of an understanding of the complex, diverse, any varying meristematic patterns of the stem tips of seeds plants. Providing a basis and a method, it has stimulated intensive study, and detailed information has been obtained for a considerable number of plants. The position, number, and behaviour of the initiating cells in seed-plants stems is now in some measure known, and early stages in the development of the primary body of the shoot are much better understood. Although, as with the histogens of the histogen theory, a distinction of tunica and corpus has littler or no morphological significance, it is of topographical value in studies of detailed development. Caution is necessary last morphological interpretations accompany use of the theory: the boundaries of tunica and corpus are often obscure; where the limits are fairly distinct, the two regions may be inconstant in structure and function, varying with seasonal conditions, vigour of growth, age of plant, position on plant, and morphology of the stem tip concerned. For example, in some plants the innermost tunica layer of a primary branch becomes an outermost part of the corpus on secondary branches. The number of tunica layers varies even in apices of the same type at the same time; it is commonly less on the less vigourous and the lateral branches than on strong, leading stems.

The differences found, even on the same plant, in the thickness and distinctness of the tunica at different times and under different morphological conditions–as in main and lateral branches, deciduous and persistent twings, intermediate vegetative tips and the determinate apices of flowers and thorns are related directly to growth status and to the morphological nature of the developing organ. A flexible basis is necessary for all descriptions of meristems. Tunica and corpus should be recognized as dynamic and fluid, not functionally or morphologically distinct or constant regions. The lateral organs of the stem–leaves, branches and floral organs–arise near the apex and studies of tunica and corpus have added greatly to a knowledge of the origin and early development of these organs.

The Floral Apex

The structure of the floral apex differs in no fundamental way from that of vegetative stems. Such differences as can be distinguished are those inherent in the determinate nature of the axis, the telescoping of internodes, and the crowding of appendages.

It has been claimed that the floral apex differs markedly from the vegatative apex; that the apical initials build up the "central core" in the vegetative axis, but that both apical and flanking tissues build up the floral apex; that the tunica of the floral apex has no definable initiating zone and that periclinal divisions occur at any depth within it. Such differences may occur, varying greatly in degree with the form and structure of flower concerned, but they are not morphological differences. They are associated directly with the function of an apex that is developing vegetative growth lose their identity, and growth activity is restricted to the peripheral region where many appendages are arising at compacted nodal levels; the numerous periclinal divisions at the depths are associated with the broadening of the receptacle and with the origin of the floral appendages which cover it. Distinction of outer and inner zones varies in the floral apex as it does in the vegetative apex. The number of layers in the tunica of the floral apex is more in some plants, less in other, then in the vegetative apex. In ontogency a floral apex develops by the gradual or abrupt transformation of a vegetative apex.

The Root Apex

The apical meristematic region of the root, lacking developing appendages and segregation into nodal and internodal regions, is simpler in gross structure than that of the stem but is complicated at the tip by the *root cap* and different methods of formation of this structure. The root cap is a terminal portion of the root tip, more or less definitely set off from the tissues beneath, covering and protecting the initiating apex of the root proper. It is formed by the same initials as those that build the root proper in all types of root tip excep those of the monocotyledons where it has independent origin. Its formation is necessarily discussed with that of the root apex.

The apical meristem is short compared with that of the stem; specialization into zones is concentrated, with the early stages lying within the terminal millimeter beneath the cap. Growth proceeds in cap and root proper in opposite directions, with sequence in development towards the tip in the cap and away from the tip in the root itself. In number, the initials range from one to many. Where the initials are more than one, they are arranged in one to four fairly distinct, uniseriate groups. In each group there are one to several initials. Since the initials are the central members of uniseriate layers and can be distinguished from their recent derivatives only by position and restriction in division to anticlinal planes, the number can often be determined only approximately.

Furthermore, the number of initials in a group seems to vary with diameter and rapidity of growth of the root. In slender roots it may be reduced to one–as in some grasses–but the zonation remains clear. Lines of zonal segregation are most obsecure in root tips of large diameter. Where there is more than one group, the groups lie adjacent to one another on the longitudinal axis of the root. Each of these groups quickly develops one or more growth zones which are usually more clearly marked than are similar zones in the stem apex. In many plants these zones appear to represent "the histogens", and interpretations of the root apex have long continued under the histogen theory. Although the terms dermatogen, periblem, and plerome are no longer in general use in descriptions of stem ontogeny, they have been continued for convenience to indicate general zones in studies of root development. A fourth histogen, the *calyptrogen*, is added where the cap has an independent origin.

In published descriptions there is much confusion as to method of origin of the zones and no arrangement as to the number of types of development. The distinctness of the zones varies, and forms intermediate between the types have been described. There seem, however, to be basic patterns for the major plant groups. This pattern is determined by the number of initials, the number of groups of these initials, the zones formed by each group, the morphological nature of the cap and the degree of independence of the cap. The vascular cryptogams that have a soliditary apical cell in the stem–the horsetails, most of the ferns, some species of *selaginella*–have a similar solitary apical cell in the root. This one cell forms the entire root and the cap which is usually sharply distinct structurally. The origin of the cap from the large apical cell is clear.

In many gymnosperms there are two groups of initials. The inner forms the plerome; the outer forms the periblem and the cap. No line can be drawn between these two regions; the cap appears as a distal poliferation of the periblem.

2

MERISTEMATIC TISSUES

The profuse and inconsistent terminology in the voluminous literature on apical meristems reflects the complexity of the subject matter. Most commonly the term *apical meristem* is used in a wider sense than merely with reference to the initials and their immediate derivatives; the term also includes variable lengths of shoot or root proximal to the apex. Yet, when determinations of the dimensions of the apices of shoots are made, only the part above the youngest leaf primordium, or youngest node, is measured. Usually *shoot apex* and *root apex* are employed as synonyms of apical meristem.

DELIMITATION

The wide meaning of apical meristem is adopted in the discussions to follow, but when it is important to differentiate the most distal part of this meristem the term *protomeristem* is used; it refers to the least determined part of the meristem and includes the initials and their most recent derivatives. The delimitation of this protomeristem is arbitrary, but the term is useful for referring to the distal part of the apical meristem, which is given much attention in the literature. The promeristem of Clowes (1961a) includes only the initials and thus does not coincide with the protomeristem. Johnson and Tolbert's (1960) metrameristem, on the other hand, refers to the same group of cells as the protomeristem.

Apical meristem and its synonyms are appropriate substitutions for the somewhat inaccurate term growing point. Growth in the sense of cell division, which is so characteristic of the meristematic state, is not restricted to the so-called growing point but occurs abundantly – and is even more intense – at some distance from the apical meristem.

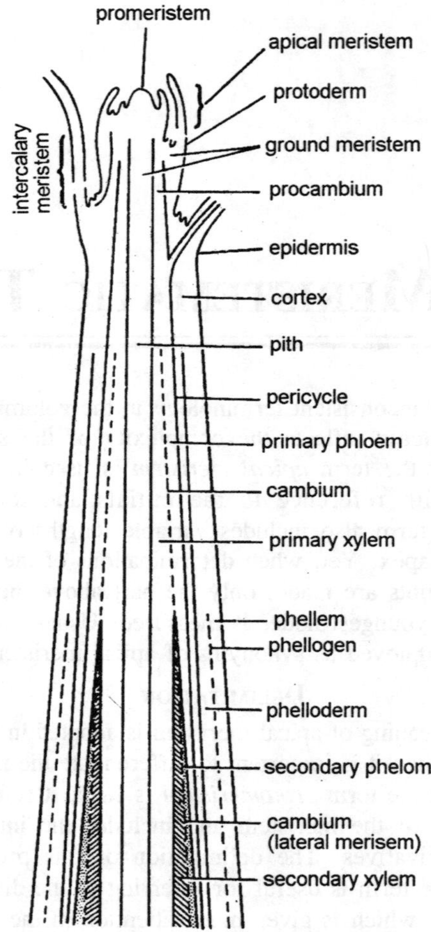

Fig. 2.1. Diagrammatic representation of the location and differentiation of various meristems in plant body.

Similarly, growth in the sense of increase in size of cells, tissues, and organs is most pronounced not in the apical meristem but in its derivatives.

Initials and Derivatives

An initial, or initiating cell, is a cell that divides into two sister cells one of which remains in the meristem and the other is added to the meristematic tissues that eventually differentiate into the various tissues characteristic of the plant. The cell remaining in the apical meristem functions as an initial like its precursor. Investigators visualize the involvement of polarity, and a consequent cytologic differentiation, in the division into an initial and derivative; at the same time they agree that the status of a cell as an initial depends on its position in the protomeristem, and that the initial may be displaced by another cell and then differentiate into a body cell.

The inference about the existence of apical initials is generally based on microscopic views and on theoretical considerations, but there is also experimental evidence regarding this matter. By treatments with colchicine it is possible to change the number of chromosomes in individual cells. When cells occupying the position of initials in the shoot apex are thus affected, the change becomes detectable and is perpetuated developmentally in more or less extended parts of the plant body that develop after the treatment, and the alterations may be traced directly to the cells in the apical meristem. These cells thus fit the definition of initials. Changes in growth may cause a shift in the relative position of the modified cells in the apical meristem so that an initial ceases to act as such. This observation supports the concept that a cell is an initial, not because of its inherent characteristics but only because of its particular position in the meristem.

The number of initials in root and shoot apices is variable. In many vascular cryptogams a single initial cell occurs at the apex; in other lower vascular plants, as well as in the higher, several initials are present. The single initial is morphologically rather distinct from its derivatives and is customarily spoken of as the *apical cell*. If the initials are more or less numerous, they are called *apical initials*, although considered semantically it would be appropriate to call them apical cells also. The recognition of apical initials under the microscope, in contrast to that of the single apical cells, is uncertain.

The apical initials may occur in one or more tiers. If there is only one tier, all cells of a plant body are ultimately derived from it. In the alternative situation, different parts of a plant body are derived from different groups of initials. The existence of more than one independent layer of initials in certain plants has been clearly demonstrated in the previously mentioned experiments with colchicines. The treatment may induce polyploidy in one or more superficial layers of the apical meristem and thus convert the plant into a cytochimera. Induced and spontaneous chimeras showed that polyploidy could be perpetuated ontogenetically if any one of the three superficial layers in the apical meristem were polyploid, and that these three layers behaved independently in the transmission of their characteristic chromosome numbers. These plants obviously had three tiers of initials, that is, three self-propagating layers.

Induced polyploidy has also served to demonstrate the presence of more than one initial cell in each tier. In addition to periclinal chimeras, sectorial polyploidy was observed in *Vaccinium*. The restriction of polyploidy to individual sectors of the stem is possible only if the initials occur in groups, with each component cell capable of becoming polyploid independently of the others.

Evolution of the Concept of Apical Organization

As has been discussed by several writers, the view concerning the number, the arrangement, and the activity of the initial cells and their recent derivatives in the apical meristems has undergone profound changes since the shoot apex was first recognized by Wolff (1759) as an undeveloped region from which growth of the plant proceeded.

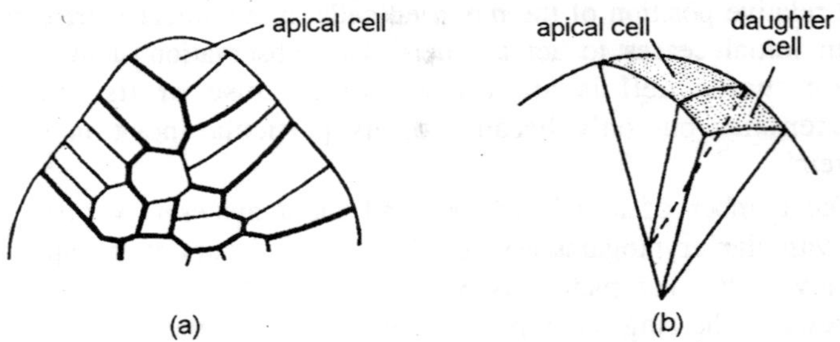

Fig. 2.2. Schematic representation of growth of the shoot apex interpreted in terms of apical cell concept. (a) Longitudinal section of shoot apex of Equisetum showing a single apical cell; (b) A pyramidal apical cell with its recent derivative.

The discovery of the apical cell in cryptogams led to the concept that such cells exist in phanerogams as well. The apical cell was interpreted as a constant structural and functional unit of apical meristems governing the whole process of growth. Subsequent researches refuted the assumption of a universal occurrence of single apical cells and replaced it by a concept of independent origin of different parts of the plant body. The *apical-cell theory* was superseded by the *histogen theory*.

The histogen theory was developed by Hanstein (1868, 1870) on the basis of extensive studies of angiosperm shoot apices and embryos. Its basic theses are, first, that the main body of the plant arises, not from superficial cells but from a mass of meristem of considerable

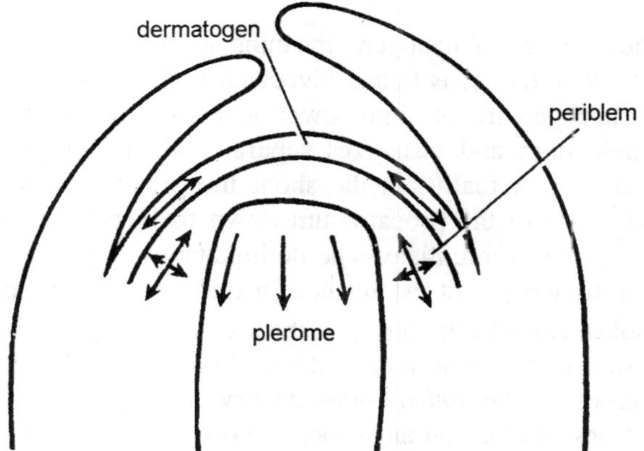

Fig. 2.3. Schematic representation of growth of the shoot apex of a dicotyledon interpreted in terms of Hanstein's histogen concept. Arrows indicate direction of growth.

depth, and, second, that this mass consists of three parts, the ***histogens***, which may be differentiated by their origin and course of development. The outermost, the *dermatogen* (from the Greek words meaning skin and to bring forth), is the primordial epidermis; the second, the *periblem* (from the Greek, clothing), gives rise to the cortex; the third, the *plerome* (from the Greek, that which fills), constitutes the entire inner mass of the axis. The dermatogen and the periblem form mantle-like layers covering the plerome. The dermatogen, each layer of the periblem, and the plerome begin with one or several initials distributed in superposed tiers in the most distal part of the apical meristem.

Hanstein's "dermatogen" is not equivalent to Haberlandt's (1914) "protoderm." The protoderm refers to the outermost layer of the apical meristem, regardless of whether this layer arises from independent initials or not, and regardless of whether it gives rise to the epidermis only or to some subepidermal tissue also. In many apices the epidermis does originate from an independent layer in the apical meristem; in such apices the protoderm and dermatogen may coincide. The plerome and periblem in the sense of Hanstein are discernible in many roots but are seldom delimited in shoots. Thus the subdivision into dermatogen, plerome, and periblem has no universal application. But Hanstein's histogen theory is criticized chiefly because it contains an assumption that the destinies of the different regions of the plant body are determined by the discrete origin of these regions in the apical meristem. The prevalent view is that histogenesis and organogenesis have no obligate relationship to the segmentation and layering of cells in the apical meristems.

A modified use of histogen, meaning an already determined but still meristematic tissue, is being advocated by Guttenberg (1960). He places the histogen initials into lower levels of the apical meristem than Hanstein does and visualizes separate initials for procambium, pith, and cortex. Actually, in the shoot the ground meristem of the cortex adds cells to the procambium down to levels where vascular elements begin to differentiate. The delimitation between vascular and nonvascular tissues is not established in the apical meristem.

The apical-cell and the histogen theories have been developed with reference to both the root apex and the shoot apex. The third theory of apical structure, the *tunica-corpus theory* of Schmidt (1924), was an outcome of observations on angiosperm shoot apices. According to this theory, two tissue zones occur in the apical meristem; the *tunica*, consisting of one or more peripheral layers of cells, and the *corpus*, a mass of cells overarched by the tunica. The demarcation between these two zones results from the contrasting modes of cell division in the tunica and the corpus. The layers of the tunica show anticlinal divisions; that is, they are undergoing surface growth. The corpus cells divide in various planes, and the whole mass grows in volume. Each layer of the tunica arises from a small group of separate initials, and the corpus has its own initials located beneath those of the tunica. In other words, the number of tiers of initials is equal to the number of tunica layers plus one, the tier of corpus initials. In contrast to the histogen theory, the tunica-corpus theory does not imply any relation between the configuration of the cells at the apex and histogenesis below the apex. Although the epidermis usually arises from the outermost tunica layer, which thus coincides with Hanstein's dermatogen, the

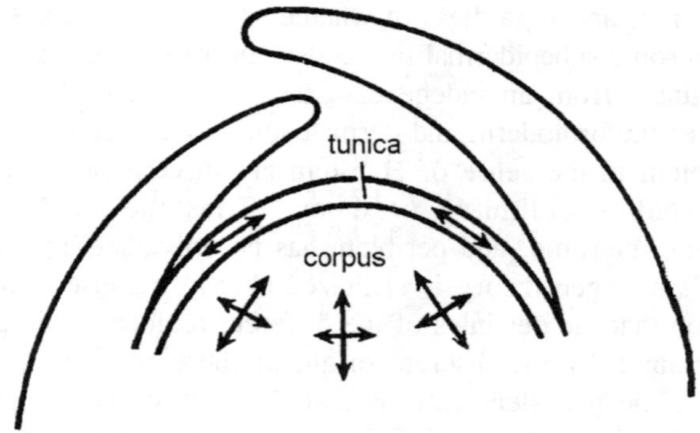

Fig. 2.4. Schematic representation of growth of the shoot apex of a dicotyledon interpreted in terms of tunica-corpa concept. Arrows indicate direction of growth.

underlying tissues may have their origin in the tunica or the corpus or both, depending on plant species and the number of tunica layers.

Interest in the tunica-corpus theory was strongly stimulated by the work of Foster and his students and has dominated studies on shoot meristems for two decades. As more plants came to be examined, the concept underwent some modifications, especially with regard to the strictness of definition of the tunica. According to one view, tunica should include only those layers that never show any periclinal divisions in the median position, that is, above the level of origin of leaf primordial. If the apex contains additional parallel layers that periodically divide periclinally, these layers are assigned to the corpus and the latter is characterized as being stratified. Other workers treat the tunica more loosely and describe it as fluctuating in number of layers; one or more of the inner layers of tunica may divide periclinally and thus become part of the corpus. The term *mantle* has been proposed for tunica in the loose sense; it overarches a body of cells called the *core*. Still others reject the tunica-corpus concept entirely because it does not relate the apical activity to the origin of tissues. Nevertheless, the tunica-corpus theory remains useful for characterizing growth in the shoot apex of angiosperms. It is used in this book with the assumption that during the vegetative growth the tunica has a characteristic number of layers, which may be attained in steps during the development of the plant and may change during the transition to the reproductive stage; and that the corpus may vary between stratified and nonstratified configurations.

As was mentioned before, the tunica-corpus concept was developed with reference to the angiosperms; it proved to be largely unsuitable for the characterization of the apical meristem of gymnosperms. Shoot apices of only a few gymnosperms have an independently propagating layer that could be interpreted as tunica; in others, the outermost layer divides periclinally and thus is ontogenetically related to the subjacent tissue. Studies of gymnosperm apices, stimulated by Foster (1941), led to the recognition of a zonation based not only on planes of division but also on cytologic and histologic differentiation and degree of meristematic activity of component cell complexes. Similar cyto-histologic zonation has since been observed in many angiosperms. The concept of zonation in Foster's sense has considerably advanced the understanding of growth in shoot apices. It has also related the apical organization to that of the underlying derivative shoot parts without reintroducing a formalized concept of histogen initials. But efforts to bring about such reintroduction are not lacking.

The cytologic zones that may be recognized in apical meristems vary in degree of differentiation and in details of grouping of cells. As a result, the pertinent terminology is constantly growing and changing. Succinctly, the zonation may be characterized by dividing the apical meristem into a *distal axial zone* terminating the axis and two zones derived from it. One of these, the *proximal axial zone*, or the *inner zone*, appears directly below the distal zone, is centrally located in the apex, and usually becomes the pith after additional meristematic activity has occurred. The other, the *peripheral zone* or *outer zone*, encircles the other zones. The peripheral zone is also called flank meristem I the literature because of a common tendency to describe structures as seen in sections in two dimensions.

The peripheral zone typically is the most meristematic of all three zones, has the densest protoplasts and smallest cell dimensions. It may be described as a eumeristem. Leaf primordia and the procambium arise here, as well as the cortical ground tissue. The inner zone early shows its destiny differentiation into the vacuolated pith by being cytologically less dense than the outer zone. Depending on the manner of growth of the shoot, especially the degree of elongation of future internodes, the inner zone assumes more or less definitely the characteristics of a rib meristem. The distal zone is somewhat variable characteristics of a rib meristem. The distal zone is somewhat variable in appearance. All of it, or only its proximal part, may be considerably vacuolated. The term protomeristem is applicable to the distal zone in the sense that it contains the initials and their most recent derivatives.

The derivatives of the distal zone, the outer zone and the inner zone, may merge imperceptibly with the distal zone or they may be delimited from the distal zone by an additional, *transitional zone*, often compared to the cambium because of orderly seriation of cells resulting from divisions periclinal with reference to the distal zone. The transitional zone is composed of particularly actively dividing derivatives of the distal zone. The presence of the transition zone apparently depends on the rate of growth in the shoot apex, and the zone shows fluctuations in its distinctness in the same kind of apex.

The next development in the interpretation of apical meristem resulted from the efforts of Buvat and his students to obtain a unified concept of growth of this meristem. Meristematic activity drew the chief attention in this work. Counts of mitoses, and cytological, histochemical, and ultrastructural studies served to formulate the theory that the distal zone of the apical meristem is relatively inert during vegetative growth and that the real initial zone is the peripheral one, where leaf primordial arise. The distal zone received the appellation of waiting meristem (*meristeme d'attente*), because it was said to be waiting for the change from vegetative to reproductive stage before taking up meristematic activity. The peripheral zone became the initiating ring (*anneau initial*) and the inner zone the medullary (pith) meristem (*meristeme medullaire*). The concept of the inactive distal zone in the apical meristem was extended from the shoots of the angiosperms to those of the gymnosperms (called the distal zone *zone apicale*) and the lower vascular plants, and to the roots. The concept was later somewhat modified in that variations in degree of inactivity of the distal zone in relation to the size of the apex and its stage of development came to be recognized.

The revision of the concept of apical initials by the French workers stimulated a considerable amount of research in other countries and led to refinement of techniques for determining the degree of meristematic activity in the apical meristem. Extensive counts of mitotic figures; studies of cell patterns in fixed and living apices; histochemical studies; use of labeled compounds to determine the location of synthesis of DNA, RNA, and protein; experimental injury of the meristem; and theoretical discussions served to evaluate the concept of the inactive distal zone in the apical meristem. Most of the investigators outside France consider that the apparent scarcity of divisions in the distal cells of the shoot does not justify regarding these cells as being inconsequential in the construction of the shoot; these cells are the ultimate source of all other cells of the shoot and hence a re the initials. This interpretation is used in the description of the shoot apices in the subsequent sections of the present chapter.

With regard to the root apices, the occurrence of an inactive center in the meristem found confirmation in many studies, which resulted in the development of the concept of *quiescent center* by Clowes (1961a). This center is described as a nonmeristematic group of cells roughly hemispherical in shape and surrounded by actively dividing cells, the initials, or the promeristem. The center becomes quiescent during the development of the root, either the main root (taproot) or the lateral root, after the architectural pattern of the apex is established, and it remains capable of resuming meristematic activity. There is apparently a range of development of the quiescent center. The center may be larger in large roots and smaller, or even absent, in small roots.

The origin of the architectural pattern in roots and shoots beginning with the embryo has been studied in a number of species. The subject is reviewed by Guttenberg (1960, 1961). The pattern is organized gradually in terminal apices of epicotyls, in lateral shoots, in radicles of embryos or seedlings, and in lateral and adventitious roots. Moreover, the distribution of the meristematic activity in the apical meristem changes with the development of shoot or root.

Apical meristems receive much attention in connection with studies of causal relations in morphogenesis. Many efforts have been directed toward determining the role of the apical meristem in the development of form and internal organization of plant organs. Some studies have been concerned with the determination of the arrangement of leaves (phyllotaxy) and their bilateral symmetry, others with the determination of the vascular patterns in roots and shoots. Workers also consider the question whether the apex is a self-determining and dominant center of development controlling the growth of the parts derived form it or whether it is a plastic region operating under the control of stimuli sent to it from the mature subjacent tissues.

Results of experimental studies involving cultures of isolated shoot and root tips and partial isolation of apical meristems and leaf primordia by operations on growing plants have been interpreted as indicating a high degree of independence of the apical meristem. The culture studies have shown that apical meristems of roots are capable of forming vascularized roots and that the tissue pattern in the root is a product of apical activity. Apical meristems of shoots including the youngest leaf primordial may grow into entire plants, whereas the subjacent regions form only vascularized masses of cells. Operations on shoot apices show that the apex may continue to grow and form primordia after its procambial connection with the subjacent region is severed. Some experimental work indicates a considerable degree of resistance of the apical meristem to disturbances that may be caused by environmental conditions, such as variation in light, temperature, and nutrient conditions.

VEGETATIVE SHOOT APEX

Vegetative shoot apices vary in shape, size, cytologic zonation, and meristematic activity. The shoot apices of conifers are commonly relatively narrow and conical in form; in *Ginkgo* and in the cycads they are rather broad and flat. The apical meristem of some monocotyledons (grasses, *Elodea*) and dicotyledons (*Hippuris*) is narrow and elongated, with the distal zone much elevated above the youngest node. In many dicotyledons the distal zone barely rises above the leaf primordia, or even appears sunken. In some plants the axis increases in width close to the apex and the peripheral region bearing the leaf primordia becomes elevated above the apical meristem leaving the latter in a pit-like depression. Examples of widths of apices measured in microns at insertion of the youngest leaf primordia are: 280, *Equisetum hiemale*; 1,000, *Dryopteris dilatata*; 2,000-3,300, *Cycas revolute*; 280, *Pinus mugo*; 140, *Taxus baccata*; 400, *Ginkgo biloba*; 288, *Washingtonia filifera*; 130, *Zea mays*; 500, *Nuphar lutea*. The shape and size of the apex change during the development of a plant from embryo to reproduction, between initiation of successive leaves, and in relation to seasonal changes. An example of change in width during growth is available for *Phoenix canariensis*. The diameter in microns was found to be 80 in the embryo, 140 in the seedling, and 528 in the adult plant.

Attempts to classify apical structure of shoots have resulted in the assigning of shoot apices to several types, but these classifications are subject to criticism on the grounds that they do not reflect fundamental differences in structure and are not helpful in making the behavior of meristems better understood. The simple classification into three types on the basis of whether there is only one initial (many vascular cryptogams), or several initials in one cell layer (most gymnosperms), or several initials in more than one layer (most angiosperms) is convenient for descriptive purposes; but the basic pattern of growth in these three kinds of apices is the same; they all consist of a distally located initiating zone (protomeristem) and two derivative zones (the outer and the inner), in which organogenesis and histogenesis begin.

Vascular Cryptogams

In the lower Tracheophyta growth at the apex proceeds from either one or few apical initial cells. These cells are often conspicuous because of their large size and relatively high degree of vacuolation. Most commonly the single apical cell is pyramidal (tetrahedral) in shape. The base of this pyramid is turned toward the free surface, the other three sides downward. New cells are cut off roughly parallel to these three sides. In apices with a tetrahedral apical cell, the derivative cells often form an orderly pattern, which apparently is initiated by the orderliness of divisions of the apical cells; the successive divisions follow one another in acropetal sequence along a helix. Tetrahedral apical cells are found in *Equisetum* and most leptosporangiate ferns.

The eusporangiate ferns may have one or more initials. In *Botrychium*, for example, the apex bears a superficial layer of prismatic cells among which an apical cell is occasionally recognized. Some investigators suggest that, in the ferns, the apex with several initials represents a more primitive evolutionary stage than the one with a single apical cell. The opposite view, that an apex with a multicellular initial layer could evolve through loss of genetic provision for a single apical cell, has also been expressed. Single apical cells may be three-sided, with two sides along which new cells are cut off. Such apical cells are characteristic of bilaterally symmetrical shoots, as in the water ferns *Salvinia* and *Azolla*. The flattened rhizome apex of *Pteridium* also bears a three-sided apical cell.

In Lycopsida, single apical cells and groups of initial cells have been described. The Isoetaceae appear to have a relatively poorly defined group of initial cells. In *Psilotum nudum* a more or less distinct apical cell was observed in both the gametophyte and the sporophyte.

Gymnosperms

As was mentioned previously, the cytologic zones in the apical meristem were first recognized by the study of a gymnosperm, namely *Ginkgo*. The zonation found in the apex of this genus has served as a basis for the interpretation of shoot apices in other gymnosperms. In *Ginkgo*, the protomeristem has been divided in two cell groups, the apical surface initials, from which all other cells of the apex are ultimately derived, and the subjacent group of cells originating from the surface initials and termed *mother cells*. Cell division is sluggish in the interior of the mother-cell group but is active on its periphery. The products of the divisions along the periphery of the mother-cell group combine with derivatives resulting from anticlinal divisions of the apical initials. All together these lateral derivatives form a mantle-like peripheral zone of densely staining and relatively small cells which appear less differentiated (eumeristem) than the mother cells and also less so than the cells of the initiating zone. The derivatives produced at the base of the mother-cell zone become pith cells, and usually they pass through a rib-meristem form of growth. During active growth, a cup-shaped region of orderly dividing cells, the transitional zone, delimits the mother-cell group and may extend to the surface of the apical dome. The peripheral mantle of cells is the seat of origin of the leaf primordia and of the epidermis, the cortex, and the vascular tissues of the axis. Part of the pith may arise from the peripheral zone.

The details of the structural pattern just reviewed vary in the different groups of gymnosperms. The cycads have very wide apices with a large number of surface cells contributing derivatives to the deeper layers by periclinal divisions. Foster (1941, 1943) interprets this extended surface layer and its immediate derivatives as the initiation zone; others seek to confine the initials to a relatively small number of surface cells. The periclinal derivatives of the surface layer converge toward the mother-cell zone, a pattern apparently characteristic of cycads. In other seed plants the cell layers typically diverge from the point of initiation. The convergent pattern results from numerous anticlinal divisions in the surface cells and their recent derivatives - evidence of surface growth through a tissue of some depth. This growth appears to be associated with the large width of the apex. The mother-cell group is relatively indistinct in cycads. The extensive peripheral zone arises from the immediate derivatives of the surface initials and from the mother cells. The rib meristem is more or less pronounced in the inner zone beneath the mother-cell zone.

Most conifers have periclinally dividing apical initials in the surface layer. A contrasting organization, with a cell layer dividing almost exclusively by anticlinal walls, has been described in *Araucaria*, *Cupressus*, *Thujopsis*, and *Agathis*. In these plants the apices have been interpreted as having a tunica-corpus organization. The mother-cell group may be well differentiated in conifers, and a transitional zone may be present. In conifers with narrow apices mother cells are few and may or may not be enlarged and vacuolated. In such apices a small mother-cell group, three or four cells in depth, is abruptly succeeded below by highly vacuolated pith cells without the interposition of a rib meristem; and the peripheral zone is also only a few cells wide.

Coniferous shoot apices have been studied with regard to seasonal variations in structure. The basic zonation does not change, but the height of the apical dome above the youngest node is greater during growth than during rest. Because of this difference, the zones are differently distributed in the two kinds of apices with regard to the youngest node; the rib meristem occurs below this node in resting apices and partly above it in active apices. This observation calls attention to a terminological problem. If the apical meristem is defined strictly, as the part of the apex above the youngest node, it must be interpreted as varying in its composition during different growth phases.

The Gnetales commonly have a definite separation into a surface layer and an inner core derived from its own initials. Therefore, the shoot apices of *Ephedra* and *Gnetum* have been described as having a tunica-corpus pattern of growth. The tunica is uniseriate, and the corpus is comparable to the central mother-cell zone in its morphology and manner of division. The shoot apex of *Welwitschia* produces only one pair of foliage leaves and does not possess a distinct zonation. Periclinal divisions have been observed in the surface layer.

The data on the shoot apices in the gymnosperms have been used to suggest possible trends in the evolution of apical structure in this group of plants. The large apex of the cycads with its extensive initiation zone, massive core of mother cells, and generally diversified growth zones is probably primitive. Evolutionary advancement seems to have involved a refinement of the meristem in the sense that it became simpler, with less diversity in growth zones and, at the same time, with a more precise separation of zones of surface and volume growth, each derived from independent initials.

Angiosperms

The main features of the tunica-corpus organization of the angiospermous shoot apex have been discussed in a foregoing part of this chapter. One to five layers of tunica have been reported for dicotyledons, with two represented in the largest number of species; one to four layers for monocotyledons, with one and two predominating. An absence of tunica-corpus organization, with the outermost layer dividing periclinally, has also been observed (*Saccharum*). To delimit the tunica from the corpus is not a simple matter. The number of parallel periclinal layers in the shoot apex may vary during the ontogeny of the plant and under the influence of seasonal growth changes. There may also be periodic changes in stratification in relation to the initiation of leaves. As was mentioned previously, some workers treat such changes as variations in the thickness of the tunica; others interpret them as reflections of variations in the stratification of the corpus.

According to Guttenberg (1960), the tunica could consist of no more than two layers, which he terms dermatogen and subdermatogen. The subdermatogen sometimes lacks its own initials, a condition corresponding to a single tunica layer configuration. Beneath the two outer layers is the central mother-cell complex, which may or may not be stratified. Its derivatives, through intermediary meristems, are the pith, the vascular tissue, and most of the cortex. The crucial evidence for the two-layered tunica is said to be the unbroken continuity of the dermatogen and subdermatogen in the emerging axillary bud. It seems that this scheme, as well as Guttenberg's concept of histogens, implies a high degree of uniformity in the relation between apical structure and origin of tissues below.

The analysis of apical meristems in terms of tunica and corpus is usually combined with that based on cytological zonation. The characteristics of the central mother-cell group - relatively large, light-staining cells - are sometimes limited to the corpus or part of it; sometimes they also appear in the tunica layers. Thus there may be a uniformly light-staining distal zone (frequently called central zone), or there may be a light-staining core covered by a more densely staining layer or layers. The participation of tunica and corpus in the formation of the peripheral and inner zones depends on the relative proportions of tunica and corpus in the apex. The degree of distinctness of zonation varies in the angiosperms, as it does in the gymnosperms, and is usually better expressed in larger apices. As was reviewed before, the studies of zonation may include determinations of meristematic activity, especially with reference to the concept of the inactive distal zone.

Origin of Leaves

In this chapter only those features of leaf origin are considered that are related to the structure and activity of the apical meristem. A leaf is initiated by periclinal divisions in a small group of cells in the peripheral zone of an apical meristem. According to the concept of the initiating ring, the leaves arise in this ring in positions related to the leaf arrangement. Successive sectors of the ring are visualized as being partly used up in the formation of leaves. Cell division restores each sector above a newly formed primordium so that the ring moves upward and the leaves arise at successively higher levels.

In the dicotyledons the first periclinal divisions initiating the leaves occur most frequently in the subsurface layer and are followed by similar divisions in the third layer and by anticlinal divisions in the surface layer. In some monocotyledons the surface layer also undergoes periclinal divisions and gives rise to some or most of the internal tissue of the leaf in addition to the epidermis. Since the initiation of leaves in angiosperms follows a relatively consistent pattern, whereas the depth of the tunica is variable, the tunica and the corpus are variously concerned with leaf formation, depending on their quantitative relationship in a given apex.

In the gymnosperms the leaves arise in the peripheral zone. The surface layer may contribute cells to the internal tissue of the primordium by periclinal and other divisions. According to Guttenberg (1961), such activity of the protoderm is characteristic of those gymnosperms in which no independent surface layer is present in the apical meristem. In the vascular cryptogams the leaves arise from either single superficial cells or groups of such cells, one of which enlarges and becomes the conspicuous apical cell of the primordium.

The divisions initiating a leaf primordium cause the formation of a lateral prominence on the side of the shoot apex. This prominence constitutes the leaf base or the so-called *leaf buttress*. Subsequently the leaf grows upward from the buttress. The level at which a leaf buttress appears, in relation to the initiating region of the apical meristem, varies in different species. In some species the apical meristem has the form of a relatively high cone, with the divisions that initiate the leaf appearing low on its sides. In others the apical meristem is less prominently elevated above the youngest leaf buttress. In still others it appears practically at one level with such a buttress, or in a depression below it. Depending on the level at which a leaf primordium is initiated, the shoot apex may or may not change in shape and structure during the period between the initiation of two successive leaf primordia (or pairs or whorls of primordia in plants with an opposite or whorled leaf arrangement). Such a period has been designated *plastochron*.

The term plastochron was originally formulated in a rather general sense for a time interval between two successive similar events occurring in a series of similar periodically repeated events. In this sense the term may be applied to the time interval between a variety of corresponding stages in the development of successive leaves, for example, the initiation of periclinal divisions in the sites of origin of primordia, the beginning of apical growth of a primordium, or the initiation of lamina. Plastochron may be used also with reference to the development of internodes and of axillary buds, to stages of vascularization of the shoot, and to the development of floral parts. With reference to the development of the plant as a whole, plastochron is applicable for indicating the age of the plant. A refinement of such use is provided by the formula of Erickson and Michelini (1957) for calculating the plastochron index. In this formula, as developed for *Xanthium*, a leaf 10 mm long is used as reference, so that, if the plant has n leaves, it is n plastochrons old when leaf n is 10 mm long. For characterizing leaf development, this index proved to be more useful than the chronological age. Fresh weight, dry weight, chlorophyll synthesis, and oxygen uptake of individual developing leaves had a straight-line relationship to the plastochronic stage of growth of the leaf.

Successive plastochrons may be of equal duration, at least during part of vegetative growth of genetically uniform material growing in a controlled environment. The stage of development of the plant and environmental conditions are known to affect the length of the plastochrons. In *Zea mays*, for example, the successive plastochrons in the embryo lengthen from 3.5 to 13.5 days, whereas those in the seedling shorten from 3.6 to 0.5 days. In *Lonicera nitida* the duration of plastochrons varied from 1.5 to 5.5 days apparently in relation to changing temperature. The rate of production of leaves is also affected by light.

The changes in the morphology of the shoot apex occurring during one plastochron may be referred to as plastochronic changes. Such changes are showing a shoot apex of a plant with a decussate (opposite, with the alternate leaf pairs at right angles to each other) leaf arrangement. Before the initiation of a new leaf primordium the apical meristem appears as a small rounded mound. It gradually widens. Then leaf buttresses are initiated on its sides. While the new leaf primordia grow upward from the buttresses, the apical meristem again assumes the appearance of a small mound. In some plants growth of leaves overshadows that of the apex. The divisions initiating the leaves encroach upon the distal zone so that the latter appears to be almost exhausted during each plastochron and, as a consequence, the position of this zone oscillates around the apex of the axis. The other extreme is illustrated by shoots with long, slender tips in which leaves arise considerably below the distal zone and cause no plastochronic changes in the apex.

If the shoot apex undergoes plastochronic changes in size, both its volume and surface area change. To designate these changes, the expressions *minimal-area* and *maximal-area phases* – now abbreviated to minimal and maximal phases – were introduced. When the leaves are in a decussate arrangement the maximal phase is attained by a symmetrical distribution of periclinal divisions on two sides of the apical meristem. Thus two diameters of the apex, which cross one another at right angles, elongate alternately in successive plastochrons. In shoots with a helical leaf arrangement the divisions alternate in different sectors around the circumference of the apical meristem and thus the enlargement of the apex in the maximal phase is asymmetrical. Because of the lack of delimitation between the emerging leaf primordium and the stem, the determination of the maximal stage is problematical. There is disagreement whether the foliar buttress should or should not be included in the measurement of width. The best compromise appears to be to identify the maximal phase with the first divisions initiating a leaf, before the cells resulting from these divisions begin to enlarge and thus to affect the outline of the shoot apex.

Plastochronic changes in the apical meristem may also affect the cytologic zonation, the degree of stratification of the corpus, and the distribution of mitoses.

The plastochronic changes may follow a regular sequence through the successive plastochrons. In the embryo and seedling of maize, for example, both the minimum and the maximum plastochronic sizes of the apex were found to increase from plastochron 1 to plastochron 14 (latest observed). This enlargement involved an increase in the number of cells, but the size of cells remained constant. The rate of this increase, calculated as increments per unit material, were deceleratory during embryogeny and acceleratory during seedling development.

A considerable amount of research has been carried out on factors determining the emergence of leaf primordia in their characteristic arrangement, or phyllotaxy, and their development into bilateral structures. To detect the causal relationships in leaf initiation, workers use experimental methods, such as application of growth-regulating substances to the apices and the making of incisions designed to affect the development of a leaf.

According to the concept of the origin of leaves in the initiating ring, the existing primordia determine the position of the new leaves. Leaf primordia arise in contact with one another along two or more helices, each of which terminates in the initiating ring with a putative generative center, which induces cell division leading to the emergence of a new leaf. According to the opposite and more prevalent view, a leaf arises in a locus that is spatially removed from inhibitions exercised by the distal part of the apical meristem and the adjacent youngest leaf primordia. This field-effect concept has been developed mainly through experimentation with ferns.

Positions of leaves have been changed by incisions isolating potential leaf sites. Such isolations sometimes resulted in the development of a centric leaf or a bud in the place of a dorsiventral leaf, observations suggesting that the dorsiventral symmetry is imposed by the physiological environment. The dorsiventral symmetry, however, becomes fixed in older primordia. As a result, older primordia grown in *vitro* develop into dorsiventral leaves, whereas younger primordia become centric structures.

ORIGIN OF BRANCHES

In the lower vascular plants, such as *Psilotum*, *Lycopodium*, and *Selaginella*, branching occurs at the apex, without reference to the leaves. It is described as *dichotomous* when the original apical meristem undergoes a median division into equal parts and as *monopodial* when a branch arises laterally at the apical meristem. In seed plants, branches commonly are formed in close association with the leaves – they appear to originate in the axils of the leaves – and in their nascent state they are referred to as *axillary buds*. Judged from most investigations the term axillary is somewhat inaccurate because the buds generally arise on the stem but become displaced closer to the leaf base, or even onto the leaf itself, by subsequent growth readjustments. Such relationship has been observed in ferns, dicotyledons, and Gramineae. In the grasses, the lack of developmental relation between the bud and the subtending (axillant) leaf is particularly clear. The bud originates close to the leaf located above it. Later the bud becomes separated from this leaf by the interpolation of an internode between it and the leaf. A rather similar origin of the lateral buds has been observed in other monocotyledons (*Tradescantia*). In the conifers, bud development resembles that in the dicotyledons.

Axillary buds are commonly initiated somewhat later than the leaves subtending them, frequently in the second plastochron. Therefore, it is not always clear whether the meristem of the axillary bud is derived directly form the apical meristem of the main shoot or whether it originates from partly differentiated tissue of the internode. Both situations probably occur, because plants vary with regard to the number of plastochrons intervening between the origin of the leaf and that of its axillary bud.

The initiation of the bud in higher vascular plants is characterized by a combination of anticlinal divisions, in one or more of the superficial layers of the young axis, and of various divisions, sometimes predominantly periclinal, in the deeper layers. This coordinated growth in surface area and in volume at greater depth causes the bud to protrude above the surface of the axis. Sometimes the divisions initiating a bud are quite regular and result in the formation of a series of curved layers approximately parallel to each other. Because of this configuration the early bud meristem has been named shell zone. Depending on the quantitative relationships between the tunica and the corpus in the shoot apices of angiosperms, the derivatives of the two zones variously participate in the formation of the axillary bud meristem and not necessarily in the same proportions as in the formation of the leaves of the same plant, because the buds frequently arise in deeper layers than the leaves. Epidermal origin of axillary buds has also been reported. If the axillary bud develops into a shoot, its apical meristem is gradually organized-commonly duplicating the pattern found in the parent shoot apex and proceeds with the formation of leaves.

Buds that arise without connection with apical meristem from more or less mature tissues are classified as adventitious buds. No clear ontogenetic distinction exists between adventitious and axillary buds, because axillary buds also may originate in more or less differentiated parenchyma some distance from the apex. Adventitious buds occur on stems, roots, and leaves on intact plants and on isolated cuttings or leaves. In cuttings the buds usually are initiated in callus tissue, which develops before the buds. Adventitious buds may originate more or less deeply in the tissue or in the epidermis.

Axillary buds are described as arising *exogenously*, that is, in relatively superficial tissues. This appears to be an entirely appropriate interpretation when the origin of such buds is compared with that of lateral roots, which are initiated deeply in the parent axis (*endogenous origin*). The adventitious buds may be exogenous or endogenous.

Many physiologic studies have been carried out on initiation of both axillary and adventitious buds. The phenomenon is evidently a complex one and involves interactions of numerous determinants. Growth-regulating substances play a part but probably in a characteristic balance with a number of specific metabolites, and apparently the different stages of bud development depend on different sets of conditions.

Reproductive Shoot Apex

In the reproductive state in angiosperms, floral apices replace the vegetative either directly or, more frequently, through the development of an inflorescence. Flowers are borne on a wide variety of inflorescences. The structural modification that occurs in the apical meristem during the transition to the reproductive stage may become recognizable in the inflorescence apex. Thus a discussion on the reproductive apex in angiosperms should include reference to both the inflorescence and the floral apical meristem.

The change to the reproductive stage may be early detectable by the modified growth habit of the shoot. When the flowers are borne on axillary-branch inflorescences an accelerated production of axillary buds is one of the earliest indications of approaching flowering.

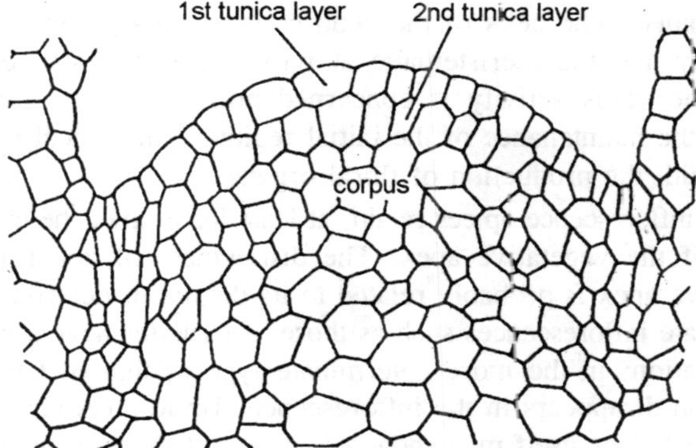

Fig. 2.5. Median longitudinal section of the shoot apex of a dicotyledon showing two-layered tunica.

Concomitantly, the nature of the foliar organs subtending the axillary buds changes; they develop as bracts more or less distinct from the foliage leaves. Growth relationships appear to change. During the vegetative stage the growth of foliar primordia is emphasized; during the reproductive stage the axillary buds appear earlier and grow more vigorously than the subtending bract primordia.

The second feature frequently revealing the beginning of the reproductive stage is the sudden increase in the elongation of the internodes. This change is particularly striking in plants that have no elongated axis during the vegetative stage, as in many grasses and rosette types of plants.

Histologically and cytologically the reproductive meristem differs from the vegetative in varying degrees. It may retain the same quantitative relationship between the tunica and the corpus as was present in the vegetative apex, or the number of discrete surface layers may be reduced or augmented. The most frequently described change is expressed in the distribution of the eumeristematic and the more highly vacuolated cells. In many species the apex of the inflorescence or the flower shows a uniform, densely staining, small-celled mantle-like zone of one or more layers enclosing a lighter-staining, larger-celled core; and such an apex may be flatter and wider than the vegetative. The mantle does not necessarily coincide with the tunica; part of the corpus may be included in it. This type of configuration is an expression of determination of growth and of a shift in its direction. Axis elongation becomes limited, and therefore the characteristic activity of the corpus, resulting in the formation of the rib meristem, is discontinued. The cells of the central tissue enlarge and vacuolate prominently and the meristematic activity becomes restricted to the mantle zone. This activity is concerned not with elongation of the shoot and the maintenance of the initial region of the apical meristem, but only with the production of floral organs.

Some inflorescence apices retain, at least for a time, the cytological zonation of the vegetative apex. The distinctness of zonation in the reproductive apex is probably related to its degree of determinateness; indeterminate inflorescences such as those of Crucifer have a persisting apical zonation; in the more determinate type as that of Compositae the zonation disappears in the inflorescence. Up to a certain stage, the apex of the flower itself may show a zonation of the vegetative type.

In the absence of internodal elongation in the axis of the flower the floral parts appear in close succession spatially and temporally. The wide meristematic surface accommodates simultaneously many centers of cell proliferation and the plastochronic rhythm that characterized the vegetative growth may become indistinguishable. If, however, the flower is less determinate and its apex has a prolonged meristematic activity-common features of flowers with numerous free parts-plastochronic fluctuations in size and configuration of the apex may be retained during floral ontogeny.

Cytologic studies on transition of the apex to reproductive state have shown that mitotic activity is increased and changed in its distribution at this time. In *Xanthium* the stimulation to increased cell division in the apex was observed within 24 hours after a single inductive dark period, before any other change was detectable. In connection with the appearance of the mantle-like eumeristem, the distinction between the highly active peripheral zone and the less active distal zone commonly seen in vegetative apices is effaced. Correspondingly, the staining that indicates presence of DNA becomes more uniform than in the vegetative stage when the distal cells stain lightly. RNA and protein are uniformly distributed in both kinds of apices but both increase in concentration in the reproductive stage.

As was mentioned previously, the proponents of the waiting-meristem concept consider that the distal part of the apical meristem, which is said to be inactive during the vegetative stage, becomes active during the development of the flower. The initiating ring still produces the sepals but may disappear immediately after this event. The formerly inactive zone assumes two roles. The upper part is sporogenous and becomes the meristem initiating the floral parts, the lower is the receptacular meristem, which produces the axis of the flower (or the inflorescence). Thus this concept implies a functional discontinuity between the vegetative and the reproductive apical meristems and, therefore, is in agreement with well-known view of Grégoire (1938) that the flower and the vegetative shoot are not related structures and that their meristems are fundamentally different.

The concept that the reproductive apex results from a more or less extensive reorganization of the vegetative apex is the prevalent one and is accepted for both the angiosperms and the gymnosperms. It is adopted in this book. The two kinds of meristem intergrade through intermediate forms and the existing differences are not fundamental; they are related to the different modes of growth of the vegetative and reproductive axes. The absence of discontinuity between the two kinds of growth is emphasized by Hillman (1962) in his review of the physiology of flowering. He suggests that floral induction represents not a sudden change in the condition of the shoot but is a process with many intermediate stages. The ontogenetic development of the reproductive apex from the vegetative is in agreement with this concept.

The change from the vegetative to the flowering state not only affects the apical meristems concerned with flower production but alters, physiologically and morphologically, other parts of the plant as well. This change is associated with a shift in the balance between meristematic activity and cell maturation in favor of the latter. It usually signifies the end of growth at the given apical meristem because of the determinate nature of the flower, and in annual plants in means the end of growth and the approach of death of the entire plant. The change is not irreversible, however, and may be interrupted or prevented by subjecting the plant to influences that favor vegetative growth. Even such a typical characteristic of the flower as determinate growth is not fixed, and the floral meristem occasionally resumes vegetative growth after the floral parts have been formed. Thus the visible change from the vegetative to the reproductive meristem is a reflection of a physiologic change in the plants and it may be discussed in terms of the ripeness-to-flower concept.

Root Apex

In contrast to the apical meristem of the shoot, that of the root produces cells not only toward the axis but also away form it, for it initiates the rootcap. Because of the presence of the rootcap the distal part of the apical meristem of the root is not terminal but subterminal in position, in the sense that it is located beneath the rootcap. The root apex further differs from the shoot meristem in that it forms no lateral appendages comparable to the leaves, and no branches. The root branches are usually initiated beyond the region of most active growth and arise endogenously. Because of the absence of leaves, the root apex shows no periodic changes in shape and structure such as commonly occur in shoot apices in relation to leaf initiation. The root also produces no nodes and internodes, and, therefore, grows more uniformly in length than the shoot, in which the internodes elongate much more than the nodes. The rib-meristem type of growth is characteristic of the elongating root cortex.

The distal part of the apical meristem of the root, like that of the shoot, may be termed protomeristem and, as such, contrasted with the subjacent primary meristematic tissues. The young root axis is more or less clearly separated into the future central cylinder (plerome) and Cortex (periblem). In their meristematic state the tissues of these two regions consist of procambium and ground meristem, respectively. The term procambium may be applied to the entire central cylinder of the root if this cylinder eventually differentiates into a solid vascular core. Many roots, however, have a pith-like region in the center. This region is regarded sometimes as potentially vascular and therefore procambial in its meristematic state, sometimes as ground tissue similar to that of the pith in stems and differentiating from a ground meristem. The term protoderm, if used to designate the surface layer regardless of its developmental relation to other tissue, may be applied to the outer layer of the young root. Usually the root protoderm does not arise from a separate layer of the protomeristem. It has a common origin with either the cortex or the rootcap.

Apical meristems of roots are analyzed on the basis of three concepts. The first is fundamentally Hanstein's histogen concept since it includes the assumption that a precise relation may exist between the initials in the distal zone and the tissue regions of the root. The second, the previously mentioned quiescent-center concept of Clowes (1961a), is a modification of the histogen concept. It places the initials of the tissue regions outside the distal region—the minimal constructional center of Clowes (1961a)—which is interpreted as being inactive. The third is the little-exploited body-cap concept of Schuepp (1917), which is comparable to the tunica-corpus concept, since it characterizes the root apex by reference to planes of division in its parts. The three concepts are not mutually exclusive. The histogen and the body-cap concepts deal with different aspects of apical activity, and the quiescentcenter concept includes the postulate that the cell pattern in the distal zone is not meaningless but reflects the past history of meristem activity when the root meristem was being organized, either in embryogeny or during the origin of the lateral root.

The cellular configuration of the distal zone has been the subject of many studies and has served for the establishment of the so-called types and for discussions of the phylogeny of apical organization of roots. In the lower vascular plants all tissues are derived either from a single apical cell or from several initials arranged in one tier (Marattiaceae). These plants usually have the same apical structure in both the root and the shoot. In some gymnosperms and angiosperms all tissue regions of the root or all except the central cylinder appear to arise from a common meristematic group of cells; in others one or more of these regions can be traced to separate initials. Guttenberg (1960) classifies the two kinds of organization as the open and the closed, respectively. He considers that both originate from a closed type present in the embryonic root or the primordium of the lateral or adventitious root. During later elongation of the root the closed pattern may be retained or replaced by an open one. In all the events of organization of the root meristem, central or connecting cells (Verbindungszellen) play the major role as initials. In position, they are periblem initials.

The structure based on a single apical cell lends itself to a study of segmentation patterns among the derivatives of the apical meristem. Since the root is normally radially symmetrical, the apical cell is tetrahedral. It cuts off cells either on four faces of the tetrahedron and thus produces the tissues of the root and the rootcap (Marsilea); or the rootcap has its own initials (Azolla). A root organization characterized by a precise segmentation of derivatives of the initial zone resembling that in the roots of ferns has been found in a monocotyledon, *Cyperus*.

An analysis of divisions in the derivatives of the apical cell illustrates the body-cap concept. The longitudinal rows of cells so prominent in roots radiate from the apical cell and many of them divide in two. Where they do so, a cell divides transversely; then one of the two new cells divides longitudinally and each daughter cell of this division becomes the source of a new row. The combination of the transverse and the longitudinal divisions results in an approximately T-(or Y-)shaped wall pattern and, therefore, such divisions of cell rows have been named T divisions. The direction of the top stroke of

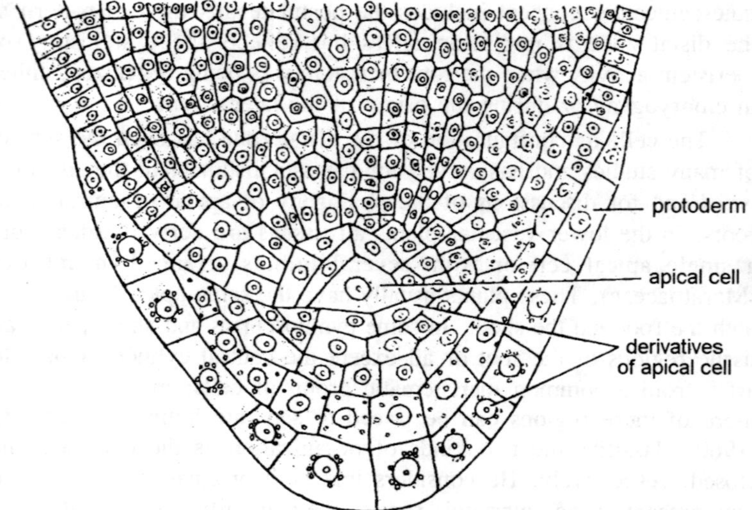

Fig. 2.6. Median longitudinal section of the root tip of Pteris showing a single apical cell with four cutting faces (one not in the plane of the section).

the T varies in different root parts. In the cap it is directed toward the base of the root, in the body toward the apex. The body and the cap are not sharply delimited if both arise from the same apical cell (*Marsilea*); the presence of separate rootcap initials causes the presence of a clear boundary between the cap and the body (*Azolla*).

The two kinds of multicellular protomeristem of angiosperms, the closed and the open in the sense of Guttenberg (1960), require separate consideration. The closed pattern is often characterized by the presence of three tiers of initials. One tier appears at the apex of the central cylinder, the second terminates the cortex, the third gives rise to the rootcap. The three-tiered meristems may be grouped according to the origin of the epidermis. In one group, the epidermis has common origin with the rootcap and becomes distinct as such after a series of T divisions along the periphery of the root. In the second, the epidermis and cortex have common initials, whereas the rootcap arises from its own initials that constitute the rootcap meristem, or *calyptrogen*. If the rootcap and the epidermis have common origin, the cell layer concerned is called *dermatocalyptrogen*.

Roots with a dermatocalyptrogen are common in dicotyledons, but occur also in monocotyledons. Roots with a calyptrogen are characteristic of monocotyledons. Sometimes the epidermis appears to terminate in the distal zone with its own initials. In some aquatic monocotyledons (*Hydrocharis*, *Lemna*, *Pistia*) the epidermis is regularly independent from the cortex and the rootcap.

An analysis of root meristems on the basis of the body-cap concept reveals the difference in origin of the epidermis. In the root with a calyptrogen the cap includes only the rootcap, in one with a dermatocalyptrogen the cap extends into the epidermis. The body-cap configuration shows other variations that elucidate patterns of growth of roots. In some roots the central core of the rootcap is distinct from the peripheral part in having few or no longitudinal divisions. If conspicuous enough, such a core is referred to as columella. The few T divisions that occur in the columella may be oriented according to the body pattern; then only the peripheral parts of the rootcap show the cap pattern.

Apices that lack a clear differentiation of initials – the open type according to Guttenberg (1960) – are difficult to analyze. One common interpretation is that such roots have a ***transversal meristem*** without any boundaries with reference to the derivative regions of the root. The other view is that the central cylinder has its own initials in this type of meristem. Analyses of body-cap configurations indicate that the limits between the two regions may be indefinite and may change during the growth of the root. A reinterpretation of meristems with indefinite boundaries in the distal zone was given by Allen (1947) for

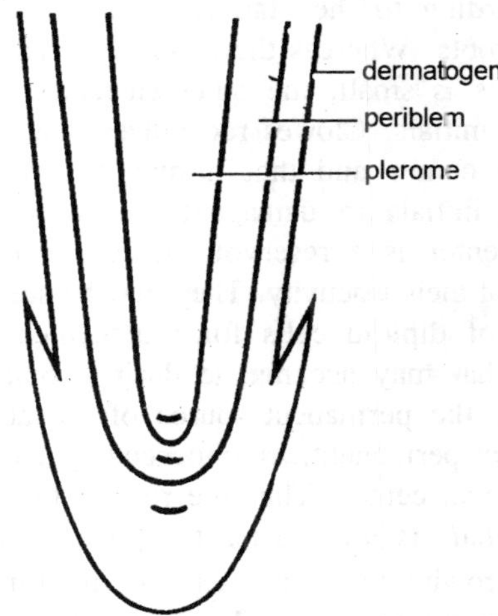

Fig. 2.7. Schematic representation of the growth of root apex of a dicotyledon inerpreted in terms of Hanstein's histogen concept.

Pseudotsuga taxifolia and by Clowes (1961a) for *Fagus sylvatica*. In *Pseudotsuga*, two kinds of initials are recognized; the "permanent", which remain in their position indefinitely, and the "temporary", which give origin to the various root regions and are from time to time replaced by derivatives from the permanent initials. *Fagus sylvatica* has a basically similar apical organization, but apparently the initials of the various regions are more independent than those of *Pseudotsuga*. Moreover, Clowes reports that the distal region enclosed by the cup-shaped group of initials is quiescent.

Apical meristems with no separate initials for the root regions have been described in dicotyledons (representatives of Proteaceae, Casuarinaceae, Leguminosae, of some Ranalean and Amentiferous families), monocotyledons (representatives of Musaceae, Palmae), and some gymnosperms. In a group of conifers, the apex is interpreted as having (1) common initials for the central cylinder and the columella and (2) a common initiating zone for the cortex and the peripheral part of the rootcap. The initiating zone 2 surrounds the initials 1 and their recent derivatives.

The quiescent-center has been studied and discussed by Clowes with consistency and imagination. After various studies on normally developing roots and on experimentally treated ones, or roots that were fed labeled compounds involved in DNA synthesis, Clowes (1961a) concluded that the inactive state of the distal zone- the zone containing the initials according to the classical histogen theory-is a general phenomenon in roots. Whereas the classical concept assumes that the number of initials is small, the quiescentcenter concept indicates a large number of initials. Clowes recognizes that occasional divisions do occur in the center and that it may become active when the previously acting initials are damaged, as, for example, by radiation. The quiescent center is a reservoir of cells relatively resistant to damage because of their inactivity. They may be sites of auxin synthesis and the source of diploid cells for replacement of polyploid and aneuploid cells that may accumulate during somatic differentiation. Finally, they are the permanent source of the active initials which themselves are not permanent, as evidenced by the fluctuations in the size of the quiescent center. Thus the role of this center may be far more important than its relative inactivity would indicate.

The tips of growing roots are frequently used in studies of growth. The zone of actively dividing cells in growing roots extends to a considerable distance from the apex, in *Zea*, for example, through 8 to 10 millimeters with a maximum at the 4 millimeter level. The distribution of meristematic activity differs in the various root regions; however, the data obtained for mitotic frequency vary probably mainly in relation to methods of analysis. At the same level of the root, the processes of cell division, cell enlargement, and cell maturation overlap not only in the different tissue regions but also in the different cells of the same tissue region, and even in individual cells. The meristematic cortex vacuolates and develops intercellular spaces close to the apex, where the central-cylinder meristem still appears dense. In the central cylinder the precursors of the innermost xylem vessels cease dividing, enlarge, and vacuolated considerably in advance of the other vascular precursors, and the first sieve tubes mature in the part of the root where cell division is still in progress. In individual cells, division, elongation, and vacuolation are combined.

3

Collenchymatous Tissues

Collenchyma is a living tissue composed of more or less elongated cells with thick primary nonlignified walls. The structure and the arrangement of collenchyma cells in the plant body indicate that the primary function of the tissue is support. Morphologically, collenchyma is a simple tissue, for it consists of one type of cell.

The presence of living protoplasts denotes a close physiologic similarity between collenchyma and parenchyma cells. In form and structure the two types of cells also intergrade. Collenchyma cells are commonly longer and narrower than parenchyma cells, but some collenchyma cells are short and, on the other hand, some parenchyma cells are considerably elongated. Where collenchyma and parenchyma lie next to each other, they frequently intergrade through transitional types of cells. The resemblance to parenchyma is further stressed by the common occurrence of chloroplasts in collenchyma and by the ability of this tissue to undergo reversible changes in wall thickness and to engage in meristematic activity. In view of these similarities and in view of the structural and functional variability of parenchyma, collenchyma is commonly interpreted as a thick-walled kind of parenchyma structurally specialized as a supporting tissue. The terms parenchyma and collenchyma are also related, but in the latter the first part of the word, derived from the Greed word *colla*, glue, refers to the thick glistening wall characteristic of collenchyma.

Position in Plant

Collenchyma is the typical supporting tissue, first, of growing organs and, second, of those mature herbaceous organs that are only slightly modified by secondary growth or lack such growth completely.

It is the first supporting tissue in stems, leaves, and floral parts, and it is the main supporting tissue in many mature, dicotyledonous leaves and some green stems. Collenchyma may occur in root cortex, particularly if the root is exposed to light. It is absent in stems and leaves of many of the monocotyledons that early develop sclerenchyma.

Collenchyma characteristically occurs in peripheral position in stems and leaves. It may be present immediately beneath the epidermis, or it may be separated from the epidermis by one or more layers of parenchyma. If it is located next to the epidermis, the inner tangential walls of the epidermis may be thickened like the walls of collenchyma. Sometimes the entire epidermal cells are collenchymatous. In its subepidermal position, collenchyma occurs in the form of continuous or somewhat discontinuous cylinders or in discrete strands. In stems and petioles with protruding ribs, collenchyma is particularly well developed in the ribs. In leaves it may differentiate on one or both sides of the veins and along the margins of the leaf blade.

In many plants the elongated parenchyma cells of the outermost part of the phloem develop thick walls after the sieve elements are obliterated, and the tissue ceases to be concerned with conduction. The resulting structure is commonly called the bundle cap. The parenchyma on the inner periphery of the xylem may be similarly differentiated. If the entire bundle is enclosed by thick-walled elongated cells, it is said to have a bundle sheath. The bundle caps and the bundle sheaths sometimes have thickened primary walls, sometimes lignified secondary walls. The tissues forming these caps and sheaths are often interpreted as collenchyma when they have nonlignified primary walls, and as sclerenchyma when they have secondary walls. The comparative characteristics of subepidermal collenchyma, on the one hand, and of nonlignified caps and sheaths, on the other, are imperfectly known. In a study of celery grown in a boron-deficient medium the collenchyma walls were found to be thinner than normal, while the walls of parenchyma cells of the phloem forming the bundle caps and of the ground parenchyma were thicker than normal. In a comparison of the strength of collenchyma and of bundle-cap tissue from the same petioles of celery, the collenchyma strands proved to be much stronger. In this book collenchyma refers only to the supporting tissue in peripheral positions in the plant. If bundle caps and sheaths resemble collenchyma, they are referred to as *collenchymatous*, an adjective implying similarity to collenchyma, but not necessarily morphologic identity.

Structure

Cell Shape

Collenchyma cells may vary in length, but typically they are considerably elongated – cells 2 mm in length have been recorded – and resemble fibers in having tapering ends. The shorter collenchyma cells are prismatic like many parenchyma cells. Both kinds of cells are polygonal in transection. Collenchyma cells may vary in shape and size in the same strand. These variations are related to the origin of the cells. A collenchyma strand is formed by a series of longitudinal divisions, which spread from a central point toward the periphery of the future strand. The longitudinal divisions are followed by elongation of the resulting cells, so that the first, that is, the innermost, cells start to elongate earlier than the more peripheral ones and attain a greater length.

The development of collenchyma was studied in considerable detail in the umbellifer *Heracleum*. In this plant the elongation of a collenchyma cell either follows immediately the last longitudinal division of a mother cell or is preceded by one or rarely more transverse divisions. In macerated preparations the products of the latest transverse divisions often remain together, enclosed by the common mother-cell wall. Such cell complexes resemble septate fibers. When transverse divisions occur before elongation, the shape of the cells is affected. The ends formed by transverse divisions may be slightly oblique or almost transverse. Without such divisions the cells taper at both ends. Peripheral cells of a collenchyma bundle are short, and their end walls taper little.

Cell Wall

The structure of the cell wall is the most distinctive feature of collenchyma cells. The thickenings are deposited unevenly in a manner somewhat variable in different groups of plants. As seen in transactions, a common form of collenchyma shows the main deposits of wall material in the corners where several cells are joined together (*Ficus*, *Vitis*, *Ampelopsis*, *Polugonum*, *Beta*, *Rumex*, *Bochmeria*, *Morus*, *Cannabis*, *Begonia*, *Pellionia*). The degree of restriction of wall thickenings to the angles varies in relation to the amount of wall thickening present on other wall parts. If the general wall thickening becomes massive, the thickening in the corners is obscured and the lumen of the cell assumes, in transactions, a circular outline, instead of an angular one. Such developmental modification is observed, for example, in the Umbelliferae. In another form of collenchyma the thickening occurs

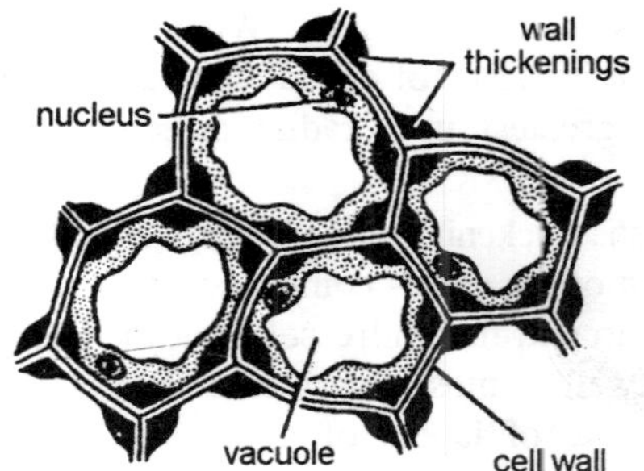

Fig. 3.1. Collenchyma showing wall thickenings.

chiefly on the tangential walls (*Sambucus*, *Sanguisorba*, *Rheum*, *Eupatorium*). Still another form is characterized by intercellular spaces and the development of the collenchymatic thickenings on the walls facing the intercellular spaces (Compositae, *Salvia*, *Brunella*, *Malva*, *Althaea*).The three forms of collenchyma have been named by Muller (1890) angular (Eckencollenchym), lamellar (Plattencollenchym), and lacunar (Luckencollenchym), respectively. The word lamellar has reference to the plate-like arrangement of the thickenings; the lacunar, to the presence of intercellular spaces. In collenchymatous bundle caps and sheaths the cell-wall thickening is sometimes most emphasized in the corners. More commonly, however, the thickening is either relatively even over the entire wall, or is uneven without being restricted to the corners or the tangential walls.

In longitudinal sections collenchyma shows thin and thick wall portions, depending on the direction of the cut with reference to the thickenings. The nearly transverse end walls are usually thin, whereas the pointed ends may appear solid because of accumulation of wall material. Primary pit-fields occur in collenchyma cells, in both the thinner and the thicker parts of the walls.

The walls of collenchyma consist mainly of cellulose and pectic compounds and contain much water. In some species collenchyma walls appear to have an alternation of layers rich in cellulose and poor in pectic compounds with layers that are rich in pectic compounds and poor in cellulose. Ultrastructurally, the thickenings of collenchyma in celery petioles show an alternation of layers of longitudinally oriented microfibrils and noncellulosic material. According to a study with polarization optics, cellulose forms transverse as well as longitudinal lamellations.

Collenchyma walls may contain over 60 per cent water, based on fresh weight: over 200 per cent, based on dry weight. Heating destroys the ability of walls to absorb water. When collenchyma walls lose their water under the influence of dehydrating agents they shrink visibly. This shrinkage is greatest in a radial direction and smallest in a vertical direction.

The characteristic thickening of collenchyma walls begins to develop before the extension of the cell is completed. Apparently the successive layers are formed around the entire cell, but the individual layers are thicker where the wall is most massive in final state. In the electron microscope a merging of layers of microfibrils in the thinner wall parts was recognized.

As was previously mentioned, collenchyma may or may not have intercellular spaces. In the absence of spaces the corners where several cells meet frequently show prominent accumulations of pectic substances. These accumulations may not fill the space completely but protrude into it in the form of warts or coralloid structures. Similar formations may occur in parenchyma tissue.

The wall thickness in collenchyma is increased if during development the plants are exposed to motion by wind. Apparently inhibition of cell elongation occurs at the same time. The wall thickenings of collenchyma are sometimes removed, as, far instance, when a phellogen arises in this tissue or when collenchyma cells respond to injuries with wound-healing reactions. Loss of wall material in collenchyma was also induced experimentally by etiolation. The occurrence of simultaneous increase in thickness and surface area of collenchyma walls, that is, the increase of wall thickening during the elongation of cells is a noteworthy phenomenon. Because of this development the term thickened primary wall has been applied to the collenchyma wall. Ultrastructurally, too, the collenchyma has been interpreted as primary.

Collenchyma walls may become modified in older plant parts. In woody species with secondary growth, collenchyma follows, at least for a time, the increase in circumference of the axis by active growth with retention of the original characteristics. In some plants (*Tilia*, *Acer*, *Aesculus*) collenchyma cells enlarge and their walls become thinner. Apparently it is not known whether this reduction in thickness results from a removal of wall material or from stretching and dehydration. Collenchyma may develop lignified secondary walls. It thus becomes changed into sclerenchyma.

Cell Contents

As was stated previously, collenchyma cells have living protoplasts at maturity. Chloroplasts occur in variable numbers. They are most numerous in collenchyma which approaches parenchyma in form. Collenchyma consisting of long, narrow cells—the most highly specialized type—contains only a few small chloroplasts or none. Tannins may be present in collenchyma cells.

Structure in relation to function

Collenchyma appears to be a mechanical tissue particularly adapted for support of growing organs. Thick walls and close packing make it a strong tissue. At the same time, the peculiarities of growth and the structure of its walls permit adjustments to elongation of the organ without loss of strength. As has been previously stressed, collenchyma cells are capable of increasing simultaneously the surface and the thickness of the walls and, therefore, can develop thick walls while the organ is still elongating.

Collenchyma tissue combines considerable tensile strength with flexibility and plasticity. Measurements of the strength of collenchyma have been made by determining the weight necessary to break a strand of the tissue dissected out from the organ. The values thus obtained were in some instances recalculated per unit area of strand and used to express the tensile strength of the tissue. Such values obviously give a measure, not of the tensile strength of the wall proper, but of the tissue as a whole. Nevertheless, they provide useful information about the strength of a tissue. Since in the plant body the mechanical effect of a tissue is determined not only by the nature of the walls but also by the shape and arrangement of the cells.

A comparison of collenchyma with fibers is particularly interesting. Collenchyma has been found capable of supporting 10 to 12 kg per mm^2, fiber strands, 15 to 20 kg per mm^2. The fiber strands regain their original length even after they have been subjected to a tension of 15 to 20 kg per mm^2, whereas collenchyma remains permanently extended after it has been made to support 1½ to 2 kg per mm^2. In other words, fibers are elastic, and collenchyma is plastic. If fibers were to differentiate in growing organs, they would hinder tissue elongation because of their tendency to regain their original length when stretched. Collenchyma, on the other hand, could be expected to respond with a plastic change in length under the same conditions.

The importance of the plasticity of collenchyma walls for the internal adjustment of growing tissues is emphasized by the observation that much of the elongation of the internodes occurs after the collenchyma cells have thickened their walls. In one particular study on *Heracleum* plants thick-walled collenchyma cells were found in

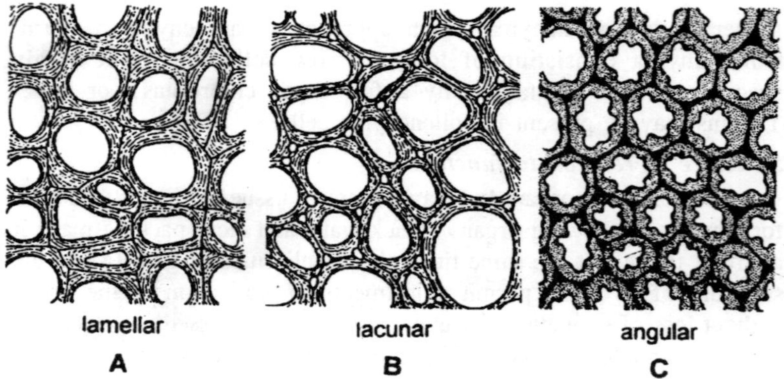

Fig. 3.2. Types of collenchyma: A–lamellar; B–lacunar; C–angular.

young internodes that were several times shorter than the extended internodes of the same axes. The individual collenchyma cells in the young internodes were markedly shorter than those in the extended ones.

The plasticity of collenchyma changes with age. Old tissue is harder and more brittle than young. As mentioned previously, in some plants collenchyma may finally become sclerified. Hardened collenchyma occurs in plant parts that have ceased to elongate.

ORIGIN

Collenchyma has been variously reported as originating jointly with the vascular tissues from the procambium and separately from these tissues in the ground meristem. This disagreement results from a difference in the interpretation of histogenetic events. Although it is proper to speak of a differentiation of the derivatives of the apical meristems into protoderm, procambium, and ground meristem, these meristems usually become delimited from each other gradually, particularly in shoots. The protoderm may be distinguished close to the initial region and may even have its own initials, but the procambium of stems and leaves is built up by longitudinal divisions involving an increasingly larger number of cells in the meristem that also gives rise to ground tissue. Thus at first it is impossible to designate a certain part of the meristem as giving rise to the procambium, another to the ground meristem. It may be said, therefore, that the cortical collenchyma and the procambium originate in a common meristem. The final delimitation of the procambium occurs later in some plants than in others, and consequently the ontogenetic relationship between the cortex and the procambium appears close in some plants

(Umbelliferae, Piperaceae, Araceae) and distant in others (Labiatae, *Clematis*, *Aristolochia*, certain Cucurbitaceae, *Chenapodium*, Compositae).

The development of collenchyma in the Umbelliferae clearly illustrates the lack of separation between cortex and procambium in the early stages of their development. In mature petioles of celery, collenchyma strands appear near the periphery in the ribs, separated by cortical parenchyma from the vascular bundles. In early ontogeny longitudinal divisions occur in the peripheral part of the petiole. Some of these initiate the procambium, others form the cortex. Subsequently, the procambium becomes distinct from the cortex because its cells are smaller in transverse diameters and greater in length. A secretory duct differentiates outside the procambium. After the procambium appears the cells between the duct and the protoderm—ground meristem cells—undergo a series of divisions giving rise to the collenchyma.

Collenchyma that differentiates early in a given organ becomes highly specialized in its morphology, whereas that formed later is more parenchymatous. This difference is also reflected in the nature of the meristem giving rise to the different kinds of collenchyma. The more specialized collenchyma has its origin in a procambium-like meristem, the less specialized in a parenchymatic ground meristem. Since Haberlandt (1914) extended the concept of procambium to include meristems giving rise to all elongated cells of the primary body, he called the collenchyma meristem having elongated cells procambium, a usage not followed in this book.

4

PARENCHYMATOUS TISSUES

The term *parenchyma* refers to a tissue composed of living cells variable in their morphology and physiology, but generally having thin walls and a polyhedral shape, and concerned with vegetative activities of the plant. The individual cells of such a tissue are *parenchyma cells*. The word parenchyma is derived from the Greek *para*, beside, and *enchein*, to pour, a combination of words that express the ancient concept of parenchyma as a semiliquid substance "poured beside" other tissues which are formed earlier and are more solid.

Parenchyma is often spoken of as the fundamental or ground tissue. It fits this definition from morphological as well as physiological aspects. In the plant body as a whole or in its organs parenchyma appears as a ground substance in which other tissues, notably the vascular, are imbedded. It is the foundation of the plant in the sense that the apical meristems and the reproductive cells are parenchymatic in nature. Furthermore, parenchyma cells are involved in phenomena of wound healing and regeneration. Phylogenetically, parenchyma is also the precursor of other tissues, as evidenced by the structure of the most primitive multicellular plants whose bodies consist of parenchyma only.

This tissue is the principal seat of such essential activities of the plant as photosynthesis, assimilation, respiration, storage, secretion, excretion – in short, activities depending on the presence of living protoplasts. Parenchyma cells which occur in the xylem and phloem tissues appear to play an important role in connection with the movement of water in the nonliving tracheary elements and with the transport of food in the sieve elements whose protoplasts lack nuclei.

Developmentally, parenchyma cells are relatively undifferentiated. They are unspecialized morphologically and physiologically, compared with such cells as sieve elements, tracheids, or fibers, since, in contrast to these three examples of cell categories, parenchyma cells may change functions or combine several different ones. However, parenchyma cells may also be distinctly specialized, for example, with reference to photosynthesis, storage of specific substances, or deposition of materials which are in excess in the plant body. Whether they are specialized or not, parenchyma cells are highly complex physiologically because they possess living protoplasts.

It was pointed out previously that living cells are not fixed in their characteristics and that they possess, in varying degrees, the ability to resume meristematic activity. Parenchyma constitutes the principal category of tissue showing such developmental plasticity resulting from its relatively low level of differentiation. Apparently the potentiality to divide may be retained by parenchyma cells for many years, as evidenced by callus development from pith cells (medullary sheath) of an approximately 50 year-old stem of *Tilia*. This development, however, was possible only after removal of the pith tissue, by tissue-culture techniques, from the correlative inhibitions to which the cells are subjected in the plant.

Delimitation

Parenchyma cells may occur in extensive continuous masses as parenchyma tissue. They may also be associated with other types of cells in morphologically heterogeneous tissues. The pith and the cortex of stems and roots, the photosynthetic tissue, or mesophyll, of leaves, the flesh of succulent fruits, the endosperm of seeds are all examples of plant parts consisting largely or entirely of parenchyma. As components of heterogeneous tissues parenchyma cells form the vascular rays and the vertical files of living cells in the xylem and the phloem. Sometimes an essentially parenchymatic tissue contains parenchymatic or nonparenchymatic cells or groups of cells, morphologically or physiologically distinct from the main mass of cells in the tissue. Sclereids, for example, may be found in the leaf mesophyll and in the pith and cortical parenchyma. Laticifers occur in various parenchymatic regions of plants containing latex. Sieve tubes traverse the cortical parenchyma of certain plants.

The variable structure of parenchyma tissue and the distribution of parenchyma cells in the plant body clearly illustrate the problems involved in the proper definition and classification of a tissue. On the one hand, parenchyma may fit the most restricted definition of a tissue as a group of cells having a common origin, essentially the same structure, and the same function. On the other hand, the homogeneity of a parenchyma tissue may be broken by the presence of varying numbers of non-parenchymatic cells, or parenchyma cells may occur as one of many cell categories in a heterogeneous tissue.

Thus the spatial delimitation of the parenchyma as a tissue is not precise in the plant body. Furthermore, parenchyma cells may intergrade with cells that are distinctly nonparenchymatic. Parenchyma cells may be more or less elongated and have thick walls, a combination of characters suggesting specialization with regard to support. A certain category of parenchyma cells is so distinctly differentiated as a supporting tissue that it is designated by the special name of collenchyma. Parenchyma cells may develop relatively thick lignified walls and assume some of the characteristics of sclerenchyma cells. Tannin may be found in ordinary parenchyma cells and also in cells basically parenchymatic but of such distinct form (vesicles, sacs, or tubes) that they are designated as idioblasts. Similarly, certain secretory cells differ from other parenchyma cells mainly in their function; others are so much modified that they are commonly treated as a special category of elements (laticifers).

The present chapter is restricted to a consideration of parenchyma concerned with the most ordinary vegetative activities of the plant, excluding the meristematic.

Structure

Cell Contents

The variation in contents of parenchyma cells is intimately related to the activities of these cells. The cells of the photosynthetic parenchyma have variable numbers of chloroplasts. At certain times during the day chloroplasts may contain assimilation starch. Because of its high content in chlorophyll the photosynthetic parenchyma is sometimes called *chlorenchyma*. The most distinctly specialized chlorenchyma is represented by the leaf mesophyll, but chloroplasts occur also in the cortex and sometimes more deeply in the stem, even in the pith. Cells not concerned with photosynthesis have no chloroplasts or have chloroplasts with a weakly differentiated internal lamellar system. Cells devoid of chloroplasts may have leucoplasts. Actively synthesizing cells are commonly conspicuously vacuolated.

Many different food substances are synthesized and stored by parenchyma cells. The same protoplast may store one or more kinds

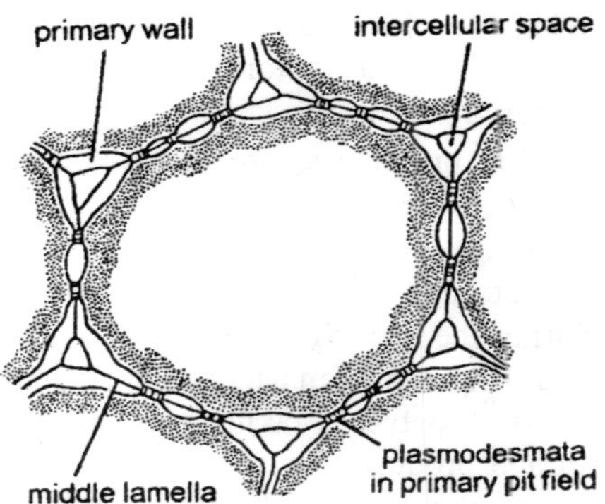

Fig. 4.1. A parenchyma cell showing plasmodermata in primary pit fields.

of substances. These substances may be dissolved in the vacuolar sap, or they may be discrete solid or fluid bodies in the cytoplasm. Bodies or masses of material may consist of such ergastic substances as starch grains, granules and crystalloids of protein, and globules of fats and oils. The cell sap may be a depository of sugars and other soluble carbohydrates and of nitrogenous substances in the form of amides and proteins. Some examples of plant organs and their storage products may be cited by reference to Netolitzky (1935). Amides, proteins, and sugar are dissolved in the cell sap of the fleshy beet root and the bulb scales of the onion. The parenchyma of the potato tuber and of the rhizomes of many other plants contains amides and proteins in the cell sap and starch in the cytoplasm. Protein granules and starch grains occur in the cytoplasm of parenchyma cells in the cotyledons of the bean, the pea, and the lentil; protein granules and oil occur in the endosperm of *Ricinus* and the cotyledons of *Glycine* (soy bean). The most widely distributed storage product is starch. It occurs in parenchyma of the cortex and the pith; in parenchyma of the vascular tissues, that is, xylem and phloem parenchyma and ray parenchyma; and in parenchyma of fleshy leaves (bulb scales), rhizomes, tubers, fruits, cotyledons, and the endosperm of seeds. In the leaves, starch predominates as the storage carbohydrate in the dicotyledons, sugars in the monocotyledons.

The physiological activity of protoplasts varies in different kinds of storage parenchyma. In stems and roots of woody species starch accumulation undergoes seasonal fluctuations. It is deposited at one time and is removed at another. Such periodic changes indicate that the storage cells have active protoplasts. The specialized storage organs, such as tubers, bulbs, and rhizomes, may serve for storage only once, their protoplasts dying after the removal of the reserves to the growing organs. During the development of storage tissues the cells can divide in the presence of starch.

In seeds the living protoplasts are directly concerned with the accumulation of storage products, but their relation to the subsequent mobilization of the storage products, but their relation to the subsequent mobilization of the storage material is not always clear. Storage cotyledons that, during the growth of the seedling, emerge above the surface of the ground (epigeous germination) and become green evidently have active protoplasts capable of taking part in photosynthesis after the removal of storage products. In contrast, cotyledons that remain below the surface of the ground during germination (hypogeous germination) usually die after they release the food reserved to the growing parts of the seedling. In both types of cotyledons the storage cells themselves probably control the mobilization of the stored food. There is some evidence, however, that the epidermis of cotyledons may be the seat of production of enzymes that digest the food materials. The protoplasts of the endosperm of some seeds are reported as being actively engaged in the process of dissolution of starch and other reserves. In other seeds the endosperm protoplasts die after the accumulation of the storage material is completed. In such seeds, the digestion of the food reserves is initiated and regulated by the enzymic activity of the embryo, alone or in conjunction with parts of the endosperm. In the Gramineae, for example, the digestion of starch is brought about by the scutellum of the embryo and by the outermost layer of the endosperm, the aleuron layer.

Water is abundant in all active vacuolated parenchyma cells so that the parenchyma plays a major role as a water reservoir. In a study of species of bamboo, the variations in moisture content of the different parts of culms were found to be clearly associated with the proportions of parenchyma cells in the tissue system.

Parenchyma may be rather specialized as a water-storage tissue. Many succulent plants, such as the Cactaceae, *Aloe*, *Agave*, and *Mesembryanthemum*, contain in their photosynthetic organs chlorophyll-free parenchyma cells full of water. This water tissue consists of living cells of particularly large size and usually with thin walls. The cells are often in rows and may be elongated like palisade cells. Each has a thin layer of parietal cytoplasm, a nucleus, and a large vacuole with watery or somewhat mucilaginous contents. The mucilages seem to increase the capacity of the cells to absorb and to retain water and may occur in the protoplasts and in the walls.

In the underground storage organs there is usually no separate water-storing tissue, but the cells containing starch and other food materials have a high water content. Potato tubers may start shoot growth in air and provide the growing parts with moisture for the initial growth. A high water content is characteristic not only of underground storage organs, such as tubers and bulbs, but also of buds and of fleshy enlargements on aerial stems. In all these structures the storage of water is combined with that of the ergastic substances serving as food reserves.

Many parenchyma cells accumulate phenol derivatives, including tannin. Tanniniferous cells may form a connected system in the plant body, or they may occur singly or in groups. In leaves they are often distributed in continuous zones with no relation to the structural characteristics of the cells occurring in these zones. In stems there may be a concentric zonation of tannin cells. Frequently tanniniferous cells are conspicuous in the outermost zone of the pith, the so-called medullary sheath. Cells containing tannins may accompany the vascular bundles or be included within them. Tannins commonly accumulate in cells located near injuries or infections.

As visible deposits, tannins occur in the vacuoles. The metabolism of carbohydrates and tannins are interrelated, and, according to some studies, starch and tannins mutually exclude each other, except when both are produced in large amounts. Tannin-containing cells may be as readily stimulated to growth and division as cells free of tannins. Cells with tannins, for example, may divide in callus cultures, initiate phellogen, and produce tyloses – proliferations of parenchyma cells into lumina of vessels – or divide with the rest of the ground-parenchyma cells during stem elongation.

Parenchyma cells also store mineral substances and form the different kinds of crystals. Some cells giving rise to crystals retain their protoplasts; others die after the development of crystals.

Cell Walls

Chlorenchyma and many kinds of storage cells commonly have thin primary walls. Some storage parenchyma, however, develops remarkably thick walls. The carbohydrates deposited in these walls, notably the hemicelluloses, are regarded by some workers as reserve materials. Thick walls occur, for example, in the endosperm of date palm (*Phoenix dactylifera*), persimmon (*Diospyros*), *Asparagus*, and *Coffea arabica*. They become thinner during germination. Although the thick-walled storage cells appear to have living protoplasts, the removal of the walls is not necessarily an independent activity of such protoplasts but may be regulated by the embryo.

Relatively thick and often lignified secondary walls also occur in parenchyma cells, particularly in those of the secondary xylem.

Cell Arrangement

Mature parenchyma tissue either is closely packed or is permeated by a more or less prominent air-space system. The storage parenchyma of fleshy axial organs or fruits has abundant intercellular spaces. In contrast, the endosperm of most seeds contains none or only small intercellular spaces. During germination, however, the cells are gradually separated from each other. This structural peculiarity seems to support the notion, mentioned previously, that the mobilization of the reserves in the endosperm is stimulated and regulated, not by the activity of the storing cells themselves, but by the embryo and perhaps also by the peripheral layers of the endosperm.

Chlorenchyma is a familiar example of a tissue having a well-developed aerating system. This structural detail is particularly characteristic of the leaf mesophyll, where the proportion of air by volume may be between 77 and 713 parts in 1,000. Intercellular spaces are abundant in the photosynthetic parenchyma of stems too. In general, they characterize such parenchyma in all groups of land plants from the liverworts and mosses to the angiosperms. Parenchyma removed from light, as in the pith and roots, also has more or less prominent intercellular spaces. On the basis of studies on penetrability of plant organs by gases under pressure, the concept has been introduced that plants possess two kinds of intercellular-space systems, continuous and discontinuous.

The intercellular spaces of vascular plants commonly arise schizogenously or lysigenously. The schizogenous method may give rise to very large spaces, particularly if the cells divide with reference to these spaces. In stems and leaves of *Elodea* and other monocotyledons cells divide parallel to the longitudinal axis of the stem or petiole and perpendicular to the surface of the initial air spaces so that these spaces become bounded by an increasingly large number of cells. Other large air spaces may arise lysigenously or *rhexigenously* (by mechanical rupture; from the Greek *rhexis*, a rending). For example, cortical cells break down in the roots of some Gramineae, Cyperaceae, and other families leaving large lacunae arranged radially or tangentially. Parenchyma tissue with large and abundant intercellular spaces is termed *aerenchyma*.

Air spaces reach a particularly high development in the aquatic angiosperms, both in individual size and in combined volume. In these plants, the aerenchyma forms an elaborate system that appears to be continuous from the leaf to the root. The significance of the development of aerenchyma in aquatic plants has been much discussed in the literature. The continuity of the system through the plant suggests a provision for aeration. The air also gives buoyancy to the plant. But these roles may be incidental to those that are determined by the primary requirement met with in an aquatic environment: a structure that for a given diameter provides strength with the least possible amount of tissue. A honeycomb structure answers this double requirement.

Cell Shape

Parenchyma cells are commonly described as having a polyhedral shape, with the various diameters differing relatively little from each other, but they vary considerably even in the same plant. Many kinds of parenchyma cells are more or less elongated and merge imperceptibly with the so-called prosenchyma cells (elongated cells with tapering ends) in shape and dimensions. Furthermore, parenchyma cells in the mesophyll and other plant parts may be variously lobed, folded, and armed.

Parenchyma cells have served as an object for intensive studies on cell shape, involving special techniques for isolating cells, constructing models of cells, and subjecting the models to statistical analyses. These studies show that, in general, parenchyma cells in relatively homogeneous complexes, with small intercellular spaces or none, have the shape of polyhedra with an approximate average of 14 faces. A geometrically perfect, 14-sided polyhedron with 8 hexagonal and 6 quadrilateral faces has been named the orthic tetrakaidecahedron. This ideal figure is extremely rare among plant cells but is approximated more commonly and more closely than the 12-sided figure with 12 rhombic faces (the rhombic dodecahedron) which the early botanists considered to be the fundamental form of cells in undifferentiated parenchyma. From the beginnings of botany, cells in tissues were regarded as assuming shapes giving the greatest economy of space – minimum surface with maximum volume – and consequently they were interpreted as potential spheres that had a polyhedral shape because of mutual contact and pressure. The rhombic dodecahedron was at first assumed to fit this interpretation best; later, it was found that the orthic tetrakaidecahedron satisfies more conditions in liquid films and

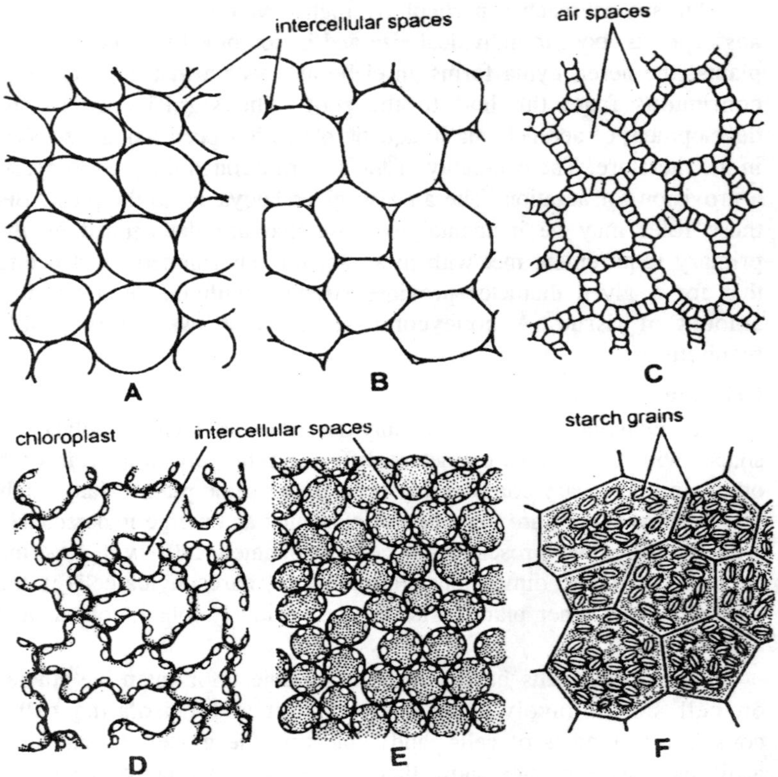

Fig. 4.2. Types of parenchyma: A, B, thin walled parenchyma; C, aerenchyma; D-E, chlorenchyma; D, spongy parenchyma; E, palisade parenchyma; F, storage parenchyma.

also has a greater economy in the surface-to-volume relationship. The rare occurrence of the ideal tetrakaidecahedron among plant cells is understandable. Even in the most homogeneous tissues cells are not of equal volumes, and are not regularly spaced.

The approximation to a 14-sided figure was observed, for example, in parenchyma of various vegetative parts in monocotyledons and dicotyledons, of the carpel vesicles in citrus, and of petioles of a fern. Presence of intercellular spaces, particularly of large ones, reduces the number of contacts. If a tissue contains large and small cells, the number of faces is correlated with size. The small cells have fewer than 14 facets, the large cells more than 14. In *Elodea* cells, the number of faces rises to almost 17 during preparation for cell division, but each new cell has at first less than 13 faces.

By studies of nonliving systems attempts were made to determine some of the possible factors that influence cell shape. In one system - lead shot placed in metal cylinders and subjected to pressure - pressure was the main factor determining the shape. In the other - soap foam bubbles placed in a container and left to adjust themselves - the surface tension played the major role. Plant cells occupied an intermediate position between lead shot and soap bubbles in the characteristics of the three-dimensional configurations. These observations suggest that pressure and surface tension both might play a part in shaping the cells. But there must be other factors as well.

The forces operating in the growth of folded cells (arm-palisade cells) or of cells with internal ridges, as in *Pinus* mesophyll, are obscure. In the ontogeny of stellately armed parenchyma cells lateral tension appears to be one of the factors determining the final shape. Ultrastructural studies of growing stellate cells of *Juncus* indicate that the arms are elongating through their entire length and that the growth of the cell wall is of the multinet type. Certain developmental phenomena, such as increase in length of cells and division of cells, violate the principle of least surfaces; and in cell division the usual position of a new wall indicates no relation to surface tension phenomena.

Origin

The parenchyma tissue of the primary plant body, that is, the parenchyma of the cortex and the pith, of the mesophyll of leaves, and of the flower parts, differentiates from the ground meristem. The parenchyma associated with the primary and secondary vascular tissues is formed by the procambium and the vascular cambium, respectively. Parenchyma may also arise from the phellogen in the form of phelloderm, and it may be increased in amount by diffuse secondary growth.

5

SCLERENCHYMATOUS TISSUES

The term *sclerenchyma* refers to complexes of thick-walled cells, often lignified, whose principal function is mechanical. These cells are supposed to enable plant organs to withstand various strains, such as may damage to the thin-walled softer cells. The word is derived from the Greek and is a combination of *sclerous*, hard, and *enchyma*, an infusion; it emphasizes the hardness of *sclerenchyma* walls. The individual cells of sclerenchyma are termed *sclerenchyma cells*; collectively, sclerenchyma cells form the sclerenchyma tissue. In terms of the entire mechanical system of a plant, collenchyma and sclerenchyma are combined in the physiological concept of *stereome*. The plastic, highly hydrated primary walls of collenchyma, however, distinguish this tissue from sclerenchyma with its hard, elastic secondary walls.

Sclerenchyma cells show much variation in form, structure, origin, and development, and the different types of cells are intergrading. A division of such a graded series of forms into a limited number of categories can be only arbitrary, and the value of such a division depends on the clearness of definition of the criteria upon which it is based. Judged by the variety of systems that have been proposed for the classification of sclerenchyma cells, precise criteria for the separation of the forms have not been established.

Most commonly, the sclerenchyma cells are grouped into fibers and sclereids. Fibers are described as long cells, sclereids as relatively short cells. Sclereids, however, may grade from short to conspicuously elongated, not only in different plants but also in the same individual. The fibers, similarly, may be shorter or longer. Although the pitting

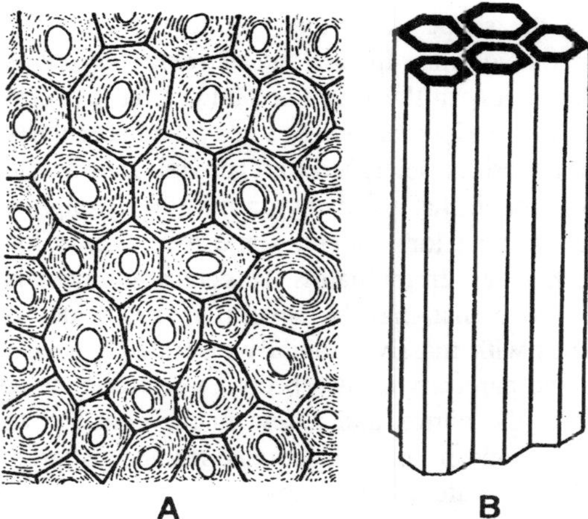

Fig. 5.1. Sclerenchyma: A, in transverse section; B, in longitudinal plane.

is generally more conspicuous in the walls of sclereids than in those of fibers, this difference is not constant either. Sometimes the origin of the two categories of cells is considered to be the distinguishing characteristic: sclereids are said to arise through secondary sclerosis of parenchyma cells, fibers from meristematic cells that are early determined as fibers. But there are sclereids that differentiate from cells early individualized as sclereids (*Camellia*, *Monstera*), and in certain plants phloic parenchyma cells differentiate into fibers when the tissue becomes old and ceases to be concerned with conduction. When it is difficult to assign sclerenchyma cells to one or the other category, the compound term *fiber-sclereid* may be used.

Sclerenchyma cells frequently possess no living protoplasts at maturity. This characteristic, combined with the presence of secondary walls, distinguishes sclerenchyma from parenchyma and collenchyma. But ground parenchyma cells may develop secondary walls (*sclerotic parenchyma*) and fibers and sclereids may retain their protoplasts at maturity. Thus, parenchyma and sclerenchyma are not sharply delimited from one another.

Occurrence and Arrangement in Plant

Fibers occur in separate strands or cylinders in the cortex and the phloem, as sheaths or bundle caps associated with the vascular bundles, or in groups or scattered in the xylem and the phloem. In the stems of monocotyledons and dicotyledons the fibers are arranged in several characteristic patterns. In many Gramineae the fibers form a system having the shape of a ribbed hollow cylinder, with the ribs connected to the epidermis. In *Zea*, *Saccharum*, *Andropogon*, *Sorghum*, and other related genera the vascular bundles have prominent sheaths of fibers and the peripheral bundles may be irregularly fused with each other or united by sclerified parenchyma into a sclerenchymatic cylinder. The hypodermal parenchyma may be strongly sclerified. A hypodermis containing long fibers, some over 1 mm long, has been recorded in *Zea mays*. In the palms the central cylinder is demarcated by a sclerotic zone that may be several inches wide. It consists of vascular bundles with massive, radially extended fibrous sheaths. The associated ground parenchyma also becomes sclerotic. In addition, fiber strands occur in the cortex and a few in the central cylinder. Other patterns may be found in the monocotyledons and patterns may vary at different levels of the stem in the same plant. Fibers may be prominent in the leaves of monocotyledons. Here they form sheaths enclosing the vascular bundles, or strands extending between the epidermis and the vascular bundles, or subepidermal strands not associated with the vascular bundles.

In stems of dicotyledons, fibers frequently occur in the outermost part of the primary phloem, forming more or less extensive anastomosing strands or tangential plates. In some plants no other than the peripheral fibers (primary phloem fibers) occur in the phloem (*Alnus*, *Betula*, *Linum*, *Nerium*). Others develop fibers in the secondary phloem also, few (*Nicotiana*, *Ulmus*, *Boehmeria*) or many (*Clematis*, *Juglans*, *Magnolia*, *Quercus*, *Robinia*, *Tilia*, *Vitis*). Some dicotyledons have complete cylinders of fibers, either close to the vascular tissues (Geranium, *Pelargonium*, *Lonicera*, some Saxifragaceae, Caryophyllaceae, Berberidaceae, Primulaceae) or at a distance from them, but still located to the inside of the innermost layer of the cortex (*Aristolochia*, *Cucurbita*). In dicotyledonous stems without secondary growth the isolated vascular bundles may be accompanied by fiber strands on both their inner and outer sides (*Polygonum*, *Rheum*, *Senecio*). Plants having phloem internal to the xylem may have fibers associated with this phloem (*Nicotiana*). Finally, a highly characteristic location for fibers in the angiosperms is the primary and the secondary xylem where they have varied arrangements. Roots show a distribution of fibers similar to that of the stems and may have fibers in the primary and in the secondary body. Gymnosperms usually have no fibers in the primary phloem, but many have them in the secondary phloem. Cortical fibers are sometimes present in stems.

Classification

Fibers are divided into two large groups, *xylary fibers* and fibers of the various tissue systems outside the xylem, the *extraxylary fibers*. The developmental and topographic relationships of xylary fibers are usually quite precise. These fibers develop from the same meristematic tissues as the other xylem cells and constitute an integral part of the xylem. The assigning of the extraxylary fibers to their proper tissue systems is much less simple and direct. Some of the extraxylary fibers are as definitely related to the phloem as the xylary fibers are to the xylem; in many others the developmental relationship is less clear. The fibers that form continuous cylinders in monocotyledonous stems arise in the ground tissue, at varying distances from the epidermis. They might be classified as cortical fibers except that vascular bundles occur among them and that the limits of the cortex in the monocotyledons are generally vague. The fibers forming sheaths around the vascular bundles in the monocotyledons arise partly from the same procambium as the vascular cells, partly from the ground tissue. The fibers of some vine types of stem, such as *Aristolochia* and *Cucurbita*, occur inside the layer of cells characterized by abundant starch accumulation, the starch sheath, which is commonly interpreted as the innermost layer of the cortex. Thus these fibers are part of the vascular cylinder but are not related to the phloem on a developmental basis.

The fibers located on the outer periphery of the vascular cylinder are frequently classified as pericyclic fibers. The pericycle is interpreted as a tissue separate from the vascular, both topographically and developmentally. But in the stems of the majority of dicotyledons investigated ontogenetically, the phloem abuts on the cortex and there is no distinct tissue between the cortex and the phloem that could be termed pericycle in the usual sense of the word. Nevertheless, in much of the literature the primary phloem fibers are called pericyclic fibers because the developmental relation of these fibers to the phloem is being disregarded or not recognized. It would be desirable to assign all the extraxylary fibers to the tissue systems to which they belong by origin, but such a classification requires developmental studies and also a proper reevaluation of the concept of the pericycle.

The extraxylary fibers are sometimes combined into a group termed *bast fibers*. The word bast was originally applied to fiber strands obtained from the extracambial region of dicotyledonous stems. In its etymology "bast" is related to the verb "to bind." It was adopted as a name for the fiber strands because they were used for binding. The

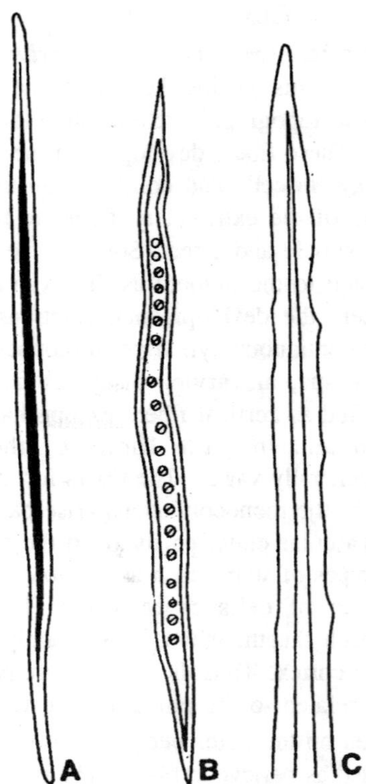

Fig. 5.2. Types of fibres: A, libriform fibre; B, fibre tracheid; C, bast fibre.

bast of the extracambial region of dicotyledonous stems is constituted, in most instances, of fibers of the phloem. In its development the concept of bast followed a double course. In one direction it was extended to cover the extraxylary fibers in various other arrangements than those in the dicotyledonous stems; in the other, it became a specific term for the phloem and was widened to include all the cells of this tissue. In the second usage, the nonsclerified conducting and parenchymatic elements of the phloem received the name "soft bast," the fibers, "hard bast". The term bast fibers is still used for phloem fibers in references dealing with the economic use of plant fibers.

In the present book, the term extraxylary fibers is used for fibers not included in the xylem, and these are grouped as follows: *phloic* or *phloem* fibers, fibers originating in primary or secondary phloem; *cortical fibers*, fibers originating in the cortex; *perivascular fibers*, fibers located on the periphery of the vascular cylinder inside the innermost cortical layer but apparently not originating in the phloem. The term perivascular has been used in the literature in a similar descriptive topographic sense.

The xylem or wood fibers have a common origin, but they are morphologically heterogeneous. They intergrade with the imperforate tracheary elements (the tracheids) and with the parenchyma cells, and certain xylem fibers resemble phloem fibers. Wood fibers are subdivided into two main categories, the *fiber-tracheids* and the *libriform fibers*. The fiber-tracheids are transitional between the tracheids and the extreme, or most specialized, libriform fibers. The libriform fibers resemble phloem fibers, hence, the name libriform. It is derived from *liber*, a Latin term for "*inner bark*," that is, phloem. Some of the xylem fibers form transverse partitions late in their development and are spoken of as *septate fibers*. Phloem fibers also may be septate.

Structure

Extraxylary Fibers

Although a long spindle-like shape is considered typical of extraxylary fibers (and fibers in general), these elements may vary in length, and their ends may be blunt, rather than tapering, or branched. Generally, primary extraxylary fibers are longer than the secondary. The bast fibers of commerce (various extraxylary fibers) vary from a fraction of a millimeter to about ½ meter (primary phloem fibers of ramie, *Boehmeria nivea*).

The cell walls of the extraxylary fibers are frequently very thick. In the phloem fibers of flax (*Linum usitatissimum*) the secondary thickening may amount to 90 per cent of the area of the cell in cross section. The pits are simple or slightly bordered. Some extraxylary fibers have lignified walls, others nonlignified. Flax, hemp, and ramie fibers have little or no lignin and their secondary wall consists of 75 to over 90 per cent cellulose. Some extraxylary fibers, notably those of the monocotyledons, are strongly lignified.

Concentric lamellations may be observed in extraxylary fibers with or without treatment with swelling reagents. In flax fibers the individual lamellae vary in thickness from 0.1 to 0.2 micron and the cellulosic layers are alternately strongly and weakly birefringent and vary in density of staining, probably a reflection of varying densities of the cellulosic matrix in the successive lamellae. In certain types of extraxylary fibers the lamellation appears to result from an alternation of cellulosic and noncellulosic layers. The orientation of the cellulosic microfibrils has also received attention and has been found to vary in fibers from different plants.

Xylary Fibers

Wood fibers typically have lignified secondary walls. They vary in size, shape, thickness of wall, and type and abundance of pitting. The variations in structural details and the corresponding divisions of this group of fibers into categories are best understood if they are considered with reference to the possible evolution of the xylem fibers. Phylogenetically, these fibers are considered as having been derived from an imperforate xylem cell combining the functions of water conduction with that of support, that is, a tracheid. A good indication that fibers and tracheids are phylogenetically related is the occurrence of almost imperceptible gradations between these two cell types in certain angiosperms such as the oak. The gradations suggest the following principal changes during the evolution of the fiber from a tracheid: increase in wall thickness, decrease in length, and reduction in the size of bordered pits. In the extreme condition the pit is simple or nearly so. Of these characteristics, wall thickness and particularly the nature of pitting are used to differentiate between the two main categories of wood fibers, fiber-tracheids, and libriform fibers. Even these criteria do not permit a circumscribed characterization of each category for use in the identification of elements in different species. The limits of the categories are best decided upon by comparing the elements of a given species among themselves. First the tracheid is identified by the resemblance of its pitting to that of the vessel members in the same tissue. Then the limits for the fiber-tracheids are established by the identification of cells with pits whose borders are reduced as compared with those in the tracheids. Finally the cells with simple pits, or essentially so, are classified as libriform fibers.

Commonly, the thickness of wall increases in the sequence of tracheid, fiber-tracheid, libriform fiber. This increase results in an increase in the length of the pit canal. In the fiber-tracheid the pit canals lead into small but evident pit chambers, and the inner apertures are lenticular to slit-like and usually extended beyond the outlines of the border. The libriform fibers also have long slit-like canals, but their pit chambers are much reduced or absent, an expression of the extreme reduction of borders. The inner apertures of the pit-pairs in the fiber-tracheids and libriform fibers are often crossed with each other.

The phylogenetic decrease in length during the evolution of a fiber from a primitive tracheid is a concomitant of a decrease in length of cambial fusiform initials. In a given sample of wood, however, the tracheids are usually shorter and the fibers longer, with the libriform fibers attaining the greatest length. The fibers become longer than the associated tracheids because they undergo a more extensive apical elongation during tissue differentiation.

Fiber-tracheids and libriform fibers may both be septate. The septa are true walls, but they are formed after the deposition of the secondary layers on the longitudinal walls of the element. The formation of septa in the fiber-tracheids of *Hypericum* has been found to involve a regular mitosis followed by cytokinesis.

Septate and nonseptate fibers may retain living protoplasts in the sapwood and serve for storage of starch, oils, and other reserve materials. Thus the living fibers intergrade with xylary parenchyma cells in function. The retention of protoplasts by fibers is an evolutionary advance and is associated with a reduction or an elimination in amount of axial parenchyma in the xylem.

In the reaction wood of dicotyledons (tension wood), the fibers-either fiber-tracheids or libriform fibers-are frequently of the *gelatinous* type. The name gelatinous refers to the appearance of a layer in the secondary wall which has a peculiar cellulosic structure and frequently lacks lignin. The cellulosic matrix has a coarse texture and in some species has been found to be highly crystalline, with the micelles oriented axially. The wall is highly hygroscopic and undergoes striking changes in volume on drying.

Origin and Development

Remarks made in the beginning of the chapter indicate that fibers arise from various meristems. Fibers of the xylem and the phloem are derived from procambium or cambium. In the cambium, the fibers arise from the fusiform initials. Extraxylary fibers other than those of the phloem arise from the ground meristem, but the cells that eventually become fibers early cease to divide transversely and elongate. In some Cyperaceae fibers are epidermal in origin. Protodermal cells divide periclinally and anticlinally and the derivatives differentiate into fibers, except for the outermost, which assume ordinary epidermal characteristics. In plants having fibrous bundle sheaths, part of the fibers may be derived from the procambium and part from the ground meristem. In the shoots of some monocotyledons the proportion of bundle-sheath fibers in a vascular bundle may be very high, or the bundles may consist of fibers only. Since such fibrous bundles may be connected with vascular bundles and since there are bundles with various proportions of fibers and vascular elements, the fibrous bundles probably should be considered as originating from the procambium.

From the developmental standpoint the attainment of great length by fibers is of particular interest. Fibers originating during primary growth have a different kind of development than those formed in the secondary tissues. Primary fibers are initiated before the organ has elongated, and they can reach considerable length by elongating while the associated cells are still dividing. To this symplastic growth may be added apical intrusive growth. In contrast, secondary fibers originate in the part of the organ that has ceased to elongate, and they can increase in length only by intrusive growth. This difference in the method of growth explains why in the same stem primary phloem fibers may attain greater lengths than the secondary. In *Cannabis* (hemp), for example, the primary phloem fibers were found to average about 13 mm, the secondary about 2 mm.

The growth of primary extraxylary fibers in unison with the other tissues in the growing organ causes longer fibers to occur in longer organs. For example, the adult length of primary phloem fibers in *Cannabis* and *Bochmeria* is correlated with the adult length of the internodes. Similarly, the longest phloem fibers in flax occur in the longest stems. In *Sanseviera*, *Agave*, and *Musa* the average length of extraxylary fibers depends on the length of the part of the leaf from which the fibers are obtained.

The great length attained by some primary extraxylary fibers cannot be explained, however, on the basis of elongation by symplastic growth only. In *Sanseviera*, *Agave*, and *Musa* the fibers become 40 to 70 times longer than the meristematic cells from which they arise. In *Luffa* the elongation of the fibers of the fruit at first keeps pace exactly with the increase in size of the fruit itself, but, after the fibers attain about 200 microns in length, their rate of extension becomes greater than that of the fruit as a whole. Thus the fibers appear to have an independent growth in addition to that correlated with the other tissues. Microscopic observations support this assumption. The apices of the fibers long remain thin walled and rich in cytoplasm. They may be serrated and forked because of adjustments to the outlines of adjacent cells. Furthermore, the number of fibers, as determined in transections of stems, gradually increases although longitudinal divisions do not occur. All these observations indicate that the apices of the fibers elongate and intrude among the associated cells. Since this growth occurs in a stem that is still elongating, intrusive growth is probably followed by symplastic growth of the new three-ply wall system formed by the apposition of the new wall of the fiber apex to that of another cell. In flax the phloem fibers have been found growing at both apices, upward and downward, and the length of the stem in which this apical growth of fibers was taking place was estimated to be about 19 mm. Although the secondary phloem fibers fail to attain the same lengths as the primary, they commonly become longer than the cambial initials by apical intrusive growth.

Apical elongation is well substantiated with reference to the secondary xylem fibers. Frequently the occurrence of intrusive growth in secondary xylary fibers may be recognized in the mature shape of the cells. The cells consist of a wider median part corresponding to the unelongated cambial cell and two slender ends that originated during intrusive growth. Pits are limited to the median part in such fibers.

When the extraxylary fibers begin to develop they cease dividing. The nuclei, however, may continue to divide so that the developing fibers become multinucleate. This phenomenon is particularly characteristic of the very long primary phloem fibers. In the same plants the primary phloem fibers may be multinucleate and the shorter, secondary phloem fibers uninucleate.

The prolonged growth in length of the primary phloem fibers results in a highly complicated method of secondary wall development. As has been explained previously, the deposition of secondary walls begins after the primary wall completes its increase in surface. While the primary fibers elongate by symplastic growth in correlation with the surrounding cells, they remain thin walled. Presumably at this stage the entire fiber wall is increasing its surface. Later, during the apical growth stage, the apices of the cells remain thin, whereas the median portions of the cells, which have completed their elongation, begin to form secondary walls. This secondary thickening of the primary phloem fibers was studied in particular detail in *Linum* and *Boehmeria*. In these two plants the secondary wall of the fibers develops in the form of distinct lamellae, each being tubular in shape and growing from the base upward. (Presumably in the earlier stage of the process there is also a downward growth of the tubule while the lower end of the fiber is still elongating. It is conceivable that this end would cease growing first, because it is imbedded in maturing tissues, whereas the upper end is advancing into a still-growing tissue). Thus, several telescoping hyaline tubules arise successively, each tubule being longer than the immediately following one. After the cell ceases to grow at the apex, some of the successively formed layers reach the apex, others stop their growth at lower levels, while new layers arise above them and complete the wall thickening in the upper cell parts. This partial interruption of wall growth results in the formation of compartments, which may be in communication with each other. Apparently the deposition of secondary walls in primary fibers may continue long past the stage at which the cell completes its elongation. In flax and in hemp the fibers of the phloem in the adult plant parts are reported to possess living protoplasts and to continue depositing secondary wall layers.

One of the striking features observed in the growth of the secondary wall in primary phloem fibers is that this wall is not cemented to the primary wall and the successive layers of the secondary wall also appear to be distinct, at least while the cell is not yet mature. In sectioned material the secondary wall of an immature fiber commonly appears detached from the primary and is often separated into two or more layers which are more or less infolded. This infolding and wrinkling might be an artifact, but it also is taken as an indication that the secondary wall layers are in a loose and relaxed state during their formation.

Economic Fibers

Fiber plants have been utilized economically since ancient times. Flax is known to have been cultivated by man as early as 3,000 years B.C. in Europe and Egypt, and hemp at approximately the same time in China. In the technical field, the term fiber usually does not have the strict botanical connotation of individual cells of a certain category of sclerenchyma. In plants in which the commercial fibers originate in the phloem (flax, hemp, ramie, jute), the term fiber denotes a fiber strand. The fibers obtained from monocotyledonous leaves commonly represent vascular bundles together with the associated fibers. Raffia consists of leaf segments of *Raphia* palm; rattan, of stems of *Calamus* palm. The epidermal hairs of cotton seed and of the kapok seed pod are also termed fibers. In still other plants the vascular system of the root (*Muhlenbergia*) or of the entire plant (*Tillandsia*) are used as fibers.

Commercial fibers are separated into hard fibers and soft fibers. The hard fibers are monocotyledonous leaf fibers with heavily lignified walls, hard and stiff in texture. Examples of plants yielding such fibers and ranges of lengths of these fibers in millimeters, according to Harris (1954), are *Agave* species (henequen and sisal, 0.8-8.0), *Musa textiles* (abaca), *Yucca*, and *Phormium tenax*. The soft fibers, that is, bast fibers, may be lignified or free of lignin, but all are soft and flexible. Here are included the phloem fibers of such plants as *Linum usitatissimum*, *Cannabis sativa*, *Corchorus capsularis*. *Boehmeria nivea*, and *Hibiscus cannabinus* (kenaf). The seed hairs of *Gossypium* (cotton) range 16-30 mm in length.

The length of the fiber strand depends on the length of the organ from which the fibers are obtained and on the degree of anastomosing of the strands within the plant. The vascular bundles and fiber strands of the monocotyledonous leaves commonly have a long, straight course with rather small, weak cross anastomoses uniting the bundles with each other. The phloem fiber strands of the dicotyledons, on the other hand, form a network in which the individual strands have no identity as such. It is assumed that the shape and length of the fiber cells, their degree of overlapping, and their connection with each other are factors in the development of strength in fiber strands.

In the preparation of commercial fiber the plants are subjected to a process of partial maceration called *retting* (technical form of the word rotting). In this process the plant material is exposed to a decomposing action by bacteria and fungi. These are allowed to act on the plant parts until the tissues surrounding the fibers are so softened that the fibers can be easily freed mechanically. In the early stages of retting only the intercellular material is affected by pectic enzymes; later the primary wall may be attacked also. An effort is made to discontinue the retting process before the fiber strands are macerated into individual cells. Lignification of cell walls, which usually involves the intercellular substance also, interferes with retting.

Sclereids

Occurrence and Arrangement in Plant

Sclereids are widely distributed in the plant body. The cortex and the pith of gymnosperms and dicotyledons often contain sclereids, arranged singly or in groups. Sclereids are also common components of the xylem and the phloem, where they may intergrade with fibers. In many plants the interfascicular parenchyma cells located between the strands of primary phloem fibers develop lignified secondary walls and differentiate into sclereids which, together with the fibers, form a continuous sclerenchyma cylinder on the outer periphery of the vascular system. The plants in which a continuous sclerenchyma cylinder is present in the primary state may show a disruption of this cylinder when the vascular system surrounded by the sclerenchyma increases in circumference through secondary growth. The breaks in the sclerenchyma cylinder are filled with parenchyma cells which later may differentiate into sclereids (*Aristolochia*).

Many species of plants, particularly in the tropics, contain sclereids in the leaves. The leaf sclereids may be few to abundant. In some leaves the mesophyll is completely permeated by sclereids. In certain species the leaf sclereids occur at the ends of vascular bundles. Sclereids are also common in fruits and seeds. In fruits they sometimes are dispersed in the soft flesh singly or in groups (*Pyrus*, *Cydonia*, *Vaccinium*). In solid layers they constitute hard coverings in the form of shells of nuts or of endocarps of stone fruits. The hardness and strength of the seed coat often result from the presence of abundant sclereids. Solid layers of sclereids occur in the epidermis of some protective scales.

Classification

Sclereids vary widely in the shape, size, and characteristics of their walls. It was inevitable, therefore, that an extensive terminology be developed in the course of study of these elements. Commonly the following categories of sclereids are recognized: *brachysclereids*, stone cells, short, roughly isodiametric sclereids, resembling parenchyma cells in shape, widely distributed in cortex, phloem, and pith of stems, and in the flesh of fruits; *macrosclereids*, elongated rod-like cells, exemplified by sclereids forming the palisadelike epidermal layer of leguminous seeds; *osteosclereids*, bone-shaped sclereids (that is, columnar

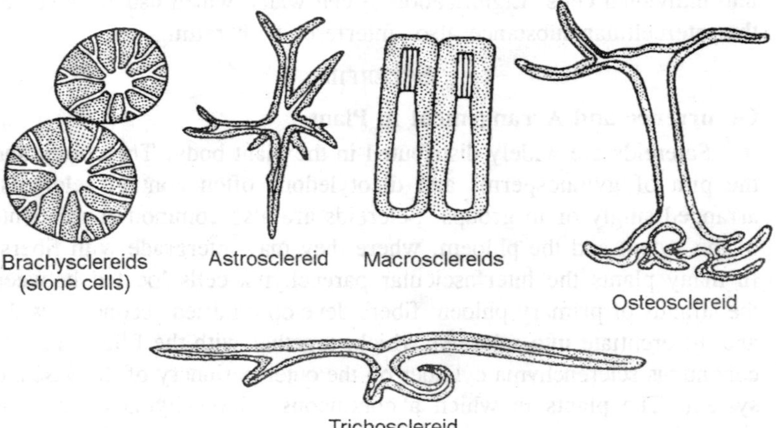

Fig. 5.3. Different types of sclereids.

cells enlarged at the ends), as those present in leaves of many dicotyledons and in seed coats; *astrosclereids*, literally star-sclereids, cells ramified to varied degrees and often found in the leaves of dicotyledons; filiform sclereids, long, slender cells resembling fibers; and *trichosclereids*, branched, thin-walled sclereids resembling plant hairs, with branches extending into intercellular spaces. This classification is rather arbitrary and does not cover all the forms of sclereids known. Its usefulness is further limited by the polymorphism of each of the established categories and by the existence of transitions among the categories. Nevertheless, sclereid forms may be characteristic of species and may therefore be of taxonomic value.

STRUCTURE

The secondary walls of the sclereids vary in thickness and are typically lignified. If the walls are relatively thin, the sclereids cannot be definitely separated from sclerotic parenchyma. The thick-walled forms, on the other hand, may strongly contrast with the parenchyma cells. In many sclereids the lumina are almost filled with massive wall deposits, and the secondary wall shows prominent pits, often with ramiform canal-like cavities. The pits are commonly simple, but sometimes the secondary wall slightly overarches a small pit chamber. The secondary wall often appears concentrically lamellated in ordinary and polarized light. The lamellation may be the result of an alternation of isotropic layers with those composed of cellulose. Crystals are imbedded in the secondary wall of the sclereids in certain species.

In some sclereids the deposition of the secondary walls is uneven. In the macrosclereids of the seed coats of the Leguminosae, for example, most of the secondary deposit is localized on the lateral walls in the end of the cell turned toward the surface of the seed. Furthermore,

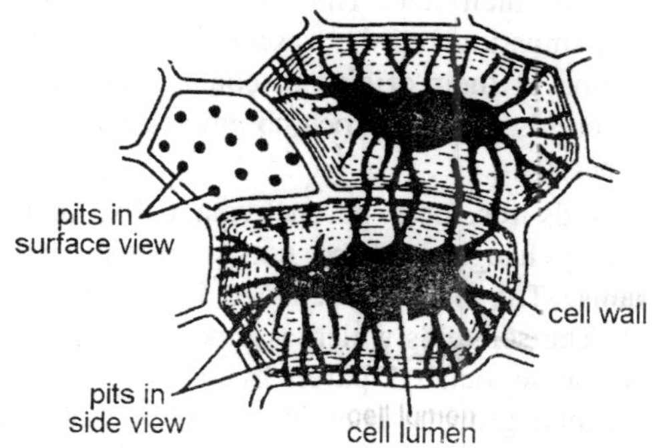

Fig. 5.4. Sclereids in surface and side view with pits.

this thickening is laid down in the form of longitudinal ridges arranged vertically or helically and constricting the lumen of the cell in such a way that it appears star shaped in sections cut at right angles to the long axis of the cell. As was mentioned before, sclereids may either retain their protoplasts on reaching maturity or become dead cells.

Origin and Development

Sclereids arise either through a belated sclerosis of apparently ordinary parenchyma cells (secondary sclerosis) or directly from cells that are early individualized as sclereid primordia. The sclerification of cells in the phloem may occur after the tissue ceases to function in conduction. The *Camellia* leaf sclereids begin their development during the final stage of enlargement of the leaf. The primordia of the sclereids in the *Mouriria* leaf, on the other hand, are clearly evident before the intercellular spaces appear in the mesophyll and while the small veins are still entirely procambial. The trichosclereids of the air roots of *Monstera* develop from cells early set aside by unequal divisions in the rib meristem of the cortex. In one and the same organ, sclereids may arise over an extended period of time, as in *Trochodendron* leaves.

Within the vascular tissues the sclereids develops from derivatives of the procambial and the cambial cells. Stone cells imbedded in the cork originate from the phellogen. Macrosclereids of the seed coats are of protodermal origin. Many sclereids differentiate from ground-parenchyma cells or, if they are individualized early, from ground-meristem cells. In some leaves the parenchyma cells developing into sclereids are part of the spongy mesophyll. In the olive leaf, the filiform sclereids originate in both palisade and spongy parenchyma cells and enlarge several hundredfold, whereas the neighboring parenchyma cells only double or triple their size. The sclereids of *Mouriria*, which are located at the terminations of the vascular bundles in the mesophyll are, from the time of their origin, in contact with procambial cells, and both the sclereids and the procambium arise from the same layer in the ground meristem.

If the sclereids resemble parenchyma cells, their development involves no striking changes in shape, compared with that of the adjacent parenchyma cells. The principal change is the development of the secondary wall. The sclereids which assume shapes strikingly different from those of the associated parenchyma cells show considerable independence in their growth. They invade intercellular spaces, intrude among other cells, and may penetrate the epidermis, sometimes between the guard cells. They becomes very much larger than the initial cells, and assume extraordinary, often grotesque shapes.

The casual relationships in the development of the sclereids constitute a challenging problem for a student of histogenesis. Auxin levels influence the development of sclereids, tending to suppress it at high levels. In some plants sclereid growth seems to be highly individualistic and uncoordinated with the growth of the other cells. In others, the origin and the development of sclereids appear to be a part of the growth pattern of the cell complex as a whole. Surgical experiments on *Camellia* leaves suggest that position may play the major role in inducing sclereid development. In some plants sclereids grow and branch in a relatively compact tissue (*Mouriria*); in others they begin development in a lacunose tissue and grow mainly by sending out protrusions into intercellular spaces (*Monstera*).

The mechanics of growth of the sclereids can best be explained as a combination of symplastic growth in the early stages of their development, when they still grow in unison with the adjacent cells, and apical intrusive growth in the later stages, when they elongate by penetrating into intercellular space and intruding among other cells.

6

VASCULAR TISSUES

The vascular system of the plant is composed of xylem, the principal water-conducting tissue, and phloem, the food-conducting tissue. As components of the vascular system xylem and phloem are called *vascular tissues*. Sometimes the two together are spoken of as the *vascular tissue*. The term xylem was introduced by Nageli (1858) and is derived from the Greek *xylos*, meaning wood.

The physiologic and phylogenetic importance of the vascular system and its prominence among the structural elements of the plant body has led to a taxonomic segregation of plants having such a system into one group, the so-called *vascular plants*, or *Tracheophyta*. This group consists of the Psilopsida, Lycopsida, Sphenopsida, and Pteropsida (ferns, gymnosperms, and angiosperms).

The terms vascular plants and Tracheophyta refer to the characteristic elements of the xylem, the vessels and the tracheary elements in general. Because of its enduring rigid walls the xylem is more conspicuous than the phloem, is better preserved in fossils, and may be studied with greater ease. It is this tissue, therefore, rather than the phloem, that serves for the identification of vascular plants.

Structurally, the xylem is a complex tissue, for it consists of several different types of cells, living and nonliving. The most characteristic components are the tracheary elements, which conduct water. Some of the tracheary elements combine the function of conduction with that of support. The xylem also commonly contains specialized supporting elements, the fibers, and living parenchymatic cells concerned with various vital activities. Fibers may retain their protoplasts in the conducting xylem and thus combine vital functions,

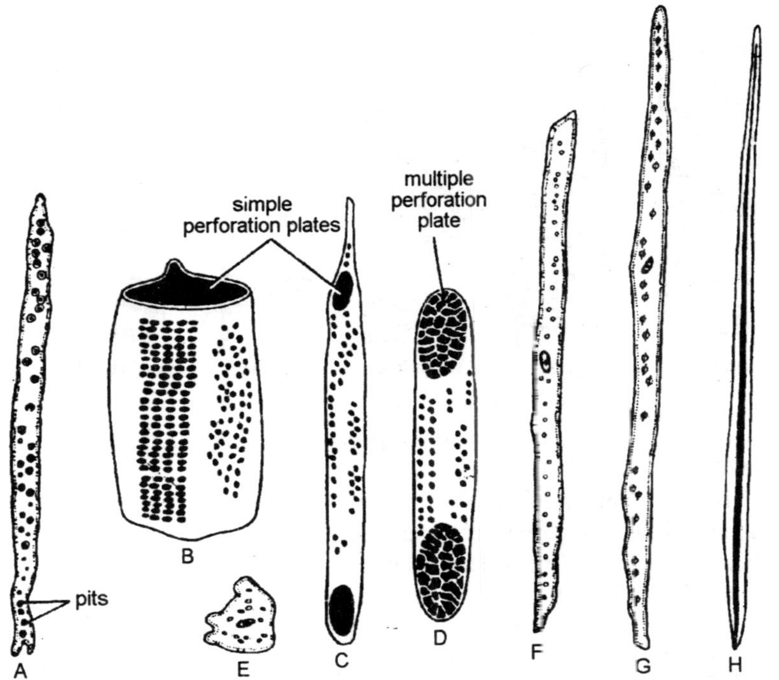

Fig. 6.1. Xylem elements: A, tracheid; B-D, vessels; E, ray parenchyma cell; F, axial parenchyma cell; G, fibre tracheid; H, libriform fibre.

as starch storage, with the mechanical one of support. In certain groups of plants, the xylem includes laticifers. Sclereids differentiated from parenchymatic elements may be present.

The common association of fibers with other xylem and phloem elements brought about the introduction of the term "*fibrovascular tissue*" with reference to the xylem and the phloem. This term is rarely employed now.

CLASSIFICATION

The first xylem differentiates during the early ontogeny of a plant– in the embryo or the post-embryonic stage–and as the plant grows, new xylem continuously develops from the derivatives of the apical meristems. As a result of this growth, the primary plant body, which is eventually formed by the activity of the apical meristems, is permeated by a continuous xylem system (together with the accompanying phloem system) whose pattern varies in different kinds of plants. The xylem differentiating in the primary plant body is the

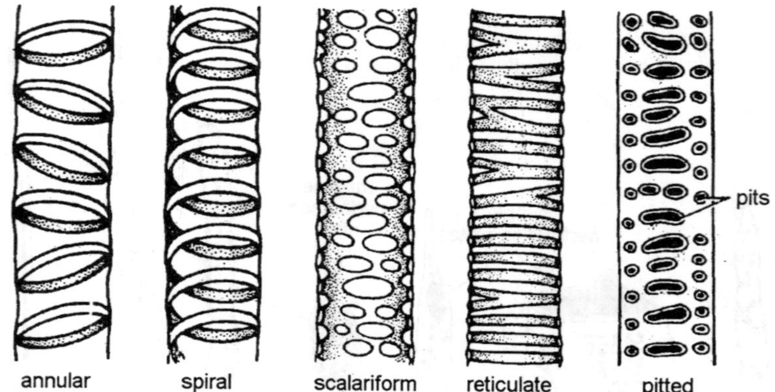

Fig. 6.2. Wall thickenings in tracheary element.

primary xylem. The immediate precursor of this xylem is the *procambium*.

If the plant is of such a nature that, after the completion of primary growth, it forms secondary tissues through the activity of the *vascular cambium*, the xylem produced by this meristem constitutes the *secondary xylem*.

The histologic characteristics of the two kinds of xylem are given later in this chapter. Depending on the nature of the plant, the primary xylem is more or less distinct from the secondary, but in many respects the two kinds of xylem intergrade with each other. Therefore, to be useful the classification into primary and secondary xylem must be conceived broadly, relating the two components of the xylem tissue to the development of the plant.

Vascular Cambium

Location in the Plant

The vascular cambium is the lateral meristem that forms the secondary vascular tissues. It is located between the xylem and the phloem and, in stems and roots, commonly has the shape of a cylinder. When the secondary vascular tissues of an axis are in discrete strands, the cambium may remain restricted to these strands in the form of strips (*Cucurbita*). It also appears in strips in most petioles and leaf veins that show secondary growth.

Cell Types

The tissues derived from apical meristems contain many cell types which differ strikingly from the meristematic cells in shape and size.

In contrast, there is a general resemblance between the cambium cells and their derivatives, and the shape and arrangement of cells in the secondary xylem and the secondary phloem are foreshadowed in the shape and arrangement of the cambial cells.

The vascular cambium contains two types of cells; elongated cells with tapering ends, the *fusiform initials* (that is, spindle-shaped initials), and nearly isodiametric, relatively small cells, the *ray initials*. The exact shape of the fusiform initials of *Pinus silvestris* has been determined as that of long, pointed, tangentially flattened cells with an average of 18 faces. The fusiform initials give rise to all the cells of xylem and phloem that are arranged with their long axes parallel to the long axis of the organ in which they occur; in other words, they form the longitudinal or axial systems of xylem and phloem. Examples of elements in these systems are tracheids, fibers, and xylem-parenchyma cells in the xylem; sieve cells, fibers, and phloem-parenchyma cells in the phloem. The ray initials give origin to the ray cells, that is, elements of the transverse or ray system of the xylem and the phloem.

These initials differ from each other most notably in length and volume, the fusiform cells being much larger than the ray initials.

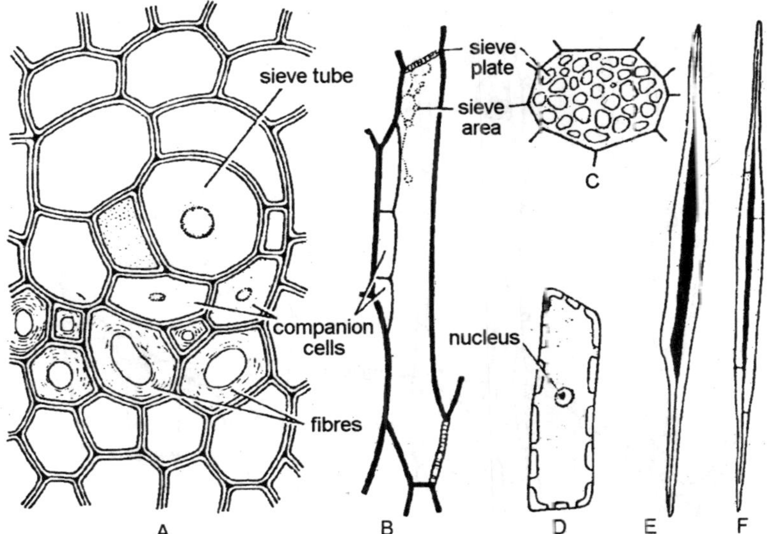

Fig. 6.3. Phloem elements: A, B, sieve tube members in transverse and longitudinal sections respectively; C, sieve plate; D, phloem parenchyma cell; E, F, aseptate and septate phloem fibres respectively.

However, in one dimension, the radial, the ray initials surpass the fusiform. In the 60-year-old stem both kinds of initials are larger than in the 1-year-old stem. The initials are uninucleate, and, although the nuclei of the fusiform initials may be markedly larger than those of the ray initials, their volumes do not increase in proportion to the cell volumes so that the ratio of nuclear volume to cell volume is much smaller in the fusiform cells.

The fusiform initials show a wide range of variation in their dimensions and volume. Some of these variations depend on plant species. The following figures, expressed in millimeters, exemplify

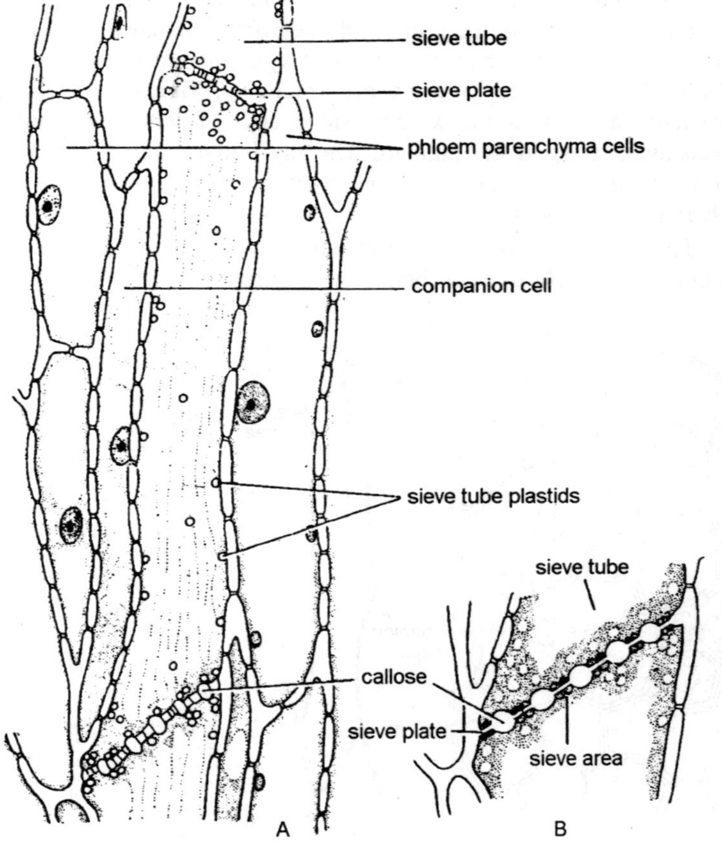

Fig. 6.4. Phloem elements: A, phleom tissue from the stem of Nicotiana tabacum showing sieve tubes, companion cells and phloem parenchyma; B, magnified view of a sieve plate.

differences in the lengths of fusiform initials in several plants: ***Pinus strobes***, 3.20; ***Ginkgo***, 2.20; ***Myristica***, 1.31; ***Pyrus***, 0.53; ***Populus***, 0.49; ***Fraxinus***, 0.29; ***Robinia***, 0.17. Fusiform initials vary in length within species, partly in relation to growth conditions. They also show length modifications associated with developmental phenomena in a single plant. Generally, the length of fusiform initials increases with the age of the axis, but after this length reaches a certain maximum it remains relatively stable. The changes in the size of fusiform initials bring about similar changes in the secondary xylem and phloem cells derived from these initials. The ultimate size of these cells, however, depends only partly on that of the cambial initials, because changes in size also occur during the differentiation of cells.

The cambial cells are highly vacuolated. Their walls have primary pit-fields with plasmodesmata. The radial walls are thicker than the tangential walls particularly during dormancy, and their primary pit-fields are deeply depressed.

Cell Arrangement

During active growth in the cambium, the initials and their immediate derivatives form a zone of similar unexpanded meristematic cells, the *cambial zone*. As seen in transections the cells in the cambial zone are arranged in radial series. On either side of the cambial zone, cambial derivatives expand and gradually assume the characteristics of the various xylem and phloem cells. Experiments with strips of bark partially detached from the stem indicate that mutual pressure of tissues is important in controlling the orderly pattern of differentiation of the cambial products. The prevailing concept is that the initials are arranged in one layer, one cell in thickness. In a strict sense, only the initials constitute the cambium, but frequently the term is used with reference to the cambial zone, because it is difficult to distinguish the initials from their recent derivatives.

In tangential views the arrangement of cambial cells shows two basic patterns. In one, the fusiform initials occur in horizontal tiers with the ends of the cells of one tier appearing at approximately the same level. Such meristem is called *storied* or *stratified cambium*. It is characteristic of plants with short fusiform initials. In the second type, the fusiform initials are not arranged in horizontal tiers, and their ends overlap. This type is termed *nonstoried* or *nonstratified cambium*. It is common in plants with long fusiform initials. Intergrading types of arrangement occur in different plants. The nonstratified type is considered to be phylogenetically more primitive than the stratified.

The former is found in fossil pteridophytes, in fossil and living gymnosperms, and in structurally primitive dicotyledons; the latter, in highly specialized dicotyledons. In primitive cambia the initials vary in length more than in the more specialized meristems.

Cell Division

The phloem and the xylem are formed by tangential (periclinal) divisions of cambial initials. The vascular tissues are laid down in two opposite directions, the xylem cells toward the interior of the axis, the phloem cells toward its periphery. The consistent tangential orientation of the planes of division during the formation of vascular tissues determines the arrangement of cambial derivatives in radial rows. Such radial seriation may persist in the developing xylem and phloem, or it may be disturbed through various kinds of growth readjustments during the differentiation of these tissues.

Tangential divisions that occur during the formation of xylem and phloem cells are not limited to the initials but are encountered also in varied numbers of derivatives, sometimes several times within the progeny of the same derivative. During the winter rest, xylem and phloem cells mature more or less close to the initials; sometimes only one cambial layer is left between the mature xylem and phloem elements. But some vascular tissue, frequently only phloem, may overwinter in an immature state in the cambial zone.

As the xylem cylinder increases in thickness by secondary growth, the cambial cylinder enlarges in circumference. The principal cause of this enlargement is the increase in the number of cambial cells in a tangential direction, followed by a tangential expansion of these cells. In stratified cambia the increase in the number of fusiform initials occurs by radial (anticlinal) longitudinal divisions. In nonstratified cambia, however, the fusiform initials divide by more or less oblique anticlinal walls (the so-called pseudotransverse walls), and then the resulting cells elongate at their apices (apical intrusive growth) until each cell is as long as, or longer than, the mother cell. Some investigators use the term *multiplicative* divisions with reference to the anticlinal divisions that increase the number of initials, as contrasted with the *additive*, periclinal, divisions that contribute cells to the xylem and the phloem. In the longitudinal divisions of the cambial initials and their derivatives, cytokinesis is a process extended in time and space. The cell plate is initiated between the two new nuclei and then spreads through the entire length of the cell, preceded by the phragmoplast fibers.

Developmental Changes

Detailed studies of the vascular cambium of conifers have shown that the increase in circumference of the meristem is accompanied by profound changes in size, number, and arrangement of cells. Both the fusiform and the ray initials are greatly multiplied in number. The fusiform initials enlarge notably along their tangential diameters, whereas the ray initials become only slightly larger in this dimension. There is also a remarkable increase in the length of the fusiform initials. The increase in number of the fusiform initials, as seen in transections, results from intrusive apical elongation following the multiplicative oblique radial divisions. Since the rays in the pine are mostly uniseriate (one cell in width), the increase in the number of ray initials, is a result, not of divisions of the existing ray initials, but of the addition of new ray initials.

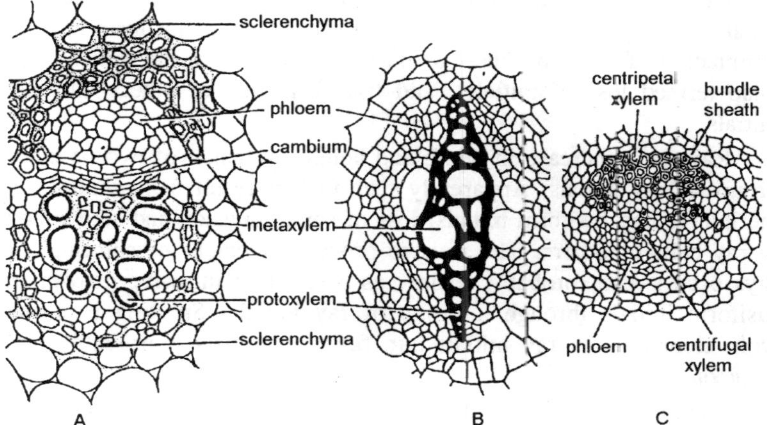

Fig. 6.5. Position of protoxylem in vascular bundles: A, endarch; B, exarch; C, mesarch.

New ray initials arise from fusiform initials or their segments. These additions maintain a relative constancy in the ratio between the rays and the axial components during the increase in the circumference of the vascular cylinder. New rays have fewer cells than older. A ray may be one cell wide and one cell high in the beginning; later, the initial divides or more initials are added to the first. The ray thus increases in height and may increase in width if multiseriate rays are characteristic of the plant. Some investigators report that new ray initials may be cut off the apices or cut out of the sides of fusiform initials. In a herbaceous species of *Hibiscus* rays were found to be derived by transverse divisions of one of the two fusiform cells resulting

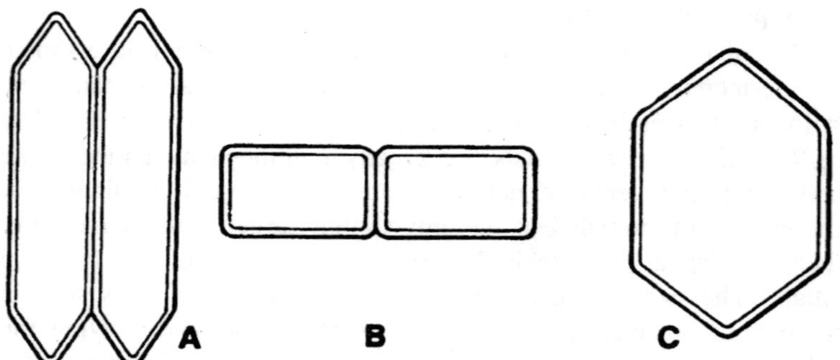

Fig. 6.6. Vascular cambium: A-B, two fusiform initials in tangential and longitudinal sections respectively; C, a ray initial in tangential section.

from an anticlinal division of a fusiform initial. Studies on certain conifers and *Liriodendron* show that ray initiation in these plants is usually a complicated process involving subdivision of fusiform initials, elimination of some products of these divisions from the initial layer (also termed loss of initials), and transformation of others into ray initials.

Rays may increase in width and height by fusion of two or more groups of ray initials. Apparently such fusions result from changes in the intervening fusiform initials, loss of some, division and conversion to ray cells of others. The reverse process, division, or splitting, of rays occurs also, probably mainly as a result of intrusive growth of fusiform initials through a group of ray initials. Splitting resulting from elongation of ray initials into fusiform initials is probably less common.

The phenomenon of loss of initials has been studied extensively in the conifers; less so in the dicotyledons. The method employed is commonly that of following the changes in radial files of cells in the xylem or the phloem as seen in serial tangential sections and reconstructing from these changes the past events in the cambium. Transections are used for confirmation since they reveal loss of initials by discontinuities in the radial files of cells.

The loss of fusiform initials is usually gradual. Before a cell is eliminated from the initial layer, its precursors fail to enlarge normally possibly even diminishing in size through loss of turgor – and become abnormal in shape. Periclinal divisions separate such cells into smaller and larger derivatives, the smaller of which remains in the initial layer. Thus, gradually the cell in the initial position is reduced in size, particularly in length. Some of the short initials are lost from

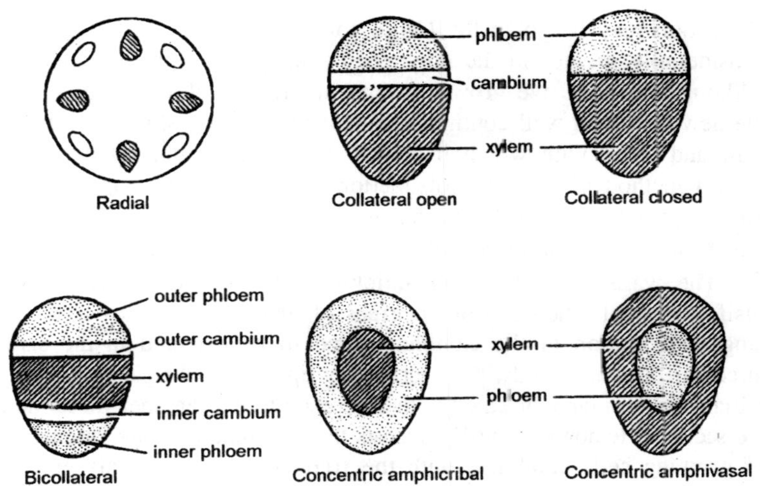

Fig. 6.7. Vascular bundles.

the initial layer by maturing into xylem or phloem elements; others become ray initials with or without further divisions. Ray initials, too, may disappear from the cambium. The space released by a declining initial is filled by the intrusive growth of the surviving initials.

The elimination of fusiform initials is associated with the anticlinal divisions giving rise to new initials. These divisions apparently would result in an overproduction of initials if they were not accompanied by the extensive loss of cells. The loss appears to be related to vigor of growth. In *Thuja occidentalis* the survival rate was found to be 20 per cent when the annual xylem increment was 3 millimeters wide, whereas at the lowest growth rates the rate of loss and that of new production were almost equal. The accommodation to the increase in girth probably occurred through elongation of cells. In *Pyrus communis* the loss was calculated to be 50 per cent among the newly formed fusiform initials. Among some 300 radial files of cells of the axial system of *Liriodendron* phloem, examined in serial sections through a layer of tissue about 400 microns in radial depth, the loss of their initials by maturation and by conversion into rays nearly equaled the addition of new tiers by anticlinal divisions of fusiform initials. Considerable evidence indicates that in both conifers and dicotyledons longer initials tend to survive and that extensive contact of these initials with the rays increases the chance of survival.

As was mentioned previously, anticlinal divisions are followed by intrusive elongation of the resulting cells. The direction of this elongation may be polar. In *Thuja*, for example, it was found to be considerably greater in the downward than in the upward direction. Although the intrusive growth occurs at the tips of cells, apparently the newly formed wall continues to expand so that a slip between this wall and those with which it comes in contact is not ruled out. In such a method of growth a distinction between intrusive growth and gliding growth can hardly be made. The tips of growing cells have thin walls and contain cytoplasmic accumulations.

The walls formed during anticlinal divisions in relatively long fusiform initials show various degrees of inclination but, as seen in a tangential section of the cambium, tend to be oriented in the same direction. In other words, the overlapping tips of the elongating fusiform initials are similarly oriented with reference to one another throughout the section. Hejnowicz (1961) suggests that this unidirectional orientation of growing cells combined with the frequent loss of initials may be causally related to the spiral arrangement of the cambial cells and of the derived vascular cells.

All the studies reviewed above deal with the vascular cambium of stems. In *Larix europea* the cambium of the root was found to have more limited elimination of initials, weaker intrusive growth, and more variable orientation of the anticlinal walls than that of a stem of similar age.

Loss of fusiform cambial initials might be less typical of herbaceous than of woody species. In *Hibiscus lasiocarpus*, a perennial herb, the loss of such initials was limited to that associated with ray formation.

Seasonal Activity

The secondary growth originating in the vascular cambium is closely connected with the activities of the primary parts of the plant body and shows fluctuations in relation to the changes in the physiologic state of the plant. Annual or biennial herbaceous plants commonly have a regular sequence of vegetative stage, reproductive stage, somatic death, and seed dispersal. Between the vegetative and reproductive stages, the plant body may attain various dimensions and its vascular tissues may be increased in amount by secondary growth. This growth ceases, however, during the transition to the reproductive stage, since the cambial activity appears to be closely associated with the vegetative stage. In perennial plants there is a repetition of vegetative and reproductive phases, without somatic death of the whole individual. As is well known, in woody species growing in temperate regions periods of growth and reproduction alternate with periods of relative inactivity during the winter. The seasonal periodicity finds its expression in the cambial activity also. Production of new cells by the vascular cambium slows down or ceases entirely during the rest period, and the vascular tissues mature more or less closely to the initial layer.

In the spring the winter rest period is succeeded by a reactivation of the cambium. From the anatomic aspect, reactivation may be divided into two stages: (1) expansion of the cambial cells in the radial direction ("swelling" of the cambium) and (2) initiation of cell division. The radial enlargement is accompanied by a weakening of the radial walls so that a slight external force applied to the stem will cause these walls to break.

The separation of the bark from the wood resulting from such a break is commonly called slipping of the bark. Slippage may also be induced later, during cell division and tissue differentiation in the cambial zone. At this time, however, the break occurs most commonly through the young xylem where the tracheary elements have attained their maximum diameters but are still without secondary walls. In evergreen species, as in *Citrus*, the histologic aspects of slippage appear to be less definite.

The cell divisions occurring during the second stage or reactivation are the additive periclinal divisions. Information on the exact sequence of these divisions is meager, especially with regard to the timing of xylem- and phloem-cell formation. In *Thuja occidentalis* periclinal divisions were found to be concentrated first in the xylem mother cells; then they occurred in the initial layer. Formation of phloem cells began when the divisions in the xylem mother-cell zone were at maximum and continued till cambial activity was terminated. In *Pyrus* additive cambial divisions began when the overwintering phloem mother cells were differentiating. Most of the first new cells were added to the phloem. Xylem cells were formed later, when a considerable amount of differentiated phloem was already present. In both conifers and dicotyledons, the annual xylem increment is characteristically wider than the corresponding phloem increment.

The resumption of cambial activity in the spring has often been found to be related to the new primary growth from buds. In many dicotyledons cambial activity of the stem begins beneath the emerging new shoots and spreads from here basipetally toward the main branches, the trunk, and the root. As an example, the data obtained with *Acer pseudo-platanus*, growing in England, may be cited. In this tree 9 to 10 weeks elapsed between the inception of xylem differentiation in the twigs (late in April) and that in the roots (early in July). Activity ceased in the same order. The formation of xylem stopped in the twigs in late July, in the roots, in later September. Thus, 8 to 9 weeks elapsed between the cessation of cambial activity in the branches and that in the roots. *Acer* exemplifies cambial behavior in dicotyledons with diffuse - porous wood (with vessels of similar width distributed throughout annual increment).

The degree of development of the bud associated with cambial reactivation is variable; the bud may be still closed, or just opening, or obviously growing. Many conifers and the dicotyledons with ring-porous type of wood (characterized by aggregation of numerous wide vessels in the early wood) show an early rapid spread of cambial reactivation throughout the trunk in the presence of little or no bud growth. Cessation of cambial activity follows approximately the same order as the reactivation. The inception of cambial reactivation beneath the new shoots and its basipetal progress explains why, in a dicotyledon, the portion of a twig that might be left in pruning above the uppermost bud dries up and forms a "snag".

The initial stimulation of cambial activity has been repeated related to the transport of growth substances in basipetal direction from the growing buds. The maintenance of cambial activity, however, appears to be independent of the extension growth of the new shoot. In *Robinia pseudoacacia* continuation of cambial activity was found to be dependent on exposure of the leaves to long-day conditions. The vascular cambium can be stimulated into activity by wounding also, possibly in relation to formation of wound hormones as a result of injury.

Intensity and amount of cambial activity varies in different seasons. Some of these variations are induced by environmental conditions, whereas others depend on an inherent rhythm of growth. At the beginning of secondary growth in pine trees, for example, the mean width of annual increments first rises then falls from one season to the next in the same internode. Other variations in ring width that may be present are overshadowed by the basic pattern. A common feature is the decrease in the rate of growth in thickness with increasing age of the tree.

In some studies the multiplicative anticlinal divisions in the initial layer were found occurring toward the end of the growth season when the cambial zone was of minimum width. Through the years, these divisions may occur more or less frequently in the same initial position. In *Thuja* the intervals between successive divisions were 1 to 8 years with an average of 3.7 years, and the frequency was reduced with increased age of the tree. The elongation of the new surviving initials resulting from anticlinal divisions begins directly after the divisions and continues for several years. In *Thuja* this elongation follows a familiar growth pattern in that it occurs rapidly at first, then at a decreasing rate.

The restriction of anticlinal divisions to the last part of the season of growth is not a constant feature. In *Picca*, these divisions occurred throughout the growth period during the earlier years of growth of the stem but became limited to the terminal part of the season in later years, when narrower annual rings were being produced.

7

Anatomy of Stem

The essential character of an epidermis is that it is a protective layer, or in other words, that it is a barrier between the internal organization of the plant and the environment. This protection is not exclusively concerned with water loss, though that is a very important factor, and the surface layer of some submerged aquatics may be truly classed as an epidermis even though it is not a protection against desiccation. But the emphasis on the barrier function excludes from this category secretory surfaces such as those of glands, and absorptive surfaces such as those of the young root.

Furthermore, a true epidermis is a primary layer, developed from the apical meristem, and the name does not apply to the exodermis of roots or the periderm of older stems, which are both secondary. On the other hand, certain specialized cells of the surface layer, such as hair cells and sclereids, which may not be primarily protective in function, are none the less morphologically part of the epidermis and we shall treat them as such. Epidermal cells are usually described as tabular in form which means somewhat flattened radially, but when viewed from the outer surface they are seen to be nearly always vertically elongased, and they are by no means always radially flattened. In outline they vary greatly, but the variation is much less in the epidermis of the stem that in the leaf. Growth of the primary stem is chiefly elongation, and the epidermal cells in the young state are consequently elongate and narrow.

The radial walls may be somewhat wavy, which is apparently the result of continued expansion of the cells after growth of the stem has ceased. The same factor may result in the outer surfaces being convex,

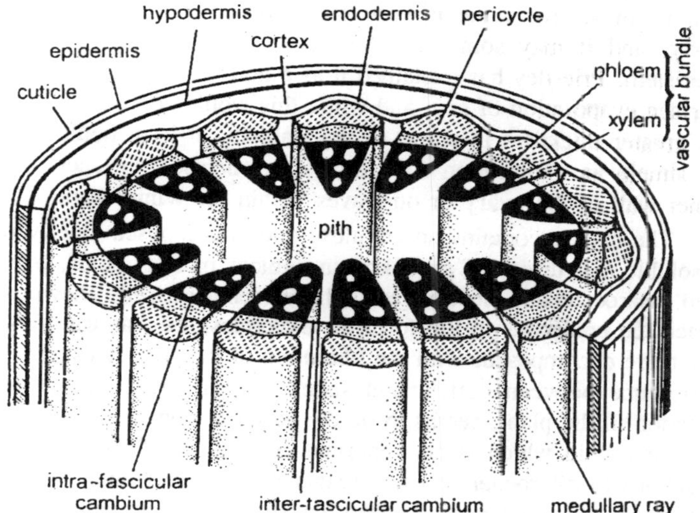

Fig. 7.1. Three-dimensional view of a typical dicotyledonous stem showing the arrangement of primary tissues.

domed or even papillate. The radial walls are not only undulating, when viewed from the exterior, but are thin and have numerous large pits. This implies that lateral movement of water in all directions in the epidermis must be relatively easy, and it is probable that a dangerous loss of water at any one point of the surface may be compensated by the tangential flow of water towards the threatened spot from other parts of the epidermis, which thus acts as a first line of defence, before the radial flow from the xylem can become effective. That epidermal cells do, in fact, part with considerable amounts of water at times, is shown by the great changes of volume they can withstand without injury, especially in leaves. When this occurs the radial walls shrink and the cells become flattened, but they rapidly expand when equilibrium is restored. The epidermal cells of some plants (e.g., Erica and Daphne) have greatly thickened, mucilaginous inner walls, thus providing for increased water storage.

The physiological as well as the mechanical cohension of the epidermal cells is increased by the absence of any spaces at the angles between them. The outer surface of the epidermis is covered by the cuticle. This layer, which varies in thickness in general relationship to the moisture conditions in the environment, is extra-cellular and continuous and therefore helps to bind the epidermal cells together. It is tough and elastic, only slightly permeable to either water or gases, and its protective value is very great. Its development begins very early, and it may sometimes be traced even on the tunica in the meristem. Priestley has explained it as a non-volatile residue from the surface evaporation of sap, and while this would serve to account for its greater thickness in plants of dry habitats, it is difficult to accept so simple an explanation for cuticles on enclosed surfaces, such as inner wall of the ovary or on leaves within the winter bud.

The cuticle contains no cellulose and is composed chiefly of the insoluble anhydrides of an unknown number of fatty acids, of which two, stearocutic and oleocutic acids, have been described. They probably reach the epidermis in the form of glycerides, i.e., as true fats, and are there decomposed, with the liberation of the free acids. The chief difference between cuticle and suberin, apart from their different location in the plant, seems to be the absence from the former of the phellonic acid which is an important constituent of the latter. Their reactions to microchemical test reagents are very similar. Continued growth of the epidermis, besides causing the wavy outlines of the cells, also throws the cuticle covering into wrinkles, which often form a minute pattern over the outer surface of each cell. The cuticle often extends downwards for some distance along the radial wall, forming distinct wedges between the cells, but it never completely separates them. Below the cuticle proper lies the *cutinized layer*, which is formed by the cellulose outer walls of the cells, more or less impregnated with cutin. It is sharply separated from the true cellulose wall next to the cell lumer.

The cutinized layer is not invariably present, but when it occurs it is frequently much thicker than the true cuticle and may be the chief protective layer. It often accompanies the cuticle in forming prominent wedges between the radial walls of the cells. Beneath the epidermis there is often, in plants of dry or exposed habitats, a second, or even a third, specialized layer, the walls of which are not cutinized but may be greatly thickened with cellulose or lignin, or else may be quite unthickened. It may augment both the protective and the water storage functions of the epidermis, and it is referred to either separately as the *hypodermis*, or both layers together may be termed a *multiple epidermis*.

Strictly speaking, the latter name should not be used unless it is known that both layers of cells have originated from the dermatogen by tangential cell divisions. As this cannot often be proved it is best to retain the general term hypodermis. The epidermis of young, green stems contains a number of *stomata*, which resemble those on the leaves. They are not normally so numerous as on the leaves, but in plants with reduced or abortive leaves they may be the only means for gas exchanges with the atmosphere, and in such cases they have the same importance in photosynthesis as those on normal leaves.

Epidermal Outgrowths

Outgrowths of the epidermis take many forms and are included under the general term *trichomes*, which covers not only true hairs, but other modified structures such as glands and prickles. They are, of course, not confined to the stem, but no distinction can be drawn between organs in this respect, since identical trichomes may be produced at all points on the surface of the shoot. Their distribution is, however, often restricted to definite lines or areas and may be a mark of distinction between related species. Despite the immense variety among trichomes, a complete anatomical series may be traced between the simplest, which are merely prolongations of a single epidermal cell, and massive structures which arise from groups of cells and involve also the sub-epidermal tissues and may even receive one or more vascular bundles. These extreme cases have sometimes been distinguished as *energences*, but in spite of their apparent differences no sharp line can be drawn between them and the simpler trichomes. Hairs can be classified under three main types: simple, branched, and peltate. Simple and branched hairs may consist of one or of many cells.

The branched types are sometimes quite elaborate, like small bushes; or they may be flattened, the branches forming a rosette around a short stalk-cell. The latter type is related to the peltate or scale hairs, which are similar to flattened branched hairs but with the branches cohering into a disc. These discs, if they overlap, as in the Eleagnaceae, from a complete protective armour over the surface of the epidermis. Secretory hair glands may be simply swollen cells which are usually water stores; or they may be compound, with long or short stalks bearing a head consisting of either one large secretory cell, or a more or less peltate group of cells. Prickles are classified as trichomes, no matter how massive they may be, whenever it is clear that they arise from the epidermis and are not modifications of any other organ.

As a matter of fact, the large prickles of *Rosa* and the vascular prickles on the fruit of Horse Chestnut are connected with simple hairs by all gradations of finer prickles. Prickles of all grades may even be present together on the same stem, as in the Burnet Rose (Rosa spinossissima), and it is notable that in some Roses the prickles at the nodes are often much larger and stronger than the others. The so-called "endogenous prickles" on the stems of some Palms are, however, really modified adventitious roots and are therefore properly classified as thorns. The very large and prominent spines of the Cacti are of doubtful nature, having been interpreted both as modified leaves and as trichomes. They grow little humps, called areolae, from which the lateral buds, if any, usually arise, and the balance of evidence points to the spines being really trichomes.

Cortex

Between the epidermis and the stele lies the zone of the parenchymatous *cortex*. In stems it is rarely as broad as in the root, for the vascular tissues, probably for mechanical reason lie much nearer to the periphery than in roots. The underground stems of Monocotyledons usually have a fairly broad cortex, but monocotyledonous aerial stems, such as that of the Maize so often used as an anatomical type, are as a rule only temporary flowering shoots and may have a very narrow cortex or none at all. Such stems may be regarded as being either entirely stelar or else as having no stele. There is at any rate, in many cases, no stelar boundary visible. Where a cortex is present it often contains numerous and small marginal leaf traces. The outer zones of the cortex frequently form a *collenchyma* either in a continuous band or, in angular stems such as those of Labiatae and Umbelliferae, developed only below the ridges.

Collenchyma is definitely a mechanical, strengthening tissue. It has this great advantage over sclerenchyma, that its cells are living and growing and hence it is able to adapt itself to the growth of young organs. It is chiefly characteristic of such organs, nevertheless in short-lived structures such as petioles or the herbaceous stems of Dicotyledons it may remain as the permanent mechanical tissue. The aerial stems of Monocotyledons, on the other hand, depend for mechanical support more on the development of the sclerotic sheaths around their vascular bundles than on collenchyma. As collenchyma cells remain alive it follows that their cell walls cannot be uniformly thickened, for this would cut them off from all external supplies. The thickening material, which is of cellulose, is accordingly distributed in such a way that thin walls are left at certain parts of the cell to allow of inter-communication. Frequently the thickening appears only at the angles of the cells, while in other cases it may be all on one side or on two

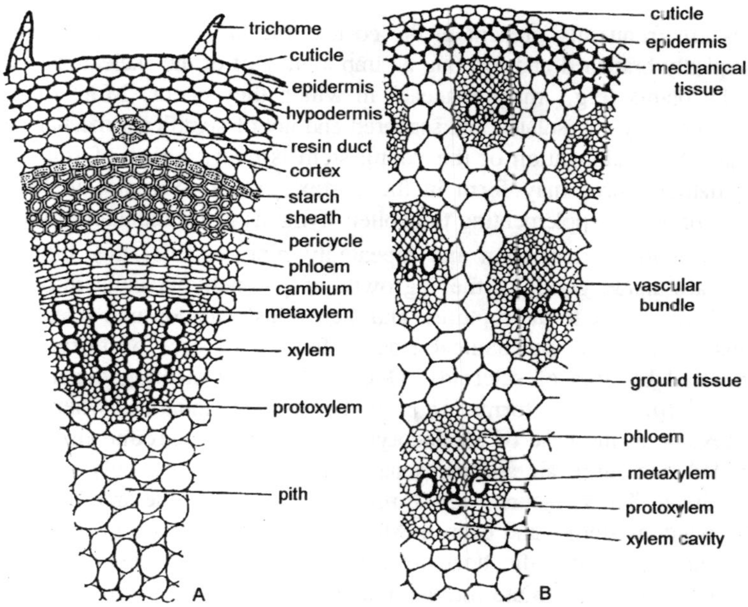

Fig. 7.2. Portions of transections of young stems: A, sunflower stem; B, maize stem.

sides, an arrangement which also has the advantage of increasing the plastic and yielding character of the tissue as a whole. The cells are somewhat elongated vertically, but never to the extent of woody fibres, and they may be regarded as mainly parenchymatous. Cortical cells usually contain functional chloroplasts since light can penetrate a distance up to 100 μ from the surface. True palisade tissue, comparable with that in the leaf, may be found in the cortex of some xerophytes, either in succulents of the Cactus type or in switch plants, like the Broom in fact, wherever the stem has taken over the function of photosynthesis from the leaves, either owing to the reduction of the latter to mere scales or to their complete disappearance. Where *chlorenchyma*, i.e., chloroplast-containing tissue, is present, whether of the palisade form or not, there will usually be some water storage tissue closely adjacent to it in the cortex. The cells of such a tissue are large and thin-walled, and they not infrequently contain mucilage. An it is nearly always in xerophytes that chlorenchyma occurs in the cortex the advantage of a water reservoir close at hand is obvious.

Indeed, in the Cacti, where the stele is often relatively slender, the greater part of the stem tissue is water-storing cortex, only a narrow outer zone being chlorenchymatous. Internal glands and sclereid cells are frequently present in the cortex, the latter being sometimes complexly branched and forming an important part of the skeletal system of the plants (e.g., magnoliaceae) in which they occur. More often, however, they are isolated and scattered and not of significant mechanical value. After elongation of the young stem is over, a ring of sclerenchymatous fibres may form in the cortex, as for example, in Oak, Birch or Ash, supplementing the collenchyma as a strengthening tissue.

Growth in girth soon and repeatedly ruptures this ring, but the gaps are made good by the ingrowth of parenchyma cells into the radial spaces due to the ruptures, and their transformation into sclereids. Thus a composite mechanical ring is formed of arcs of fibre cells separated by arcs of sclereids. One of the most striking departures from normal cortex formation is to be seen in aquatics, where the cortex, or in the case of Monocotyledons the whole ground tissue, is divided into large air spaces, separated by membranes only one the cell thick. These lacunae are limited in length to one internode in Dicotyledons stems, and even in Monocotyledons they are not of great vertical length, but the horizontal membranes which separate them are perforated by minute intercellular passages, which form at all the angles of the cells. These provide for the movement of air from one lacuna to another, but they are too small to allow water to penetrate, if it should chance to get into one lacuna through an injury to the epidermis. They are in effect waterproof bulkheads and prevent the accidental flooding of the whole internal air system.

Periderm

The epidermis is usually only retained for a short time while the shoot is young, and it is later replaced by a secondary covering. This is necessary because the thick walls and the cuticular layer of the epidermis inhibit the division of its cells. The epidermis can therefore only accommodate itself to the increasing girth of growing stems by the tangential stretching of its cells, and that is limited in extent. A few cases exist of woody plants which retain their epidermis for a number of years, e.g., *Laurus*, *Aucuba*, *Rosa*, *Acer*. The cuticle is soon ruptured by the growth of the stem, but the cutinized cell wall increases enormously in thickness and is progressively regenerated from within, while the outer layers crack and crumble away.

Most smooth, stemmed shribs do, however, eventually form a bark. One exception is the Mistletoe, in which the primary epidermis is supplemented by further cutinized layers of cells formed successively inwards, so that a multiple epidermis of great thickness is build up.

Only a growing tissue can retain permanently its equilibrium with other growing tissues, and thus the secondary covering of the stem is initiated by the formation of a cambial layer known as the *cork cambiun* or *phellogen* the chief product of which is the *phellem* or *cork*. The phellogen is a true secondary cambium formed by the rejuvenation of tissues already fully differentiated, which may be situated anywhere between the secondary phloem and the epidermis itself, although in most cases the first phellogen lies in the cortex near to the outer surface. It is a single-layered meristem, and by tangential divisions it cuts off a succession of cells outwards, which are arranged in radial rows, corresponding to the cells of the phellogen from which they have arisen.

There are a few examples of intraxylary phellogens, e.g., *Ariemisia*, where cork formation begins in the wood of the current year, forming a sleeve over the wood and ending upwards in a dome in the pith near the stem apex, thus isolating all new growth from the old. This may be related to the desert conditions under which the plants concerned chiefly live. Priestley has shown an association betwen phellogen formation and the existence of some anatomical barrier to outward sap diffusion. When a functional endodermis is present this acts as such a barrier and the phellogen commences in the pericycle. In the absence of such an endodermis there may be no corresponding barrier until the hypodermis or the epidermis is reached, and it there that phellogen appears. Intermediate cases may be accounted for by the presence of an internal cuticle at some level in the cortex or by the existence of an impermeable sclerotic layer.

According to Priestley's view the accumulation of sap on the inner side of a diffusion barrier provides the physiological stimulus to increased cell division. As the cell sap always contains diffusible fatty substances the materials for the formation of suberin in the cork layers are also accumulated at the level of the phellogen. The suberization of the walls of phellem cells consists in the deposit of a layer of suberin, unmixed with cellulose, between the middle lamella and the inner wall of pure cellulose. This layer varies greatly in thickness in different cases, and in the massive "cork wings" on the branches of species of some trees, such as Ulmus and *Acer*, it may be practically absent, which has led to the application of the term *phelloid* to such unsuberized cork tissues.

The suberin itself is apparently a mixture of the anhydrides of fatty acids, the chief of which is *pehllonic acid*, $C_{22}H_{43}O_{30}$ and it is

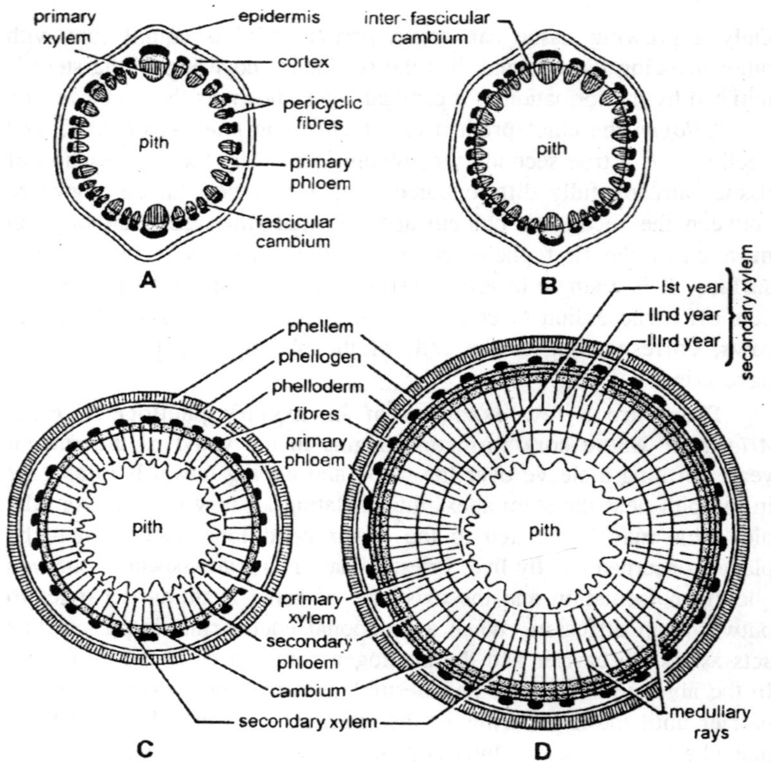

Fig. 7.3. Secondary growth in dicotyledonous stem. A, primary tissues only; B, formation of inter-fascicular cambium; C, secondary xylem and phloem are formed; D, cross secton of a three year old stem.

thus closely similar to cutin. Both substances seen to be dependent on contact with air for their development. There are no pits through the suberin layer, and as it is highly impermeable the suberized cell soon dies. Tannin and related substances often accumulate in the cells before suberization, and these give the cork its dark colour and its value for tanning leather.

Haberlandt showed that the protective value of two-year-old cork layers against evaporation is about the same as that of the primary epidermis. That it is not greater is probably due to the cracking, caused by growth pressure, which fissures the outer layers of the cork, and would destroy its protective value if fresh layers were not constantly added from within by the phellogen. The heavy cork layers of old stems are highly protective. The formation of a cortical or other superficial phellogen begins below the stomata, and is preceded by cutinization of the cells lying beneath the stomatal openings. The phellogen formed at these spots is highly active and cork formation is in progress there before the phellogen has spread round the stem as a continuous layer. The excess of cork cells formed, ruptures the epidermis around the stoma, and protrudes above the general surface level as a spongy mass called the *complementary tissue*. Such a spot is clearly visible to the eye on the stem surface and is called a *lenticel*. Blockage of diffusion through the lenticels, due to the formation of a closing layer of compact phellem in the autumn, reacts upon the phellogen and stimulates increased divisions, thus bursting through the block in the spring.

The life of a lenticel seems to be an alternation of blocking and bursting phases. Permanent lenticels are not in all cases formed at every stoma, the number being related to the vigour of the shoot. In deep-seated phellogens, both primary and secondary, the lenticels are obviously not related to stomata and their distribution appears to be related rather to the broad medullary rays which are associated with the leaf traces. Lenticels have been interpreted as ventilating apertures, but their causation has nothing to do with the assumed necessity for this function.

Suberin is, however, exceedingly impermeable to air, and the rounded and irregularly arranged cells at the lenticel undoubtedly do allow access to the underlying tissues for air, which cannot penetrate at any other point. The physiological value of this is reflected in the bad effects on the health of the plant which result if the lenticels become covered with external growths of moss or lichens. While the chief activity of the phellogen is in forming phellem outwards, there is usually also a slow production of cells inwardly, which are added to the cortex and are known as *phelloderm* or secondary cortex. These cells are not suberized and are only distinguishable from the cortical parenchyma by their arrangement in radial files and sometimes by the presence of starch grains. Their walls are often thicker than those of the cells of the primary cortex and may even be lignified, as in *Canelia*. We thus see the elaboration of a secondary protective system consisting of phellem, phellogen and phelloderm which goes by the collective name of *periderm*. This slowly develops into the *bark* of old stems, but in this development other tissues become involved. In smooth-barked trees the first phellogen remains active for some years and is only slowly replaced, if at all.

The bark of such trees therefore consists of a continuous cork film. In rough-barked trees, however, the first phellogen is soon replaced by a series of others, lying successively deeper in the tissues, and all later phellogens, after the first few, are formed in the secondary phloem, much of which is destroyed in this way and its remains incorporated in the bark, along with dead cortex, collenchyma, sclerenchyma, etc. Bark is thus a composite structure. Where the later phellogens form continuous zones the bark may show successive annual rings, like the annual rings of the wood, each marking the formation of a new phellogen, but in many other cases the later phellogens form a series of discontinuous arcs and the bark is thus cut up into distinct areas, between which deep cracks appear.

The various bark patterns on old tree trunks which are characteristic of different species, are thus traceable to variation in the arrangement of the phellogen layers. An abnormal formation of cork is commonly associated with injuries, when it is known as *wound cork*. Wounding or almost any sort of lesion of the integrity of the plant body may call it forth, and any living tissue may produce it. The first reaction to the wound is the suberization of the outer surfaces of exposed but intact cells. Suberization then spreads inwards for several cell layers. Below this barrier cell division starts and a cambium is organized which produces a few layers of regular cork cells, forming and impermeable barrier against desiccation, parasites and other inimical influences.

Endodermis

The innermost layer of the cortex, surrounding the stele, is called, as in roots, the *endodermis*. A typical endodermis, with Casparian bands, recognizably the same as that in roots, is not common in stems. The chief exceptions are among water plants, such as *potamogeton* and *Hippuris*, in which the vascular tissues are concentrated into a relatively slender axial strand, as is normal in roots. Many other exceptions occur which are not, however, related to special environmental features, e.g., in some Compositae, species of *Primula*, and in the rhizomes of some Monocotyledons.

The endodermis, in most cases where it occurs in the stem, is a common investment round the stele as a whole, but there are a number of instances among Angiosperms where each vascular bundle has an individual endodermis, which is not infrequently continued into the leaves. Species even within the same genus may differ in this character as, for example, in *Ranunculus*, where *R. lingua* and *R. flammula* are peculiar in having individualized bundle sheaths. The significance of the endodermis and of its variations cannot as yet be fully interpreted on physiological grounds, but some interesting points have been elucidated. It first appears outside the procambial zone at the stem apex, simultaneously with the first appearance of xylem elements. Materials from the breakdown of the xylem protoplasm probably diffuse out radially, and may supply the fatty acids which condense to form the Casparian Band.

It has been suggested that the condensation of fatty acids at this level is due to the penetration of air to this zone through the intercellular spaces of the cortex, but while we may have here the foundation for an explanation it is plainly not the whole story. The stele is not the only organ of the stem to possess a sheath, since many glandular passages are similarly protected, nor does the stelar sheath always have a true endodermal character. In a great many stems the layer surrounding the stele is marked by the presence of large starch grains, although the cells may not be otherwise different from those of the rest of the cortex, which may contain plastids but no large starch grains. This "starch sheath" corresponds morphologically to an endodermis, though it will avoid confusion if we do not apply this term to it, since it lacks the Casparian thickening and is physiologically different. Its morphological identity with an endodermis is shown by the fact that a starch sheath in a young stem may sometimes lose its starch and become thickened as an endodermis as a later stage.

Cases also occur where the starch sheath is dissected into separate arcs or strands, divided by ordinary parenchyma, as in the Nettle and at the nodes of Grasses. These large starch grains are movable, sinking always to the lowest side of the cell vacuole, and they have been called *statoliths*, and assigned a function in connection with geotropism. In underground stems of Monocoty-ledons, such as *Convallaria*, there is a pronounced endodermis, which may be double or even triple.

In the aerial stems, however, the boundary of the stele is usually a sclerotic or collenchymatous zone which grades insensibly into the cortex. It is evident that the stelar sheath in the steam is by no means uniform, and Strasburger has coined the useful name *phloeoterma* to apply to the innermost layer of the cortex, which must be recognized to be an important anatomical boundary, whether or not it has the character of a true endodermis.

Primary Vascular Tissues

Everything within the endodermis is denominated the *stele*. The question of the applicability of this concept, derived from the study of the stem in the Vascular Cryptogams, to the stems of the Spermatophyta, we have discussed previously. It is at any rate certain that in the absence of an endodermis, and particularly in some Monocotyledons, it is impossible to give it the precise connotation which it has, for example, in the *Ferns*; but it is a handy descriptive term and we shall continue to apply it, without prejudice, to the complex of tissues lying within the cortex and of which the vascular tissues form so important a part. Its outer boundary is usually called the *pericycle*. The pericycle in the stem is very different from the meristematic zone which goes by the same name in the root. The idea of a pericycle is that of a cell layer formed from the outer zone of the procambium, within the endodermis and immediately surrounding the phloem. This position in the stems of Dicotyledons is often occupied by a zone of sclerotic fibres or of sclereids, sometimes continuous, sometimes broken up into groups which lie outside the vascular bundles.

At the present day it is doubtful whether this fibrous zone is really distinct in all cases from the phloem, for it has been shown that it may arise from protophloem in which the sieve tubes and other soft elements have been obliterated and only phloem fibres remain. Whether this is always so cannot yet be stated. The vascular bundles of Monocotyledons, whether they have an individual endodermis or not, are frequently surrounded by sclerotic sheaths, complete or partial, which are quite independent of the phloem but are otherwise analogous to the general pericyclic fibre-sheath in Dicotyledons. It is difficult to see why both types of sheath should not be classed under the same name, as the bundle sheaths in Monocotyledons are particularly characteristic of those stems in which no general pericycle exists. Each *vascular bundle* is a complex of tissues, partly conducting elements, partly storage elements and partly strengthening elements. Haberlandt gives the following analysis of the components:

While *xylem* and *phloem* denote the whole of the two main tissue systems in the bundle, Haberlandt applies the terms *hadrome* and *leptome* to the two sets of conducting elements respectively, to distinguish them from the fibres present. The most general arrangement of tissues in the bundle is the *collateral*, in which both tissues lie on the same radius, the phloem constituting approximately the outer half of each bundle and the xylem the inner half. We have described above the first appearance of protoxylem at the inner side of the procambial strand and of protophloem at the outer side. The further differentiation of each tissue from procambial cells is centrifugal in the case of the xylem and centripetal in the phloem. In Monocotyledons generally, and in some reduced herbaceous types among the Dicotyledons, the whole of the procambium is thus differentiated and no further addition of vascular tissue is possible. Such bundles are called "closed." Among Dicotyledons, however, it is usual for a zone of cells about half-way across the procambium to become *cambium*, which remains meristematic and thereafter adds cells continuously to both xylem and phloem in radial files. This change marks the beginning of what is called *secondary thickening*.

As we have previously pointed out, the organization of cambium may begin so early that every part of the vascular tissue is formed from it and the radial arrangement rules from the beginning. Such cases show that a clear distinction of primary, or procambial growth, from secondary, or cambial growth, is not always possible. Even in plants with a meristematic ring which gives rise to a woody stem, differentiation normally begins at a number of separate points on the ring, so that separate primary bundles are for a short time distinguishable. The few exceptions, such as *Vinca*, in which a continuous ring of xylem is present from the beginning, may really be analyzed into a ring consisting of a very large number of uniseriate bundles, placed very close together and showing a radial arrangement of the elements from the start. The residual meristem between the original bundles may organize directly into *interfascicular cambium*, continuous with that in the bundles, or it may differentiate into parenchyma.

In the first case secondary xylem and phloem are differentiated in the spaces between the original bundles and secondary growth goes forward both in the bundles and between them, forming continuous zones of xylem and of phloem, as in timber trees. In the second case a layer of the interfascicular parenchyma may later differentiate into cambium, but such secondary cambium forms only or mainly parenchyma, and the original bundles remain permanently distinct, i.e., secondary vascular tissue is confined to the bundles. This is not uncommon in woody climbers like the Vine. On the other hand, the spaces between the bundles may remain permanently parenchymatous, with no interfascicular cambium at all, and secondary growth of the bundles themselves is very limited. This is the extreme herbaceous type of stem.

The parenchymatous intervals between the original bundles are the *primary medullary rays*. They are usually broad at first, and in stems of the Vine type they grow radially by additions from the interfascicular cambiums, so that they retain their initial breadth permanently, but in timber trees the original broad rays are soon bridged and closed by the interfascicular xylem. Certain interfascicular cambial cells, however, produce only parencyma cells, both inwards and outwards, so that at these places narrow rays are built up, which may be uniseriate, or at most two to three cells broad. These narrow rays continue outwards through the whole zone of vascular tissue, no matter how broad it may become, from the pith to the pericycle, interrupted only by the cambium itself. Medullary rays which originate after secondary thickening has begun are called *secondary rays*. Their point of origin is an initial parenchymatous cell, cut off, usually terminally, from one of the fusiform cambial cells. Similar narrow rays are formed in the original bundles as soon as radial growth from a cambium is established, and they may be similarly permanent. The details of secondary growth we will leave until later, while we consider some characters of the individual bundles.

Two types besides the simple collateral bundle deserve mention. The first is the *bicollateral*, in which there is a second phloem region, usually smaller than the first, inside the xylem zone. This internal or *medullary phloem* may be confined to the primary bundles, or it may develop in the interfascicular region as well, so that a more or less continuous zone of internal phloem results. It is a character found only in certain families (e.g., Cucurbitaceae and Solanaceae), sometimes in members with all types of growth, sometimes only in climbing species of the family. The second type is the *centric bundle*. This is usually, though not invariably, a closed bundle and is commoner in Monocotyledons, especially in the underground stems, than in Dicotyledons. It may have the xylem in the centre and the phloem outside in which case it is called *periphloic* or *amphipholic*, or else the reverse when it becomes perixylic or amphixylic.

The closed bundle which is characteristic of the aerial stems in Monocotyledons is small and has relatively few elements. It is sometimes described as the Y-type from the characteristic grouping of the xylem elements. The protoxylem, which is often crushed and obsolete at maturity, occupies the tail of the Y, and the arms are formed by, or end in, two very large vessels, between which lies a group of small lignified tracheids. Above this is the compact phloem, consisting of regularly arranged sieve-tubes and companion cells with a few parenchyma cells at the sides.

Another monocotyledonous type has a still simpler xylem, formed of one or two narrow protoxylem elements and one extremely large vessel in the centre of the bundle. This type of bundle is formed mostly in monaxial Monocotyledons with very large, rapidly growing leaves, e.g., Banana, and the extra large vessels are probably correlated with the need for quick development of the water-conducting capacity of the stem. The concentric type of bundle is considered to the most advanced type in Monocotyledons and has probably been derived from one of the collateral types. As is well known, the arrangement of the vascular bundles in Dicotyledons is typically in a single, wide ring, while in Monocotyledons they are typically dispersed, seemingly at random, across the transverse section of the stem.

In the Dicotyledon the orientation of the vascular tissues is constantly with the protoxylem innermost but in Monocotyledons, although the general tendency is the same, there is less uniformity, especially in the bundles near the centre. The parenchymatous tissue enclosed by the ring of bundles in Dicotyledons is the *medulla* or *pith*, but in monocotyledonous stems there is no corresponding region, and the parenchyma among the bundles is generally known as the *ground tissue* or *conjunctive tissue*. Departures from these typical arrangements will be dealt with later. We must now consider the bundle systems in three dimensions in order to get some idea of vascular architecture.

In Dicotyledons the leaves have usually a narrow base of insertion on the stem. The number of *leaf traces* is therefore nearly always small and may be only one. These traces differentiate downwards from the leaf base and intercalate themselves between the traces of lower leaves. In this way the trace of a young leaf may reach downwards through several internodes. Eventually, however, at a node, it either becomes united laterally to the adjacent bundle in the ring, or it forks and unites to the bundles on each side of it, producing "synthetic" bundles. The distance it reaches downwards before this happens depends on the leaf arrangement on the stem, since it appears that the traces to which a given leaf unites itself are those of the leaf vertically below it is in the leaf spiral.

At every node some anastomosis of traces takes place, so that there is at these levels a considerable amount of vascular linkage, while in the internodes the bundles remain separate. The lateral traces of the leaf, where such exist, are close to the median trace at the leaf base, but as they go downwards they are separated by the intercalation of traces from higher leaves, until they may lie far apart.

Where there are several traces to each leaf they will thus come to be spaced out where they enter the ring, and in types with many traces, such as *Liriodendron*, they may occupy points all round the ring before they finally join the ring bundles. The lateral traces join the ring at a higher level than the median trace, which means, in terms of apical development, that they do not differentiate until several plastochrons later than the median trace, which is always the strongest and longest and goes farthest down the stem. While the trace differentiates downwards, the period of growth available for differentiation gets shorter in each successive internode and the trace will therefore contain less and less primary xylem and more and more secondary xylem, until it is finally merged in the general mass of secondary xylem.

The phloem generally accompanies the xylem, though peripheral portions of phloem may branch off independently and join up with others far above the level at which as a whole finally joins a synthetic bundle. Comparative studies give good grounds for the opinion that the primitive leaf trace system in Dicotyledons is one with three trace bundles leaving three separate gaps in the vascular ring of the stem, the so-called *trilacunar* type. Reduction to one is due to some cases to abortion of the lateral traces, in others to fusion into one, which, however, may be three-lobed at the base. The many-bundled type of trace is also derivative from the three-trace type, which is frequently found in the seedlings and youngest leaves of plants with complex traces.

Even in Monocotyledons the three-trace type is found in many small and slender species and in the young leaves of others, and it appears to be primitive in this group also. In Monocotyledons the leaf base is broad and surrounds the stem, even in the embryonic state. It may contain thirty to forty trace bundles, which enter the stem simultaneously all round the periphery. Each leaf primordium is quickly pushed outwards away from the apex by the new encircling primordia which follow it, with the consequence that its traces become acutely fixed at the node, bending outwards from a central position into the leaf base.

In Monocotyledons with extended internodes the bundles follow a course downwards which is similar to that in Dicotyledons, tending gradually outwards towards the point of anastomosis with older bundles, which appear to be usually those of the next leaf vertically below the leaf considered, i.e., on the same orthostichy. At the nodes the number of anastomosing bundles is so great that it creates a nodal plexus of intermingled bundles which may extend right across the stem, and is increased by the insertion of numerous traces from the axillary buds. It must be admitted that the course of the bundles in even a simple Monocotyledons is not yet known with sufficient accuracy to warrant positive statements, while in those species with contracted stems the complexity and irregularily defines analysis. The whole question of vascular development is indeed in a somewhat hazy state. For example, although basipetal differentiation of vascular elements within the procambium seems to be the rule, it is by no means certain that the differentiation of the procambium itself, from the primary meristem, is not acropetal or continuous. The above account must therefore be regarded as provisional. The development of the meristematic cells in a young leaf axil into an axillary bud complicates the dicotyledonous nodal structure further by its special system of traces. The axillary bud has an apex which is a miniature of the main apex and has the same procambial arrangement, giving rise to a duplicate of the stem stele.

As this differentiates downwards into the stem, the ring or group of traces opens out fanwise, forming two bundles which pass into the main ring between the bundles which flank the median leaf gap and join them above the level at which the median trace of the subtending leaf itself joins an older trace in the ring. Exogenous adventitious buds such as those on the corms of *Cyclamen* are apparently connected secondarily to the stele of the parent stem. Procambial strands come from the bud and differentiate inwards through the cortex towards the stele, and the formation of vascular tissue in this procambium is likewise downwards from the bud.

Cambium

It has been previously pointed out that the organization of cambium from procambium marks the beginning of radial growth, commonly called secondary growth, which is capable of indefinite expansion outwards and will, if continued, build up a massive woody stem. In many herbaceous stems, although secondary growth begins as in woody plants, it does not continue beyond one year. The structure of such herbaceous stems is therefore similar to that in the one-year old stems of woody plants. The *Cambial initials* are distinguished in the procambium by their repeated tangential, longitudinal divisions, giving rise to a zone of cells which appear narrow in transverse section. This occurs in the part of the stem apex where active growth is still going on, and the cambial initials undergo considerable elongation and

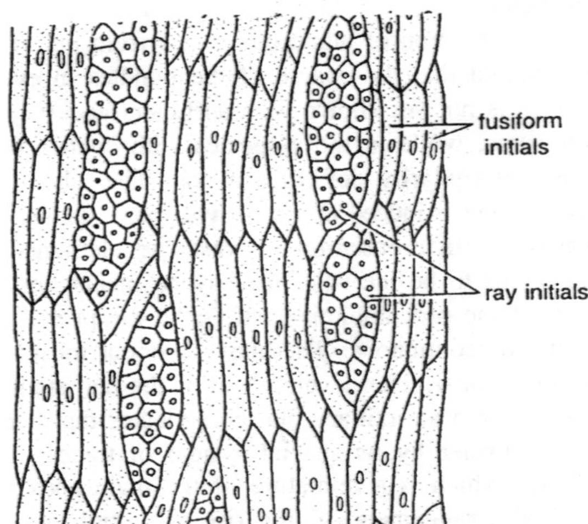

Fig. 7.4. Vascular cambium showing fusiform initials and ray initials.

become radially compressed by the expanding tissues of the pith and cortex. In spite of these changes the cambial cell remains a true meristematic cell, and its repeated tangential divisions cut off new xylem cells centripetally and new phloem cells centrifugally.

It is difficult to distinguish these new tissue elements in their early stages of development from the true cambial cells, and it is not possible to say in every case whether the fully developed cambium consists of only one layer of cells or of more. The position at which the cambium forms in the procambium may be physically determined by the gradient between acid sap from the differentiating wood and alkaline sap from the phloem. This theory, which is due to Priestley, maintains that the cambial layer lies at the pH on the gradient which corresponds to the mean isoelectric point of the cell proteins and that the protoplasm of the cambial cells remains unvacuolated because at the isoelectric point colloids have their minimum affinity for water and their maximum density. The same theory extends to the formation of interfascicular cambium from fully differentiated parenchyma cells and perhaps also to the appearance of the phellogen. Both these charges involve the alteration of cells with large vacuoles and little protoplasm into meristematic cells which have no vacuoles and abundant protoplasm.

The isoelectric point of the cell proteins is the condition which would seem most favourable to such a change. The cambium does in fact lie between acid xylem (pH 3.4 to 5.0) and alkaline phloem (pH

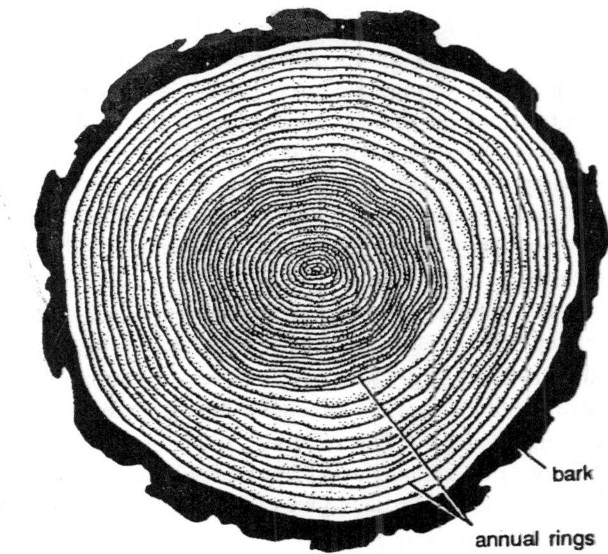

Fig. 7.5. Transverse section of an old dicotyledonous stem showing annual rings.

7.8), while the phellogen lies between the strongly acid cork (pH 3.0) and the less acid cortex (pH 5.5 to 6.5). Protoplasm at its isoelectric point will lose water to vacuolating cells and the condensed non-vacuolate condition produced is that most favourable to protein synthesis. When cambium is viewed in tangential longitudinal section, two types of cell are seen, the long, pointed fusiform cells which give rise to the vascular elements, and the small, isodiametric cells in groups, which produce the medullary rays. These latter are much more numerous than in Conifers and may account for up to 55 per cent of the cells in a cambial ring. The long, fusiform elements are thin walled and without pits except in the dormant season. They are often flexed into curves as they pass round the groups of ray cells. Their divisions are mostly longitudinal-tangential, and it is doubtful whether they divide radially.

More probably the appearance of radial division is created by transverse walls which become more and more inclined in a tangential direction until they are vertical, thus increasing the number of cells in the ring, as it expands with growth. In radial section the long cells are very narrow, with chisel edges at both ends, which overlap those of the next cells above and below. This overlapping of the cell ends has been attributed to "sliding growth," in which the the walls of neighbouring cells actually slip along each other, but thus is doubtful, for many reasons, and it seems more probable that it is produced by local growth of the young, platic cell-walls, above and below each new transverse division wall, at opposite ends respectively, in such a way that the transvese wall is tipped into the vertical without any mutual slipping of the cells. The cambial ray cell, on the contrary, divide in all three planes. Bailey has shown that in Dicotyledons there are three main types of cambium, corresponding to three types of wood structure.

1. Initial cells averaging 890 microns in length, producing long, narrow vessel segments with scalariform pitting of the end walls.
2. Initial cells averaging 410 microns, producing short, broad vessel segments with cross walls which are simply pitted or else absent.
3. Initial cells averaging 250 microns, producing vessels similar to type 2, but the cambial cells are arranged in regular horizontal rows (Etagen cambium) and give rise to a corresponding "storied" structure in the wood.

These three types are regarded as marking advancement in specialization of the vascular structure. The cambium in the resting state, in winter, is sharply marked off from the vascular tissues, with fully formed vessels and sieve tubes lying in immediate contact with the cambial cells from which they have been formed.

During the summer activity, however, the cambial zone seems much broader and less clearly delimite, because tangential divisons follow one another so rapidly that the true cambial zone is flanked on both sides by a layer of still undifferentiated cells, and the transition to the mature vascular elements is thus spread over a broader zone of development. It has been shown that the cambial growth in spring begins in the swelling buds and spreads downwards from them. It begins simultaneously all over the tree, but the rate of spread is much slower in the older branches and its stops sooner in them than in the young shoots. There is good evidence that the stimulus to cambial growth is the diffusion of hormones from the leaf initials in the budl. Auxin-α and heterauxin can both start strong cambial growth when they are supplied to a shoot, and auxin has been found in naturally active cambium and is known to be produced by leaf initials.This direct relation of leaf development to cambial activity is probably the reason for the proportionality between the extent of leaf surface and the amount of vascular tissue formed in the stem, which observation has revealed.

The hormone explanation may hold good not only in growth of the normal cambium but in the differentiation of new vascular tissue from old permanent tissues, which occurs in regeneration growth after wounding. Such new developments show a definite directional guidance from point to point, often around obstacles, which strongly suggests the action of a diffusing hormone. This is also probably true of new, adventitious, vascular differentiation, as in the case of *Cyclamen*. The cambium, like other meristematic tissues, does not itself undergo any marked change throughout its existence. The only visible indication of age is a gradual increase of the average length of the cambial cells and of the tracheids formed from them, an increase which may continue for a number of years from the beginning of cambial growth.

Eventually, however, a steady state is reached and no further change in length occurs. In the most pronouncedly herbaceous plants there may be no interfascicular cambium, and the separate bundles themselves have little or no cambial growth, while in the most extreme herbaceous types there is not even a continuous ring of procambium. Indeed the annual, monocapric habit of growth, limited to a single growing season, is probably brought about by the absence of early disappearance of cambium and the consequent lack of the power of extended growth. Monocotyledons are generally supposed to have no cambium in their bundles, but the metaxylem elements are often in radial rows, which is characteristic of cambial growth, and they probably resemble the annual Dicotyledons in having cambia that are very short lived.

The specialized cambia of some Monocotyledons which undergo secondary thickening are dealt with. In connection with the cambium we may briefly mention the abnormal tissue formation known as *callus*, which is associated with wounds and similar injures. The name is applied to irregular tumour-like outgrowths, due to a resumption of meristematic activity called out by the injury. Any living tissue may contribute to this development, but where the cambium is involved, its contribution is much the greatest. Thus a callus ring arises on the cut ends of branches or old roots, which is almost entirely due to growth from the exposed end of the cambium and which spreads until the wounded surface is completely covered. This new tissue is almost entirely parenchymatous, and there is no regular differentiation of tissues, as in normal cambium growth, At the most a few irregular scattered tracheids or sclereids may be formed, embedded in the mass, while the outermost cells may become corky.

Secondary vascular tissue—Phloem

The *metaphloem*, including the secondary phloem, if any, consists of sieve tubes, their companion cells and phloem, parenchyma. The companion cells, originally sister cells of the sieve tube cells, usually divide transversely more than once, so that they are individually much shorter than the sieve tube segments. The phloem parenchyma cells are variously shaped, and may be so long and narrow that they scarcely justify their name. Mingled with there elements in many plants are phloem fibres, either scattered or, as in *Tilia*, in tangential bands alternating with the conducting tissues. These fibres closely resemble those which form the fibres of the so-called pericycle, external to the phloem, and are often classed with them as *libriform fibres*, from the old name "liber" (on "bast") applied to the inner bark, in which the phloem was distinguished as "soft bast" and the fibres ad "hard bast."

Fibres of the secondary phloem, being produced from the cambium, show the same radial arrangement as other phloem cells, but the pericycle fibres, coming direct from procambial cells, do not. Libriform fibres are narrow and very highly thickened, and their pits are very small, so that the pit opening shows little or no border. The pit openings are slit-like and the openings on the inside and the outside of the wall are usually at right angles to each other, so that the pits appear like minute crosses on the cell wall. They are longer than the xylem fibres, but flexible and tough rather than brittle. Jute, and several other important fibres consist of this type of cell.

Secondary vascular tissue—Xylem

The term *metaxylem* includes all the secondary xylem. The cell structrue of secondary xylem is very variable, and each type of wood has its recognizable characteristics, by which it may be identified. The usual constituents are vessels or tracheae, tracheids, xylem parenchyma and fibres of various types. In true vessels the successive elements open into one another in longitudinal succession by membraneless portals so that the lumen forms a continuous passage. This involves the alteration or disappearance of the transvese walls which separate the young elements from each other. Comparative studies have shown that in the wood of certain families which are predominantly arboreal in habit the vessel elements are narrow and long and have steeply oblique end walls with numerous scalariform perforations.

In this they resemble the vessels of the Gnetales. Example are : *Magrolia Betula*, *Salix Myrica*. From this unspecialized condition there has been a progression towards vessels with short elements, as broad as or broader than their length, and with transverse end walls which have either one very large central perforation, or are, in the final stage, completely absorbed, leaving only a ring on the vessel wall to mark where they had been. Examples of this are common among herbaceous plants. The lateral walls of vessels in the secondary wood may bear pits which are either scalariform or circular, the latter being either side by side or alternating with each other.

Spiral vessels also occur in the wood of many families, but not in those which are regarded as being the most primitive. Movement of water upwards in the xylem under the influence of transpiration tension follows the lines of least resistance and large, fully open tubes must obviously afford an easier passage than narrow elements with many cross walls, even if the latter are perforated. Climbing plants which have very long, narrow stems are those in which the resistance to hydraulic flow is greatest, and they are notable for the large average diameter of their vessels, Poiscuille's Law of fluid flow expresses the velocity thus:

$$V = \frac{\pi p r^4}{8 \ell \eta}$$

where, V = velocity of flow

p = pressure of difference.

r = radius of tube.

l = length of tube

η = coefficient of fluid viscotiy.

The velocity is directly proportional to the fourth power of the radius and inversely proportional to the length of the path of flow. Velocity is thus such more sensitive to differences of radius than of length, and the resistance offered by a long stem can be readily compensated by the increased width of the vessels.To double the radius of a vessel will increase the flow capacity by sixteen times, but to halves the length of the path will only increase the flow twice.

Owing to the irregularites of their structure the vessels may have conducting capacity which is no more than 30 to 40 per cent of the theoretical efficiency but their comparative capacity is nevertheless governed by the hydraulic laws. The flow of water upwards proceeds through very numerous fine channels, branching and rejoining which are formed of single vessels or or small groups of vessels islanded among tracts of non-conducting parenchyma and fibre cells. Among herbaceous plants the leaf trace bundles serve for conduction only to the leaves to which they are connected, and there is some evidence that this physiological segregation may persist downwards in the synthetic bundles of the lower stem. Whether this applies to the massive secondary wood in trees is doubtful.

There is no positive evidence of the anatomical segregation in the mass of wood of distinct conducting tracts associated with individual branches. The selective flow through limited channels, which may be observed when a single leaf or branch draws a coloured solution through the main stem, is probably conditioned by the downward transmission of the water tension, which is the motive force of the flow, through the physically most direct path available. Vessels anastomose freely both tangentially and radically, so that the water currents, even if confined to the vessels, may shift their paths readily in either direction, even from one annual ring to another.

The force of transpiration is not, however, sufficient to draw water across the medullary rays, in the absence of a vascular connection. In most woody plants the vessels only remain functional for, at most, a few years after their formation, and frequently only for one year. The outer most annual rings of wood are often lighter in colour and softer in texture than the older wood. They form the *alburnum* or sap wood, while the older rings compose the *duramen* or heart wood. Only the outermost portions of the sap wood may be active in conducting water and the heart wood takes no part whatever in conduction.

Water storage in the duramen is confined to the living cells of the medullary rays and xylem parenchyma, but as the conducting elements pass out of action they are often filled with resinous or gummy substances and tannins, and the walls may be impregnated with colouring matters, an example of which is Haematoxylin, which oxidizes to the well-known dye, Logwood. Minerals substances, such as Calcium carbonate in *Fagus* and *Ulmus*, Silica in *Tectona grandis* (Teak) or Calcium oxalate in Chloroxylon (satinwood) may also impregnate the heart wood, adding greatly to its weight and durability and enhancing its value as a strong core in the tree trunk.

Vessels which are passing out of action often become partially or wholly filled by sac-like in growths from the living parenchyma cells which are in contact with them. The pits between parenchyma and vessels are not bordered onthe side of the former, so that they present a large surface of pit membrane, which is expanded by turgor pressure into the cavity of the vessel. Such ingrowths are called *tyloses*.They are ballon-like, cylindrical or even branched, and have thin walls, but occasionally they may be lignified, either like stone cells or like the walls of the vessel in which they lie. What determines these differences is unknown. The young vessel elements usually expand laterally very rapidly, reaching their fully diameter even while still in contact with the cambium, and creating considerable tissue pressures, while tracheids do not thus expand; yet there is no invariable distinction of size, though normally the expanded vessels are larger than the tracheids of the same wood.

While vessels are normally the channels of long distance transport, tracheids serve chiefly for abort distances and for water storage. Offshoots of the main bundles, including the leaf traces the leaf veins, consists almost entirely of tracheids. This distinction of function implies an advanced state of specialization and is characteristic of the Dicotyledons. Fibres play a large part in the make up of the secondary wood. Firstly there are the *fibre-tracheids*, thicker walled than the true tracheids but with the same bordered pits, and in some cases not sharply distinguishable. These grade through intermediate elements, increasingly narrow, thick-walled and elongated, to true *libriform fibres*. Libriform fibres in the wood are often as much as seven times as long as the cambial initials. Measurements show variations between 0.3 and 1.3 mm. Nevertheless this is small compared with the bast-fibres, which are seldom less than 1.0 mm. long and may reach 7.7 mm. in Urtica, or even 40 cm., i.e., 16 in., in Boehmeria nivea. In both cases, however, the elongation during differentiation is very considerable.

As it takes place in regions where general growth in length has ceased, it has been held that the elongation of the individual cells must imply slipping or sliding of the cells on each other, and this has been called *sliding growth*. The idea is open to some anatomical objections, but it cannot be said that the alternatives suggested are very convincing. One counter-argument, usually regarded as weighty, is that the pits in the walls of adjacent fibres show a normal degree of correspondence, which is considered to be impossible if the walls have slipped during growth. It is not impossible, however, that this correspondence may be secondarily established, since it is also found in the pits between adjacent tyloses in wood venica, in which cane the cells have manifestly only come into contact at a late stage of development. The last class of fibres is that called *substitute fibres*, which are living cells, often containing starch, but with lignified walls. They are not clearly separated, except in point of length, from xylem parenchyma, and indeed they are sometimes divided by transverse walls forming a short chain of typical parenchyma.

The xylem parenchyma cells arise by repeated transverse divisions of cambial initials. Their walls are lignified but not heavily, and they are richly supplied with simple pits on the radial and horizontal walls. They are important as stores of reserve substances, either starch, in most hard-wooded trees, or oil in soft-wooded trees. They also form important links between the cells of adjacent medullary rays. Careful investigation shows that they are seldom isolated, but occur in irregular tracts, which contact one or more of the medullary rays, usually two, between which they form a tangential bridge.

In this way they serve to make an intercommunicating system of living cells in wood, providing for tangential transport, as the medullary rays themselves do for radial transport. The distribution of the xylem parenchyma is highly characteristic of particular species of wood and is one of the features by which they may be identified. There are two main types:

1. *Apotracheal*, which includes those occuring in tracts which are independent of the vessels, i.e., metatracheal, and also the rarer cases where they are diffuse or *dispersed*.
2. *Paratracheal*, which includes all cases where they are grouped around the vessels.

In Conifers the xylem parenchyma usually occur only in the last row of elements in each annual ring. This arrangement, called *terminal*, is very rare in Angiosperms, being known only in *Magnolia* and *Salix*.

Medullary Rays

The majority of the medullary rays in secondary wood are uniseriate, that is one cell in breadth as seen in transverse sections of the wood, but broader, *compound rays* occur in some species, while *aggregate rays*, formed by the close grouping of a numer of uniseriate rays, also occur, especially in association with leaf of branch gaps in the woody cylinder. All rays which originate from cambial cells are classed as secondary rays, and new rays, intercalated between the previously existing ones, are commenced in each year of growth, so that the total number steadily increases with the age and girth of the tree, the average distance between rays, tangentially, remaining roughly uniform in successive rings of growth. Once a ray has been started it is invariably carried on through all subsequent growth rings. Outside the cambium the ray is continued in the phloem, so that each ray has a xylem and a phloem portion.

The ray in the phloem may be both longer and sider than in the xylem and in plants such as *Tilia*, where the phloem rays are V-shaped, the widening is due to serial cell divisions resembling cambial growth. Medullary ray initials, which are much shorter than other cambial cells, cut off cells both inwards and outwards by successive tangential divisions, though many more are added to the xylem rays than to the phloem rays, corresponding to the more rapid formation of xylem. The cells cut off are directly converted into parenchyma and added to the rays without further divisions. The xylem ray cells become lignified like the other xylem parenchyma, the phloem ray cells are thin-walled and pitted, like phloem parenchyma.

During the winter, when the cambium is dormant, the ray initials in the cambium take on the character of phloem ray cells, so that there is then no break in the radial continuity of the two portions of the ray. New initial cells are formed in spring by the tangential division of the inner-most phloem ray cell. Most rays are comparatively short in the vertical direction, some no more than one cell in height, and the tallest are not usually more than twenty cells.The cells are, with few exceptions, uniform in character, and all contain living protoplasm, nuclei, and reserves of food material. They remain living without growth or division for as long as the plant exists, and the ray cells of heart wood in old trees are the oldest living cells nown. They may persist for an indefinite period without growth and without division.

Medullary rays in woody Dicotyledons have been classified into a number of types on the following characters:

1. Whether both uniseriate and multiseriate rays are present, or only one kind.
2. Whether the cells composing the rays are all alike short, or whether some are vertically elongated.

The heterogeneous type, with both kinds of rays and both kinds of cells, is regarded as the most primitive, and the homogeneous type with only multiseriate rays of short cells is probably the most advanced. Where the xylem ray cells about on vessels, the ray field, i.e., the area of contact between ray cell and vessel, has simple, not bordered pits, which are sometimes single and very large, but more often small and either circular or elliptical.

Types of Wood

At the resumption of activity in spring the cambium produces comparatively large, thin-walled, rapidly expanding xylem cells, the *spring wood*, which meets the need for active conduction of water to the developing buds. At the height of summer cambial growth is reduced, but it again increases to a secondary maximum in autumn,

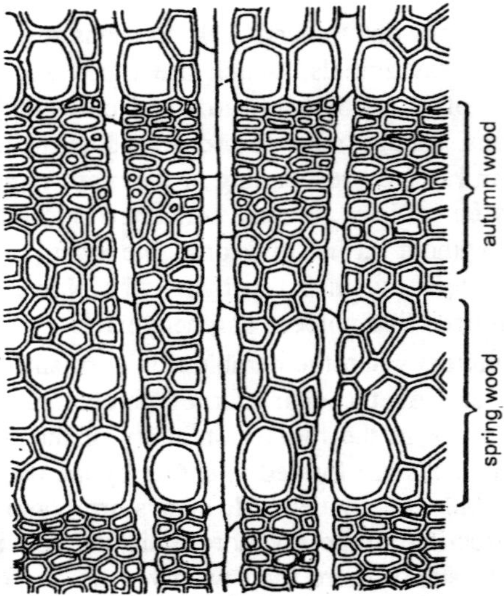

Fig. 7.6. A portion of transverse section of an old stem showing spring wood and autumn wood.

associated with increased rainfall and an abundance of photosynthetic carbohydrates. The *autumn* wood is, however, much denser than the spring wood and contains a higher percentage of fibres. It is the dense zones of autumn wood which mark the "grain" of commercial timbers. Tracheids, vessels and fibres may be formed at any time during the year's growth, and the pattern of their distribution is often on index to the identity of the wood and decides its physical characteristics.

Two alternative patterns affecting the vessels are so well marked that woods may be classified by them at a glance. In one class the formation of vessels is almost or entirely confined to the spring, so that each annual ring begins with a zone of large vessels. This is called *ring-porous* wood. In the other class vessels are formed throughout the year. Such wood is called *diffuse-porous*.

The vessels in these woods are nearly always smaller than in the ring-porous type, and are sometimes scarcely broader than the tracheids. They are also apparently much shorter. The total length of the open lumen in a single vessel is difficult to determine, but the passage of coal-gas through the wood gives us some idea of their relative lengths. Wood without vessels is completely impermeable to gas, but it will pass through a 10 ft. length of the ring-porous wood of Ash in sufficient amount to ignite, while more than 2 ft. of the diffuse-porous wood of the Maple stops it completely. A small group of genera belonging to the order Ranales, namely *Drimys*, *Trochodendron* and *Tetracentron*, is notable for having no wood vessels at all. These woods are spoken of as homoxylous, for they consist of radially arranged tracheids of uniform width, recalling the wood of *Pinus*.

The resemblance is carried further, for they have large bordered pits on the radial walls of the tracheids, like those of Conifers, though the pitting on the spring wood is scalariform. Whether this is significant from the evolutionary standpoint, as indicating a primitive character in these three genera, or whether they have lost their vessels by reduction in the course of evolution is a matter of opinion. Each year's complement of secondary thickening forms one *annual ring*. The change from the dense, small-celled autumn wood to the soft, open type of spring wood is normally well-marked and unmistakable in all temperate trees, but in trees grown in an equatorial climate growth is continuous and the annual rings may be non-existent.

The older fossil woods have either no annual rings or vague rings with no marked break between them, showing that there were no marked seasons even in temperate latitudes in Carboniferous time. A marked differentiation of climate first appears in the Mesozoic period. Annual rings are an excellent record of climatic variations, being wider or narrower according to the quality of the season and having a reduced average width in high latitudes and at high altitudes. It does not follow, as might be imagined, that narrow rings imply stronger wood.

On the contrary, unfavourable conditions limit chiefly the development of the autumn wood, and narrow rings may consist mainly of the spongy spring wood. The width of the annual ring is not uniform all over the tree but is greater in the upper part, associated with a longer period of cambial activity. Annual rings are usually symmetrical in vertical stems, but in oblique or horizontal branches, especially near the base, they become asymmetrical, the lower half of the upper half being thicker than the other and consisting of differently thickened elements. In Conifers the lower wood is reddish, the upper white, the former having much thicker cell walls.

The difference is attributed to the mechanical effect of the weight of the branch, the upper wood being called *tension-wood* and the lower portion *compression-wood*. Vertical stems subject to consistent wind pressure from one side may show the same differences. Severe frost may cause upward or downward bending of large branches according to whether the thinner-walled tissue lies on the lower or the upper side of the branch, as is remarked in "Lorna Doone". In the contracted stems of rosette plants successive rings of wood may not be formed at all, or if they are, they are related to environmental conditions, not to age.

A constant relation to annual growth is only maintained in main stems of unlimited growth. In short branches, such as fruits spurs, annual rings are only formed in the first few years; while in thorns, unless they bear leaves, there is never more than one ring, however old they may be. The phloem rarely shows indications of annual rings. The periodicity of cambial growth is less marked on the phloem side. Activity begins later and may, in some cases, continue into and perhaps through the winter, the spring change in the phloem not being the formation of new elements, but the very marked swelling of the old elements through water intake. Towards the end of its life the phloem may, though rarely, become signified, as may be seen in old stems of the annual Sunflower.

Wood anatomy has proved a very useful adjunct to systematic Botany and is serving to supplement the ideas of classification and evolution based upon external morphology. The features principally used for comparison are: fibre types, distribution of the xylem parenchyma, type of perforation in the end walls of vessels and the presence of absence of "storied structure" in the wood. The origin of storied structure has been explained under the heading of Cambium. It implies a marked uniformity of length in each series of tracheids and vessel elements as seen in longitudinal section, so that the wood appears to be built up of vertically successive layers, or storeys. These characters are not always sharply marked and a good deal has to be left to the judgement of experience in deciding systematic affinities on wood characters.

The above characters are grouped as specialized (i.e., advanced in the evolutionary sense) or unspecialized, as follows

Unspecialized	*Specialized*
Scalariform vessel pores	Simple vessel pores
Fibre tracheids	Libriform fibres
Apotracheal parenchyma	Paratracheal parenchyma.
Diffuse-porous	Ring-porous
Unstoried structure	Storied structure

Specialized unspecialized features are widely distributed among angiospermic families and show no relation to specialized or unspecialized floral structure, but truly natural families are usually fairly homogeneous in their type of wood structure.

The Pith

The comparative anatomy of *pith* is still to be written, though it is evidently a very variable tissue. Many herbaceous plants, notably among the Umbeliferae and Compositae, have hollow stems, due to the failure of the pith to keep pace with the expansion of the outer tissues.

The pith thus becomes cracked and riven, or *fistular* as it is called, only a few scanty fragments of tissue remaining to line the cavity. Certain woody plants, of which the Walnut (Junlans regia) is the best known, have discoid pith, dut to transverse cleavage, which, consequent upon growth in length of the young shoot, separates the pith into a series of discs, 2 to 3 mm. apart. The pith often becomes sclerotic, especially at the nodes, and nodal plates of highly thickened cells may remain even when the pith has disappeared in the internodes. The outer zones of the pith are frequently smaller celled, more sclerotic and more persistent than the central portion, forming a conspicuous boundary layer, called in German "Markrande", though an international name has not been adopted.

8

Structural Modification of Stem

Stems are commonly *siphonostelic protosteles* occurring in living plants only among the fevus and some other pteridophytes and usually possess secondary growth. Stems differ greately in the amount and arrangement of primary vascular tissues and in the amount of secondary tissues. The primary vascular tissue ranges in amount from that of a solid cylinder of considerable thickness to that of a few small strands forming isolated bundles. The various conditions apparently represent stages in evolutionary progress where there has been a thinning of the cylinder of primary vascular tissue, a breaking up of the cylinder into longitudinal strands,and changes in the arrangement of the resulting bundles so that these no longer form a cylindrical series. Whatever the amount and the arrangement of the primary xylem, the secondary vascular tissues may form a solid cylinder enclosing it and it this way develop an unbroken vascular cylinder even from a series of unevenly placed strands.

The amount and the arrangement of secondary xylem also vary from that of the complete cylinder of indefinite thickness formed in typical perennial woody axes to be isolated thin strands of certain types of annual herbaceous stems and to the condition in other plants where secondary growth is absent. Secondary vascular tissues, like primary, have been reduced in evolutionary modification, the cylinder being first thinned radially and then broken up tangentially. In forms most specialized in this respect, no secondary vascular tissues are formed. All stages in these changes in primary and secondary vascular tissues are represented among living plants. Great variety is found in stem structure.

Types of Stems

Woody Stem

Perennial woody plants present apparently simple stem structure. In these an unbroken layer of secondary vascular tissue sheaths a more or less continuous cylinder of primary xylem. Variations in the structure of this cylinder, range from cylinders that are unbroken except by leaf and branch gaps to those consisting of discrete bundles often complex in arrangement. The simplicity of the secondary cylinder, concealing the basic primary structure, seems to make the entire structure simple. Even where the primary cylinder is broken only by leaf and branch gaps prominent "bundles" stand out as protoxylem-containing ridges in the otherwise very thin cylinder. These ridges represent the downward continuation of the leaf traces and other early maturing parts of the cylinder. In other plants the cylinder consists of more or less closely placed bundles of varying size–with proportionate amounts of protoxylem—without connecting primary xylem. The areas between these primary strands are ray-like projections of the pith on the inside and pericycle on the outside. They are soon closed by the cambium which builds secondary vascular tissue in these spaces. Woody, perennial stems, such as those of some palms and other large monocotyledons, may be developed without secondary growth.

Herbaceous Stem

Herbaceous plants have no distinctive anatomical structure. Annual stems are commonly supposed to have their vascular tissue—whether it is primary only, or primary plus secondary—characteristically in the form of discrete bundles arranged in a cylinder, but this condition is not typical of herbaceous plants. Among dicotyledons, the majority of herbaceous forms possess cylinders of vascular tissue that are complete, except for the presence of leaf and branch gaps. Such conditions obtain not only in coarse and stout stems, such as those of many composites, mints, and legumes, but even in slender delicate stems, such as those of species of *Veronica* and *Stellaria*. The herbaceous stem with discrete bundles is not to be considered typical of annual stems. It represents, clearly, the extreme stage of reduction of vascular tissues.

Although infrequent in occurrance, it is found in various families and in both stout and slender plants. In some forms, for example, in species of *Trifolium*, *Geum*, and *Agrimonia*, the lower part of the stem

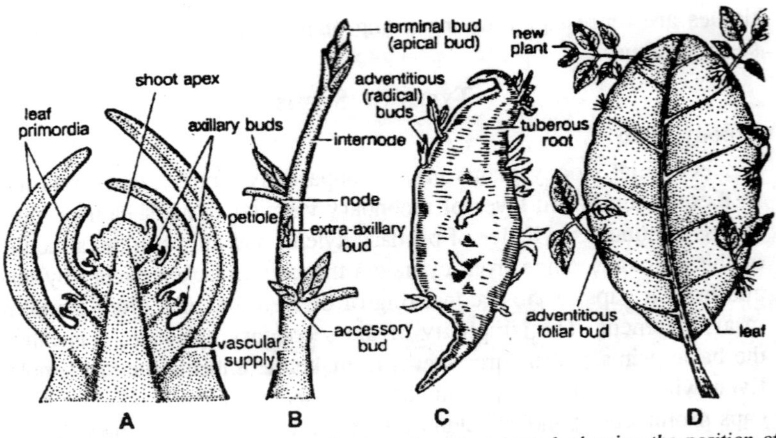

Fig. 8.1. Buds: A, a bud in longitudinal section; B, a branch showing the position of buds; C, adventitious radical buds of sweet potato; D, foliar buds of Bryophyllum.

may have a complete cylinder of vascular tissue, and the upper part separate bundles. The condition in monocotyledons is essentially the same as that in extreme dicotyledonous herbs; here the bundles may lie in a ring, but more frequently they are distributed through the stele according to a plan determined by the number of leaf traces, lead arrangement, and other factors. Herbaceous stems, except those of the extreme type where the cambium fails to unite the primary skeleton, are structurally like woody stems; differences are those of period of persistence, not of basic structure.

The herbaceous stem is one in which cambial growth is limited to one season or part of one season, or is lacking. The amount of vascular tissue, primary or secondary, is not necessarily less in a mature herbaceous stem than in the one-year-old stem of a closely related woody plant of the same type. The type of stem in a given herb depends upon the type of stele present in the woody ancestral forms. Where the vascular cylinder of the woody type possesses primary tissues in the form of a continuous cylinder, related herbaceous forms also possess unbroken primary cylinders; where the primary cylinder of woody forms is discontinuous, related herbaceous plants have a type of stele in which the cylinder consists of discrete bundles. In some details, a mature herbaceous stem with well developed secondary vascular tissue may differ from a similar woody stem. Commonly, in annuals, as seasonal growth of the stem ceases, the cambium disappears, all its cells naturing as xylem and phloem elements. All tissues outside the cambium become strongly compressed—cortical parenchyma, especially delicate, photosynthetic types, may be completely flattened; and even before flowering and seed-formation is completed, the softer cells of the phloem, including sieve tubes, are crushed.

The phloem fibres of such plants show structural evidence of this lateral compression in their "jointed," or "broken" structure. In herbaceous stems the epidermis rarely is ruptured with increase in diameter. It is maintained at first by slow division; later, the compacting and crushing of the outer tissues provide space for the developing secondary xylem and phloem. The morphological problem underlying the evolutionary changes in structure from the typical woody stem to the extreme herbaceous stem lies outside the scope of this book. From the evidence of comparative morphology and of the fossil record, it is clear that in angiosperms the herbadeous type of stem has been derived from the woody, undoubtedly independently in many families. In a few families, such as the *Berberidaceae*, woody types have been derived from herbs. Woody plants of this type have many features in common with herbs, especially structure of the primary skeleton and histological structure of the xylem.

Monocotyledonous Stem

Secondary growth of the usual type is typically lacking throughout the body of monocotyledons but vestiges of cambial activity in the bundles both of the stem and the leaves have been found in nearly all groups. The stele is broken up into bundles which are commonly distributed throughout the axis; the endodermis is lacking, and the limits of cortex, pericycle, and pith are often indistinguishable, since bundles are scattered throughout. In many forms, as in most grasses, there is present a central region—which may or may not represent the pith morphologically—in which no bundles occur; in other forms a more or less readily separable cortical region may also be seen. The vascular system of most monocotyledons is highly complex. The leaf-trace bundles are numerous and follow various types of courses in their descent, uniting in different ways with other strands.

A common condition is that where all bundles are common bundles. The traces, upon entering the stem from the leaf, penetrate deeply, the median traces more deeply than the lateral, and then in their descent return toward the periphery. The downward course may be vertical, or the traces may swing laterally and become oriented in various ways. Each common bundle sooner or later fuses with other similar bundles. The anastomoses occur chiefly at the nodes,and in some groups, such as the grasses, are abundant and largely restricted to that region.

Secondary Growth in Monocotyledons

The type of secondary tissue formed in the monocotyledons is very different from that formed in other groups. The cambium does not produce phloem on the outside the xylem toward the inside in the normal way but forms on the inner side amphivasal or collateral bundles in parenchymatous ground tissue commonly called *conjunctive tissue*. The bundles are usually without definite arrangement but may lie to some extent in radial rows. Occasional anastomoses occur. The extracambial tissues formed are small in amount and parenchymatous in natrue. In the cambium layer proper, the divisions are largely tangential, and the cells so formed are consequently radially arranged. This order is usually evident in the conjunctive tissue and aids in the separation of the secondary from the primary tissues in which the interfascicular parenchyma has no definite arrangement. It is not evident in the tissues of the bundles because of the method of development of these strands. A very small group of cambium derivatives divide longitudinally—at first anticlinally and periclinally, then haphazardly —forming a strand of cells that mature as xylem and phloem.

The cells that form the tracheids become 15 to 40 times their original length; the other cells elongate little or not at all. The tracheids may be scalariform, a type rare in secondary tissue. The mature bundles differ from the primary bundle in the small amount of phloem and the absence of annular and spiral protoxylem cells. In some genera differences in relative abundance of bundles and in wall thickness of the conjuctive cells set off weakly limited growth rings. The relation of these rings to annual growth is unknown.

Secondary tissue of this type is indefinite in amount, as is that of the normal type, but usually develops slowly, and large trunks are not commonly formed. Increase in diameter of this type occurs, in the arborescent Lilliflorae, for example, *Dracaena*, *Yucca*, *Aloe*, *Cordyline*; some of the palms; rarely in herbs, as in *Veratrum*; and in fleshy parts of some of the *Dioscoreae*. The thickening which takes place in the bases of some palm stems is not due to the activity of a definite cambium layer but is, rather the result of gradual increase in size of cells and of intercellular spaces, and rarely, of the proliferation of strands of tissue to form new fibres; it represents long continuing primary growth.

Vine Type of Stem

The stem of vines are of both structural types. Many vines, such as *Vitis*, *Celastrus*, and *Sclanum Dulcamara*, have vascular steles in the form of woody cylinders; others, such as Clematis, *Humulus*, and *Pisum*, have vascular tissues arranged in the form of a ring of bundles. These bundles are separated by rays of parenchyma, which in many forms are increased, like the bundles, by cambial growth. In such forms, as in the type of herbaceous stem with discrete bundles, the xylem and phloem are highly specialized in structure—vascular rays are commonly absent; the vessels are porous and of large diameter and great length; tracheids and fibres are proportionately few; sieve tubes are of the highest type; and phloem parenchyma and fibres are scarce or lacking. Many vines show not merely these tissue specializations, but also possess anomalous general structure. In this group fall *Aristolochia* and *Menispermum*, which, unfortunately, are often cited as examples of typical stem structure, and used to demonstrate the origin of a woody from an herbaceous stem. In the stem of the young herbaceous vine, such as that of *Pisum*, *Apios*, and *Adlumia*, the vascular bundles are separated by wide segments of parenchyma. In these stems the cambium may be restricted to the bundles, as in Adlumia, or may from a complete cylinder by extension across the interfascicular rays of parenchyma. This interfascicular cambium may be vestigial, as in the upper parts of the pea vine, forming no secondary tissue, or a very few vascular cells, as at the base of the pea vine; or it may be very active, increasing the parenchymatous rays at the same rate as the fascicular cambium builds up the bundles, as in *Clematis*. In such plants as the last, the type of structure found in the one-year stem is maintained as the stem becomes woody and perennial. Yet, since the bundles increase in tengantial extent as secondary growth continues, the stem apparently becomes a woody stem.

The bundles are, however, still separated, though by wedges of secondary rather than primary parenchyma. These radial plates or rays of tissue commonly extend unbroken from node to node, often for several internodes and constitute an important feature of vine structure. Such rays, either of primary structure, or of both primary and secondary structure, are prominent structural features of many vines, both annual and perennial. Similar parenchymatous segments are present also in some vines with complete woody cylinders, and constitute one of the prominent modifications of such steles found in lianas. Since they consist of soft tissue, they are sometimes crushed as the stem becomes older, perhaps by the "play" of the bundles upon one another during the periods of lateral stresses to which a vine stem is peculiarly subject. Apparently, this type of structure represents one type of modification related to the mechanical requirements of vines.

"Medullary Rays" of Vines and Herbs

In vines and herbs with dissected vascular cylinders, the vascular bundles are in early stages separated by sheets of parenchyma which merge internally with the pith and externally with the cortex. In these early stages there is usually little or no evidence of limitation of pith or cortex in the region of these bands, nor is there histological evidence that they belong morphologically, even in part, to the vascular cylinder. Since they appear as radiating portions of the pith, they are commonly called *medullary rays*. When these primary structures are later increased in radial extent by secondary growth, as they are in many vines, such as *Clematis*, they closely resemble broad "medullary rays" (vascular rays) of secondary wood and phloem. For this reason both these rays and true vascular rays have been termed "medullary rays," and believed to be homologous. The theory that the woody stem is developed from the herbaceous stem is based in large part upon this misconception. The broad rays of vines and herbs extend from node to node and often through several internodes and in this respect differ much from the vascular rays which are of very limited vertical extent.

Further, they clearly represent, morphologically, entire segments of the vascular cylinder, being equivalent to many vascular rays plus the tissue surrounding and included between these rays. It is obvious that the same term should not be applied to both types of rays. The term "medullary rays" as applied to rays of secondary "vascular ray." Since the broad rays of vines and herbs, when only primary, resemble projections of the pith, they may perhaps aptly be called medullary rays, especially if this term is not applied to vascular rays. That they are not homologous with the structures hitherto known commonly as medullary rays is obvious.

Internal (or Intraxylary) Phloem

In the siphonosteles of ferns the amphiphloic condition is usually prominent. Here the internal phloem forms a continuous layer and is closely similar to the external phloem. In the angiosperms, where internal phloem is also frequently present, this tissue is less conspicuous. It occurs usually as strands, large or small, more or less closely associated with the primary xylem. In amphiphloic ferns, where the cylinder is broken by leaf gaps, the internal phloem unites with the external through the gaps, and amphicribral bundles are formed.

In Angiosperms with internal phleom and broken vascular cylinder, the internal phleom forms the innermost part of bicollateral bundles, as in *Cucurbita*; where the primary xylem forms a more or less complete cylinder, strands of internal phloem also form a fairly continuous layer. Internal phloem is in most plants only primary; rarely a cambium arises inside the primary xylem and a small amount of internal secondary phloem is formed, as in *Tecoma*. The cells of internal phloem are like those of external phloem, except that fibres are few or lacking, and the sieve tubes and companion cells occur in small, restricted groups, surrounded by parenchyma. This parenchyma, together with that of the protoxylem region, forms the perimedullary zone of the pith. In ontogeny internal phloem develops later than external primary phloem. Such a layer is, of course, morphologically, a part of the vascular cylinder and not the outer layer of the pith. Where an inner endodermis is present, this layer separates the inner phloem from the pith. The internal phloem and the internal endodermis are often continuous through leaf and branch gaps with the outer phloem and the outer endodermis. Internal phloem occurs in many families among the angiosperms.

Forms of Stems

Ordinarily stems bear foliage leaves, and in such a way that the leaves are exposed to light and enabled to function as organs of photosynthesis. Stems also serve as conducting channels through which materials synthesised in the aerial parts of the plant may be translocated to roots or other underground organs, and by means of which substances absorbed by the roots from the soil can be transported to the leaves. Stems, however, may exhibit various degrees of specialization of form, correlated with special functions which they have assumed. They may be underground instead of aerial, prostrate instead of erect, serve as organs of perennation and as storage organs. It will be readily understood, therefore, that stem structures assume and immense variety of forms. Some grow *erect* and are self-supporting. Examples of erect stems are sunflower, broad bean, foxglove, in fact, this is the typical forms of stem. Erect stems may be woody or herbaceous. Herbaceous stems live generally only for a single growing season, whereas woody stems persist from year to year. Plants with herbaceous stems may be annuals, completing their life-history in a single season; others are biennial, usually growing and forming shoots in one season and producing flowers, fruits and seeds in the second year, death occurring after seed production.

Herbaceous perennials have an underground, permanent stem which each year produces a crop of aerial stems which bear foliage leaves and flowers. These aerial stems are short-lived—dying down at the end of the season. Woody stems are generally perennials. Not only are the plants long-lived, but the shoots also are perennial in character. In some woody plants many shoots arise which appear capable of only very limited growth. Such dwarf shoots occur in the jak, where they bear the flowers. Sometimes, for reasons not fully understood, a dwarf shoot or spur may assume the power of rapid and extensive elongation. Extremely short stems occur, too, in many herbaceous perennials. In *Taraxacum*, *Elephantopus* and many primulas the plant appears to consist only of a root surmounted by a rosette of leaves. In fact, we have in these plants a root and a very short stem with extremely short internodes, and it is these two characters that give the impression of a crown of leaves borne directly on a root. Aerial shoots may serve as water-storage organs, as in many cacti and other succulent plants.

The leaves are often reduced or absent and the peripheral regions of the stem are green and carry on many of the functions normally performed by the leaf. Whilst the central region of the stem consists of specialised water-storage tissue. Aerial stems may show modifications of form associated with food storage. Parts of swollen "root" of the radish and turnip and of celeriac are in reality stems, whilst in the kohlrabi the whole of the swollen storage-organ is stem and bears leaves and leaf-scars. Most stems are aerial, but some are buried in the soil, forming underground or subterranean stems.

Weak Stems

In some plants the stems are more or less prostrate or trail along the ground. The weak stems of many plants, however, which are unable to grow erect, make their way upwards by attaching themselves to surrounding objects. These are climbing and twining plants. In climbing plants the climbing is effected in various ways. *Ficus repens*, for example, climbs by means of adventitious roots; these roots, developed on the stem, fix themselves to the support on which the plant climbs. *Lathyrus*, *Passiflora*, *Vitis*, and many other plants, climb by means of special organs called *tendrils*. These tendrils, may be specialised stems, leaves, or parts of leaves. Virginian creeper (Ampelopsis) climbs by means of adhesive, sucker-like discs developed at the tips of the branches of its tendrils; *Clematis*, by means of its leaf-stalks or petioles, which act as tendrils, and, in fact, are called petiole tendrils.

In some plants, e.g., Artabotrys, and *Strychnos*, a modified inflorescence-axis forms a sensitive hook, which, after attaching itself to a support, thickens and becomes woody. The leaf-tip of *Gloriosa* is sensitive and functions as a tendril. As distinguished from these, *twining plants* achieve the same result by twining round some support, as for example, *Convolvulus*, *Thunbergia*, *Ipomoea*, and *Phaseolus*. Climbing in all cases enables plants with weak stems to attain such a position that their foliage leaves can receive adequate illumination and so be enabled to carry on the process of photosynthesis.

Modifications of the Aerial Shoot

Cladodes and Phylloclades

The term *cladode* is applied to a branch of a single internode which is flattened to stimulate a leaf, and the term phylloclade to entire shoots similarly flattened and leaf-like. There are considerably more cases in the latter category, and it is not always easy to discriminate the one type from the other, with the result that many botanists treat them as synonymous. The biological principle is the same in both, namely the replacement of reduced leaves, which are no longer functional for photosynthesis by that branches, containing chlorophyllous tissue, which function physiologically as leaves. The benefit to the plant of such a procedure is indeed puzzling, in as much as the cladopes are never as efficient photosynthetically as are true leaves. Perhaps we have here an example of the supposed "Laws of Irreversibility" in evolution, which maintains that a structure once lost cannot be regained but can only be substituted by the modification of other existing structures.

As such a substitution necessarily involves the production of new characters in the structures concerned, it is difficult to understand what should prevent the redevelopment of the lost character as the simplest response to the demand of necessity. But to these questions there are no adequate answers. A good example of a simple cladode is that of *Ruscus aculeatus* (Butcher's Broom) and of some related plants such as *Semele* and *Danae*. The cladodes are short, ovate, spiny-pointed and with parallel veins. They are deep green in colour and arise in the axils of minute scale leaves, which soon drop off. In the young shoot of *Ruscus* these leaves are much bigger and partly green; the cladode habit is only fully established in mature shoot. Each cladode bears, half-way up and in the median line, a scale-life bract, from the axil of which arises in summer a group of flowers. There has been much dispute about the morphological nature of this structure. The classical view is that the whole structure is a branch of limited growth bearing one node from which arise the flowers, subtended by the small bract.

Opposed to this is the view that only the lower half is axial in nature and that its apex (the flower bearing "node") are two opposed, bracts, one small and one large, the latter forming the upper half of the cladode, with its margins decurrent along the lower, axial portion. This latter view, though it seems farfetched is, in fact, supported by the vascular structure, which is axial only in the lower and leaf-like in the upper half, and also by the presence of opposed bracts in the related genus *Denate*, where the inflorescene is free from the cladode. A third theory interprets the whole cladode as a modified leaf, borne on an axillary shoot, to which it is united along its lower half.

The common Asparagus (A. officinalis) has cladodes which appear from the axil of each scale leaf as two little bundles of green needles. Comparison of the placing of these, relative to the scale leaf, with the arrangement of the flowers at flowering nodes, makes it very probable that they represent sterilized flower stalks arranged in two cymes. They do, in fact, occasionally bear abortive flower buds at their ends. The median axillary shoot, present at the flowering nodes, is suppressed at the vegetative nodes, but in the allied genus *Myrsiphyllum* (Climbing Asparagus) it is present and is developed into a leaf-like cladode, instead of the needles of the common Asparagus, which in this case are suppressed. As true phylloclades in our present sense, we may cite several cases in which the whole of a shoot or indeed all the shoots of the plants are flattened and leaf-like.

Perhaps the best known is the Cactus called *Opuntia* (Prickly Pear), the shoots of which form large discoid segments. The apex is, however, radially symmetrical, and in some species all the shoots are globose or cylindrical. Some other species of the Cacti are also flattened notably among those which form the genus, *Epiphyllum*, frequently grown for the sake of its large scarlet flowers. None of these Cacti has any true leaves. This is probably connected with their desert habitat, but the flattened shoots are a peculiarity of these genera, not necessarily related to the environment. In the genus *Muehlenbeckia* (Polygonaceae) and *Carmichaelia* (Papilionaceae) from the southern hemisphere, the shoots may from jointed green ribbons, but temporary leaves are produced at the nodes in spring. Certain species of *Phyllanthus*

(Eupherbiaceae) show an interesting example of what we may call pseudo-phylloclades. The main shoots be only scale leaves from whose axils arise side shoots of limited growth, which are pinnately branched and bear small true leaves in two lateral rows. The whole appearance of these shoots is exactly that of the pinnately compound leaves of such leguminous plants as *Mimosa*. In other species of *Phyllanthus* the lateral shoots are true phylloclades.

Thorns

True stem thorns are distinguished from prickles by the branch nature. Prickles are modified trichomes and spines are usually modified leaves or stipules, but thorns arise from leaf-axils and often leaves or flowers as evidence of their branch nature. The thorns of *Hippophae* and *Crataegus* are examples. They commence as axillary shoots with normal leaves and with an apical bud. Left to themselves, however, apex soon stops growth and hardens into a woody point, from which the undeveloped leaves fall away, having it naked.

The lower leaves in large thorns may persist, and from their skills may develop secondary thorn branches. In *Ulex* (Gorse) the main

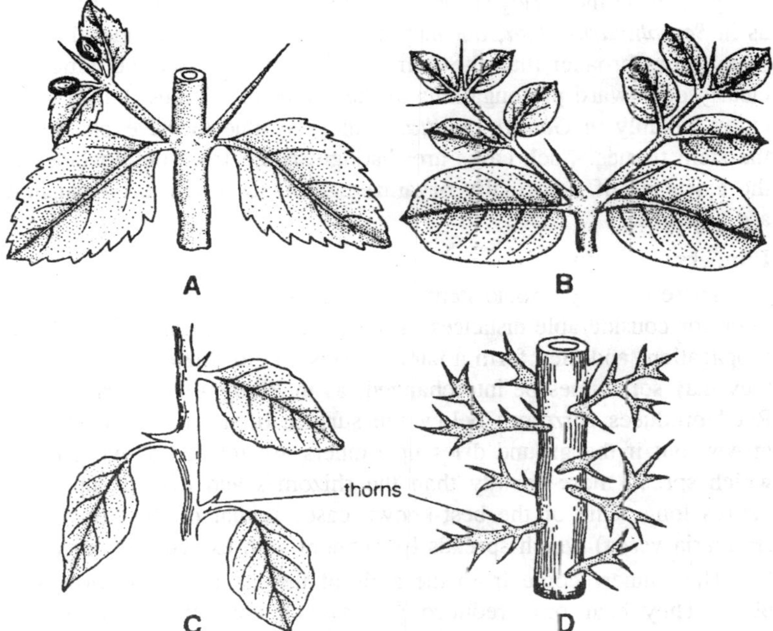

Fig. 8.2. Stem thorns: A, Duranta; B, Carissa; C, Bougainvillea, D, Flacourtia.

shoots are thickly with compound thorn branches, but in this case the leaves are also reduced to spines and the function of photosynthesis devolves on the branches including the thorns which are all green. Gorse seedlings raised in a damp atmosphere do not develops thorn or spines, but produce leafy side shoots of reduced growth.

A similar change has occurred in the Apple (Pyrusmalus) which is thorny in the wild state, but not in cultivation, where the thorns are represented by fruiting spurs. The above examples are all straight thorns, but thorns may also be curved in the form of a hook, as in the South African tree, *Acacia detinens*, from which the unwary may find it almost impossible to escape. Intermediate structure between thorns and cladodes are known. Such for example, are the remarkable flattened branches of *Colletia cruciata*, which are unlike other cladodes in being flattened in the vertical direction. They, and the secondary cladodes which grow from them, end in thorn like points.

Winged Stems

These provide another example of the stem taking over the photosynthetic functions and something of the structure of leaves. The wings may be mere ridges, as in the Broom (cytisus) or narrow frills as in *Scrophularia alata*, but they are often so developed that they are many times broader than the stem itself. In many cases the wings are clearly downward prolongations of the leaves or stipules, but in a few cases, notably in *Genista sagittalis*, they are due to the expansion of the stem tissues. Such cases are distinguishable from phylloclades by the existence of a normally organized stem, to which the expansions are attached.

Runners

These are plagiotropic stems which grow on the surface of the soil, often for considerable distances. They are a potent means of vegetative propagation, and they form a natural transition to rhizomes with which they may sometimes be interchanged, as in *Phragmites*. Normally the Reed produces rhizomes below the surface of the mud on which it grows, but if the ground dries up runners are formed above ground, which spread more rapidly than the rhizomes and may be up to 10 metres long. One of the best-known cases is that of the Strawberry (Fragaria vesca) which spreads by runners with extreme rapidity.

The runners come from the axils of the leaves on the flowering plant. They bear only reduced leaves each consisting of a pair of stipules with a small bristle between them. The runner ends at the second node, with a rosette of normal leaves. From the axial of a

plagiotropic stems which root adventitiously. It is most generally used for comparatively short, above-ground shoots which are shorter than typical runners. As it has no precise significance it had better perhaps be dropped. The same is true of the term sobole.

Droppers

Sometimes called Rhizomorphic Droppers. These are a special kind of rhizomatous branch, produced from young bulbs of Tulip, etc. An axillary growing point is covered in by the inrolling over it of a leaf initial, so that it lies inside a hollow hood comparable to a coleoptile. The base of the leaf and the stem immediately below the growing point then elongate greatly, carrying the protected apex outwards and downwards from the parent bulb, the base of the rolled leaf acting as a boring point, until a suitable depth is reached, at which the enclosed growing point may develop into a new bulb.

Tubers

These are solid, thickened stems or branches serving for storage and also, if detached, for vegetative propagation. They may be formed on rhizomes, on axillary branches or on main stems, either below or above ground. Rhizome tubers are only distinguishable as such when formed on comparatively slender rhizomes, such as those of *Circaea*, since swollen rhizomes such as those of *Iris germanica* may be regarded as tuberous throughout. In *Circaea lutetiana* and in *Cyperus esculentus* the tubers are terminal on rhizomes branches. They contain the terminal bud protected by scale leaves and become centres of new growth when the rhizomes connecting them to the parent plant has perished by decay. Those of *Cyperus esculentus*, *Stachys tuberifera* and of *Sagittaria sagittifolia* are edible. Branch tubers may be typified by the Potato. The tuber-bearing branches come from the lowest axils of the aerial stem, even, in seedlings, from the cotyledonary axils. These branches are plagiotropic or weakly geotropic, and if formed above ground they bury themselves and swell out at their apices to from the massive tubers, which bear temporary scale-leaves with buds in their axils called "eyes".

The tuber has a corky periderm and may remain dormant for some time. When it germinates, orthotropic shoots are formed from the axillary buds, which become new aerial stems. Even a single detached bud will grow and form a new plant. In some conditions axillary buds near the top of the leafy shoots may be transformed into sessile tubers. Placing the plant in the dark under a box will often stimulate this abnormal tuber formation. The tubers of the Jerusalem

Artichoke (Helianthus tuberosus) are formed like those of the Potato, but they are borne on shooter branches, are often themselves branched and have persistent, fleshy scale leaves. They contain insulin instead of starch. Similar bud more slender tubers are formed by the perennial Sunflower (Helianthus rigidus). Tubers on the main stems are formed, for example, in a climbing member of the Asclepiadaceae (Ceropegia woodii), which is often grown in green houses.

The tubers are local thickenings of the nodes, including the bases of the pair of opposite leaves home at the node, and their axillary buds. They are generally not detached, but form adventitious roots *in situ*, which attach them to the support on which the plant is climbing and so become new centres of growth. Many members of the Dioscoreaceae form stem tubers. In *Tamus communis* (Black Bryony) the first internode above the cotyledons thickens to form a parennial tuber, which may reach a very large size and may branch. From this tuber the leafy shoots are annually renewed. Another well-known example is *Testudinaria elephantipes*, a South African desert plant in which the base of the stem forms a large tuber, as much as 30 cm. across, which stands partly above ground. It is covered by a very thick, irregular periderm which gives it the name of Elephant's Foot. As in *Tamus*, delicate or trailing shoots are produced from the tuber each spring and disappear in the dry season.

Corms

The corm is simply is a special form of underground tuber, consisting of a much contracted, swollen, main stem, whose principal axis is vertical. On the top it bears an apical bud, from which summer shoots with leaves and flowers are produced, and it bears an annual crop of adventious roots, usually from the lower end. Good example are provided by *Crocus Gladiolus*. Here the corm is considerably flattened, its length being shorter than its diameter. It is covered by a tonic made up of the fibrous remains of leaf bases, which arise from nodes that form a series of ridges around the corm. At the nodes there are occassional buds, axillary to the scales, from which side shoots and new corms may arise. In *Tritonia* these buds produce rhizomes, which copiously propogate the plant, forming new aerial shoot with new corms at their bases. When flowering is over the leaves of *Crocus* remain active for some time, and the material they assimilate is stored at the base of the flowering stem, producing a new corm, on top of the old one, from a bud axillary to one of the leaves.

One or more axillary buds at the uppermost nodes of the old corm also develop into new flowering shoots and eventually into new corms, so that the number of corms increases year by year by sympodial branching. The corm scales are formed from the base of the scale leaves which envelope the lower part of the flowering stem. The old corms rot away, leaving a space beneath the new ones, and into this space the new corms are pulled down by thick contractile roots, produced in late summer from the base of the new corm. The plant thus maintains a uniform depth in the soil. In the seedling the first corm is produced by the swelling of the hypocotyl. The corm of *Arum maculatum*, often cited as an example, is however, plagiotropic and is really a very short rhizome. Although corms are not uncommon in Monocotyledons they are less common among Dicotyledons. Good examples in the latter class are the so-called Bulbous Buttercup (*Ranunculus bulbosus*) and species of *Corydalis*.

Bulbs

A bulb differs from a corm in that the stem is reduced to very small dimensions and the body of bulb is made up of thick, fleshy bulb-scales. The difference is a question of proportion. In the corm the swollen stem is the main storage organ and the scales are merely protective, while in the bulb the scales themselves have been magnified and have replaced the stem as the main storage organs. The stem is no more than a thick disc or very flat cone, but it has an apical bud on the upper side, and adventitious roots are formed in an annual crop from the marginal portion of the underside, their development following a resting period of from two to four months after the conclusion of flowering. The fleshy scales are set very close together, and their bases of insertion almost or completely surround the stem.

In *Tulipa* the bulb scales are formed independently of the foliage leaves, but in *Narcissus* they represent the persistent bases of the foliage leaves. In both cases, however, they are richly stored with food material.

The bulb of *Lilium* differs from the above examples, chiefly in the much greater number and smaller size of its fleshy scales, the outermost being loosely arranged and easily detachable. In favourable conditions these detached scales will produce adventitious roots and a bud which can grow into a plant. The flowering bud for each year develops during the previous summer. In *Tulipa* the apex of each bulb becomes transformed into a flowering axis, but in *Narcissus* the apex itself persists and forms only a few foliage leaves, the flowering axis being axillary. The branching of *Tulipa* is thus sympodial, that of Narcissus is monopodial.

The organs of the flower are completely formed in the flowering bud by the end of summer, together with the rudiments of the foliage leaves, and only require to expand when the flowering season arrives. This expansion is carried out at the expense of material stored in the scales, which consequently shrivel, the outermost becoming dead and papery. After flowering the foliage leaves persist end the food material they elaborate is sent down to bulb. In the Tulip this food promotes, the development of one or more axillary buds into new daughter bulbs, which absorb the surplus of food from the old bulb and consequently replace it.

In Narcissus the food is stored in the leaf bases, both of the current year and those persisting from the two or three previous year. An axillary bud also develops into a new flowering bud and one or more other axillary buds may develop into daughter bulbs which, however, do not replace the mother bulb, which remains active for years. The bulb of Bluebell (Scilla non-scripta) is intermediate in character between the two examples given above. The scales consist of the bases of the foliage leaves of one year only but its growth is sympodial. Among the Orchids we commonly find green bulbous structures at the base of the leaves, which are called pseudobulbs. These are really tuberous stem or branch segments, sometimes involving the whole length of the stem, sometimes only a single terminal segment, the apical bud of which aborts, but which bears one or two fully developed foliage leaves.

A new shoot arises from the axil of a scale at the base of the pseudo-bulb, and this in its turn ends in another pseudo-bulb, so that the appearance is presented of a row of green tubers connected together at their bases by a slender stem bearing scale leaves. There are numerous instances of the axillary buds on aerial shoots becoming transformed into small bulbs or *busbill*. Good examples are shown by *Dentaria bulbifera* and *Libium bulbiferum*, in which nearly all the axillary buds are so transformed. These bulbils are readily detached and form an effective means of vegetative propagation. In the genera *Allium* and *Agave* and in some alpine plants, e.g., *Polygonum viviyarum* and *Poa alpina*, the flowers are often transformed into bulbis, but the latter cases are undoubtedly related to the very short growth season available to the plants and perhaps to the scarcity of pollinating insects at high levels. Finally there is the case, described by Goebel, of

Cryptocoryne in which the embryo itself becomes a bulbil and is shed naked, detached from its cotyledon, which remains in the seed. These latter cases of modification of floral structures, though not morphologically equivalent, are included under the biological term "vivipary".

Fasciated Stems

While it is not our intention to enter upon the bewildering study of plant abnormalities known as Teratology, there is one such abnormality of stems which we shall mention because of its frequent occurrence and striking appearance. It is that called *fasciation*, which consists of the development of a strap shaped structure, often several inches broad, in place of a normal stem. Fasciated stems have a "growing line" instead of a growing point. They often carry hundreds of leaves and enormous numbers of flowers. A second type, called ring-fasciation, transforms the stem into a wide cylinder, open at the top and lined with an internal as well as an external epidermis.

A complete explanation of the morphology of these monstrosities is not available, but there are two theories, one being that there has been an enlargement of a single growing point and the other maintaining that there has been a congenital fusion of many shoots. Worsdell, in supporting the latter theory, claimed that it was due to a recrudescene of the primitive form of branching, dichotomy, otherwise scarcely known among Angiosperms, with the concrescence of the numerous shoots thus produced. Biologically there are two types of fasciation, one which is germinal and inherited and another which is somatic and nonheritable. Outwardly they are indistinguishable, but there is some evidence that the latter type is stimulated by insect damage to the growing point.

9

ANATOMY OF DICOTYLEDONS

ANATOMY OF CUCURBITA STEM

Epidermis

This single outermost layer consist of compact barrel shaped cells having no intercellular spaces. The epidermis remains covered with a thin cuticle. Some of the epidermal cells possess multicellular epidermal hairs.

Cortex

This region consists of external collenchyma, chlorenchyma and endodermis.

(a) *Collenchyma.* This lies immediately beneath the epidermis consisting of many layers of the cells in the ridges, whereas in furrows it is only two or three layered or sometimes altogether absent.

(b) *Chlorenchyma*. Just below the collenchyma two or three layers of parenchyma containing chloroplasts present which help in the process of assimilation.

(c) *Endodermis*. It is innermost layer of the cortex, lying immediately outside the sclerenchymatous zone of pericycle. This layer is wavy and contains many starch grains.

Pericycle

Just beneath the endodermis there is a multilayered zone of sclerenchymatous pericycle. The cells are lignified and appear polygonal in cross section.

Ground Tissue

The vascular bundles are found lying embedded in the thin walled parenchyma cells of ground tissue. The ground tissue extends from

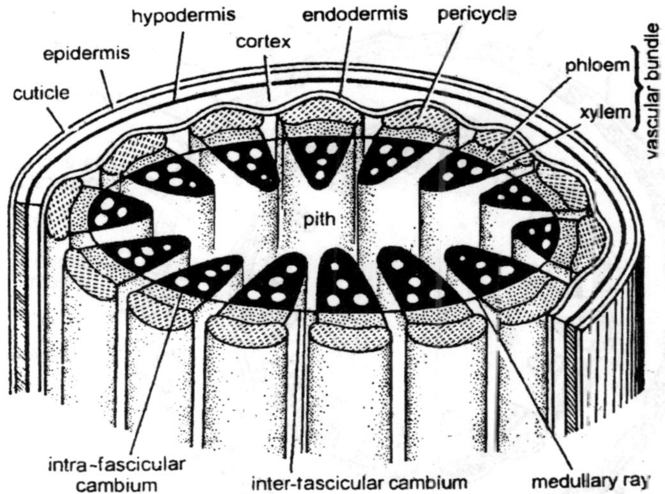

Fig. 9.1. Three-dimensional view of a typical dicotyledonous stem showing the arrangement of primary tissue.

just below the sclerenchymatous pericycle to the central medullary cavity.

Vascular Bundles

Generally vascular bundles are ten in number which are found to be arranged in two rows, those of the outer row corresponding to the ridges are those of the inner to the furrows. The vascular bundles are bicollateral each consisting of xylem, two strips of cambium and two strands of phloem.

(a) Xylem

It occupies the central position of the vascular bundle, consisting of very wide, pitted vessels towards periphery of the metaxylem, and on the inner side of narrow vessels which form the protoxylem. In the xylem certain tracheids, wood fibres and xylem parenchyma are also present. The xylem vessels are not arranged in radial rows.

(b) Cambium

In each vascular bundle two strips of cambium are found. The cambial activity remains confined within the vascular bundles. The cambial strip is found between xylem and phloem on either side of the bundle. Of the two strips of cambium it is only the external one which divides and causes growth in thickness. The cells of cambium are thin walled, rectangular and arranged in radial rows. Usually the outer cambium is many layered and flat while the inner cambium is

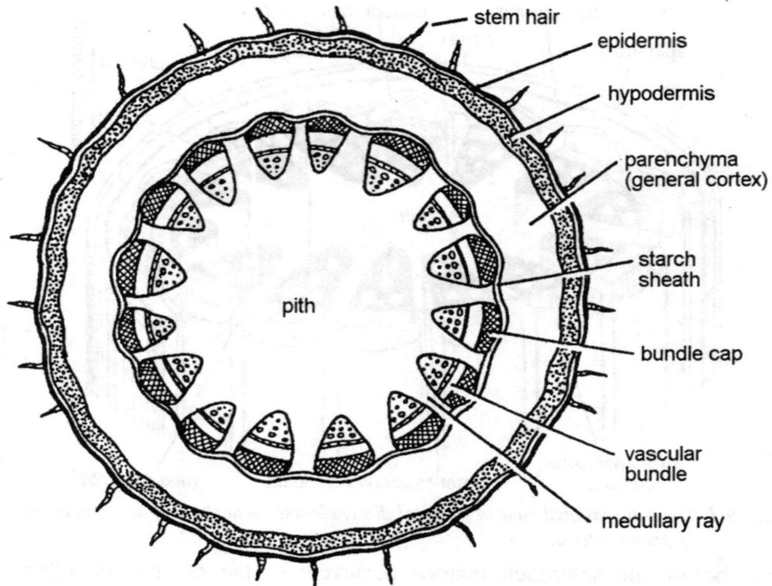

Fig. 9.2. T.S. of a young dicot stem.

few layered and somewhat curved. Only fascicular cambium is found. The stem is not woody, and therefore, the periderm and lenticels are not formed.

(c) Phloem

On the extreme ends of the vascular bundle the phloem occurs in two patches, towards the periphery, the outer phloem, and towards pith, the inner phloem. Each strand of phloem consists of sieve tubes, companion cells and phloem parenchyma. Sieve tubes are very well developed. The sieve plates with perforations are also visible. Fibres and ray cells are absent.

Special Structure

The bicollateral open vascular bundles are found each consisting of xylem (central position), two strips of cambium (outer and inner) and two patches of phloem (outer and inner).

Anatomy of Bryonia Stem

Epidermis

The single outermost layer consist of compact barrel shaped cells having no intercellular spaces. Usually the epidermis is covered with a thin cuticle. At certain places stomata are also found.

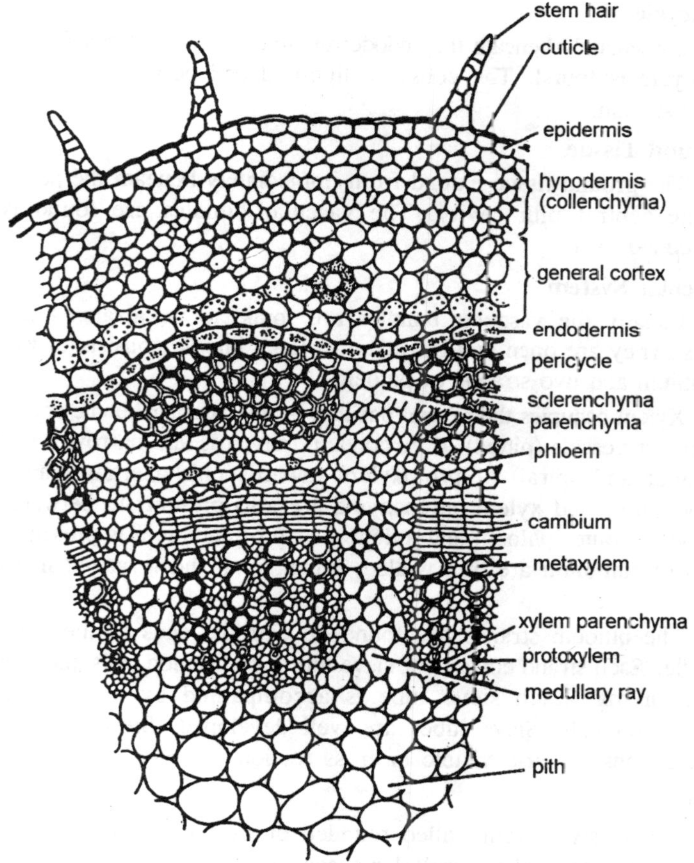

Fig. 9.3. T.S. of dicot stem.

Cortex

This consists of collenchyma, loose chlorenchyma and inconspicuous endodermis.

(a) *Collenchyma.* This lies immediately beneath the epidermis consisting of rounded or oval usually thickened laterally due to the presence of cellulose. The collenchymatous region is multilayered.

(b) *Chlorenchyma.* Below the collenchyma few layers of chlorenchyma are found. The cells are thin walled, rounded or oval, with chloroplasts and having intercellular spaces among them.

(c) *Endodermis.* The single layered endodermis is inconspicuous. The endodermal cells may contain abundant starch grains in them.

Pericycle

Immediately beneath the endodermis multilayered sclerenchy-matous pericycle is found. The cells are lignified and polygonal as seen in cross section.

Ground Tissue

The ground tissue extend from immediately beneath the pericycle to the central pith cavity. The vascular bundles are found lying embedded in it.

Vascular System

Usually the vascular bundles are found to be arranged in two rows. They are open and bicollateral. Each bundle consists of xylem, cambium and two strands of phloem.

Xylem occupies the central position of the vascular bundle consisting of bigger vessels (pitted) of metaxylem outwards and narrower vessels (annular and spiral) of protoxylem towards pith. Certain tracheids, wood fibres and xylem parenchyma are also present. In between the strands of outer phloem and xylem a strip of cambium is present. The cells of cambium are thin walled, rectangular and arranged in radial rows.

The phloem strands are found on extreme ends of the vascular bundle. Each strand consists of sieve tubes, companion cells and phloem parenchyma. Each sieve tube is accompanied by a conspicuous companion cell. Sieve tubes are well developed. Sieve plates with perforations are also visible in cross section.

Pith

It consists of thin-walled rounded or oval parenchymatous cells having well defined intercellular spaces among them.

Special Structure

The bicollateral and open vascular bundles are present. Each bundle consists of central xylem, a cambial strip and two strands (inner and outer) of phloem.

Anatomy of Monocotyledonous Stems

The monocotyledonous stems are similar to dicotyledonous stems in having an *epidermis*, a *cortex* and a *stele*. The cortex may be well developed and sharply marked off from the stele, or it may be quite narrow andinconspicuous. It is in the structure and arrangement of bundles that monocotyledonous stems markedly from dicotyledonous stems.

Stele

The vascular bundles of monocotyledonous stems, instead of being arranged in a cylinder as in dicotyledonous stems, are usually scattered throughout the stele, including the pith, so that there is no distinction between pith and pith rays. Sometimes the centre of the stele is free from vascular bundles and is occupied by parenchyma cells, which dry up and disappear at an early stage, resulting in a hollow stem, as in most grasses.

Vascular Bundles

The vascular bundles of monocotyledonous stems are like those of dicotyledonous stems in consisting of xylem towards the centre of the stele and phloem towards the periphery. The arrangement of the vascular bundles in a monocotyledonous stem is usually described as scattered, but this is not always the case, as in the stolons of grasses, sedges, rushes, and indeed in many erect stems. Generally speaking it is true that monocotyle-donous vascular bundles have no apparent regular arrangement. This is due to the behaviour of the leaf-traces bundles which enter the stem, usually from the sheathing leaf bases, and penetrate for varying distances towards the centre. They then curve outwards and ultimately coalesce with other bundles near the periphery. Meanwhile they have remained independent through a varying number of internodes. Such bundles are referred to as *common bundles* because they are common to both leaf and stem.

In addition there may be present *cauline bundles* which pertain the stem, and *foliar bundles* which join on to other bundles immediately on entering the stem. The rest of the tissue within the epidermis may consist of an undifferentiated ground tissue of parenchyma, storing starch and other food reserves. Strengthening tissue may be present in the form of sclerenchyma, variously arranged; sometimes in bundles just below the epidermis, sometimes in larger masses in which the outer vascular bundles are embedded, sometimes forming a sheath round the individual bundles. In some stems, particularly stolons and rhizomes, a central cylinder of ground tissue, in which are numerous vascular bundles, is separated from an outer parenchymatous cortex by an endodermis in which the cells have uberised walls.

Vascular Bundle

If we take the stem of maize as an example of a monocotyle-donous stem we find numerous bundles scattered throughout a ground tissue which is undifferentiated into cortex and central cylinder. Each bundle is more or less enclosed in a sheath of sclerenchy-matous fibres, and is collateral. The metaxylem consists of two large pitted vessels connected by tracheids. Internal to these is the protoxylem consisting of spiral and annular vessels associated with thin-walled xylem parenchyma. The innermost protoxylem elements, the annular vessels, may have been broken down, leaving a bysigenous cavity. External to the tracheids is the phloem, in which can be distinguished larger, apparently empty sieve-tubes, and smaller companion cells filled with contents. There is not cambium; on this account the bundles are described as *closed*. This is an important distinction from the dicotyledonous vascular bundle.

Modifications

In some monocotyledonous stems, e.g., *Asparagus*, the xylem is in the form of a V, with the point occupied by protoxylem and the arms the metaxylem with vessels of increasing diameter. The phloem lies between the arms of the V. In other stems, e.g., *Iris* and *Acorus calamus* (sweet flag) rhizomes, the xylem may even encircle the phloem.

In grasses, bundles of the maize type are arranged in more or less concentric rings, often associated sclerenchyma, round a central cavity. In the leafy stems of *Tamus* (black bryony) the vascular bundles are arranged in a ring surrounding a pith, as in Dicotyledons. Some Monocotyledons, such as *Dracaena*, *Yucca*, the screw pipes and the palms attain tree-like dimensions.

In the palms the whole of the tissues of the stout stem are derived from a massive apical meristem. This gives rise to a parenchymatous ground-tissue containing numerous, scattered, closed vascular bundles. No cambium develops and threfore there is no secondary thickening. *Dracaena*, *Yucca* and a few others possess a form of secondary growth. In *Dracaena*, for example, the primary condition of the stem shows the scattered arrangement of closed, common bundles. A secondary meristem originates in the ground-tissue outside the outermost leaf-trace bundles. This cambium cuts off cells internally and externally.

The tissue produced internally differentiates into vascular bundles, separated by parenchyma which may become lignified and pitted, or remain thin-walled. The secondary vascular bundles consist of a little phloem surrounded by xylem formed of tracheids.The small amount of secondary tissue external to the cambium is thin-walled parenchyma. Periderm may be present in Monocotyledons, but it rarely shows the regular arrangement seen in Dicotyledons. The epidermis is usually

are always collateral and closed. Each vascular bundle consists of xylem and phloem. The xylem is Y-shaped. The metaxylem vessels form of arms of Y and protoxylem, the base. Phloem consists of sieve tubes and companion cells. Bundle sheath is not found.

Anatomy of the Scape of Canna

Epidermis

It is the outermost uniseriate layer consisting of small, polygonal cells with cuticularized outer walls.

Ground Tissue Systems

Just beneath the epidermis a few layers of parenchyma occur forming small cortical region. The cells of cortex are sufficiently large and polygonal. Immediately below the cortex, a single-layered chlorophyllous tissue is found consisting of chloroplast bearing cells. The sclerenchyma patches also remain attached to the chlorophyllous tissue here and there. The rest of the portion consists of a continuous sufficiently developed intercellular spaces among them. It is called the ground tissues.

Vascular Bundles

They are many and of various sizes, lying scattered in the ground tissue. The bundles are closed and collateral. Each bundle is incompletely surrounded by a sheath of sclerenchyma called *bundle sheath*. The outer sclerenchyma patch of the bundle is more distinct and cap like whereas inner patch is not so developed. Each bundle consists of xylem and phloem. The xylem is situated on the inner side and phloem towards outer side. The xylem consists of a large spiral vessel with one or two smaller vessels of same nature. The phloem consists of sieve tubes and campanion cells.

Anatomy of the Stem of Wheat (Triticum Aestivum)

Epidermis

It is the outermost uniseriate layer, usually composed of compact tabular cells with cuticularized outer walls. The stomata are also seen here and there on the epidermis.

Ground Tissue System

Just beneath the epidermis the sclerenchyma cells occur in small patches which are not arranged in a continuous band, but are interrupted by chlorenchyma tissue here and there. The stomata are confined on the epidermis only in chlorenchymatous regions. The rest of the ground tissue consists of thin walled rounded or oval parenchyma cells having sufficiently developed intercellular spaces among them. The central region of the stem is hollow.

Vascular Bundles

The closed and collateral vascular bundles occur in two series. The peripheral series consists of smaller bundles, whereas the inner series is of bigger bundles. The vascular bundles of outer series are lying embedded in sclerenchyma band. The bundles of inner series are also surrounded by sclerenchymatous bundle sheath like that of maize stem. The bundle sheath of peripheral bundles actually touches the epidermis.

ANATOMY OF THE PHYLLOCLADE

Numerous stems are specialized for photosynthesis and take the place of leaves in the manufacture of carbohydrates. Some stems which are specialized for photosynthesis are round, others are flattened and others even have the form of leaves. Such stems as those of cacti are specialized both for photosynthesis and for water storage. In the following paragraphs, the anatomy of the phylloclade of *cocoloba* and *Ruscus* has been discussed.

ANATOMY OF THE PHYLLOCLADE OF COCOLOBA (DICOT)

Epidermis

The surfaces are bounded on both sides by upper and lower epidermal layers. A thin cuticle covers the epidermis. The epidermis is interrupted by numerous stomata on both the surfaces. Distinct guard cells of the stomata and sub-stomatal chambers are visible.

Chlorenchyma

Just beneath the epidermis there are few layers of chlorophyllous cells. The chlorophyllous cells are found only underneath the upper and lower epidermal layers. However, the chlorenchyma is not found at the edges of the phylloclade. The stomat are confined to chlorophyllous regions. There are well developed intercellular space among these cells. This is assimilatory tissue.

Sclerenchyma

The multilayered sclerenchyma tissue is found at the edges of the phylloclade. This is meant for mechanical strength. Just below the chlorenchyma there is a single layer of sclerenchyma cells, which delimits the central parenchyma and the peripheral chlorenchyma. Usually each vascular bundle is capped by a well developed sclerenchymatous patch.

Vascular Bundles

Around the central parenchyma the vascular bundles are found to be arranged in the peripheral region. The vascular budles of the two corners are bigger in size than the remaining ones. Each vascular bundle is capped by a sclerenchymatous patch, and is composed of xylem, phloem and cambium. The xylem consists of metaxylem and protoxylem groups. The xylem parenchyma is also present. The phloem strand lies towards periphery. The phloem is composed of sieve tubes, companion cells and phloem parenchyma. In between xylem and phloem strands there lies the cambium. The cambium is confined to the bundle.

Parenchyma

The central region is occupied by parenchyma. It is composed of thin-walled rounded or oval, living cells having well developed intercellular spaces. This is storage tissue.

Anatomy of the Phylloclade of Ruscus (Monocot)

Epidermis

The phylloclade possess two surfaces (upper and lower). Both the surfaces remain bounded by the upper and lower epidermal layers. The upper epidermis consists of a single row of radially elongated epidermal cells. The epidermis is interrupted by stomata at certain places. The sub-stomatal chamber and guard cells with chloroplasts are distinctly seen. The upper surface becomes somewhat bulged in the central region. The lower surface in the central region becomes somewhat angular, otherwise the anatomy of the lower epidermis is quite similar to that of upper epidermis.

Chlorenchyma

Immediately below the upper epidermis few layers of chlorophyllous cells are present. These cells are rounded or oval shaped, containing chloroplasts and having well developed intercellular spaces among them. In between upper and lower epidermis the well developed parenchyma is present.

Vascular System

It is well developed and represented by many amphivasal (phloem surrounded by xylem) vascular bundles. The phloem bundle remain surrounded by a sclerenchymatous sheath. The phloem consists of sieve tubes and companion cells. There has been some controversy on the question whether the arboreal or herbaceous habit of growth is primitive in Angiosperms. Arguments may be adduced on both sides. In favour of the arboreal habit there is the fact that the earliest known fossil Angiosperms appear to belong almost exclusively to tree families.

Further there is present predominance of woody forms in tropical climates, together with the parallels which can be drawn between woody tropical species and related herbaceous species in temperature climates; the argument being that the herbaceous habit is an adaptation to the severity of winter. Of the other side is urged the fact that the youngest stages of tree species are herbaceous, which is an appeal the theory of recapitulation. Moreover a parallel can be drawn between the great stature of trees with the large mass of their non-living tissue and the gigantism and heavy bone-structure of decadent animal groups of the past, such as the Mesozoic reptiles. There are undoubtedly many primitive characters in the structure of trees. These have been attributed to the longer life of the tree and the smaller number of generations which such species have passed through in a given time, but nevertheless the balance of evidence is generally helf to be in favour of the arboreal theory.

Two interesting anatomical theories have been propounded on the supposed mode of origin of the herbaceous habit from the arboreal, which may be thus briefly summarized:

Theory of Jeffrey

Where leaf traces enter the vascular cylinder there is a need for storage for the photosynthetic products of the leaf. This is met firstly by a concentration of medullary rays flanking the traces. Secondly, these fuse into broad compound rays and the neighbouring woody tissue becomes transformed into storage parenchyma, so that the woody cylinder is broken into separate wedges with broad rays between, through which the leaf traces pass out.

Theory of Sinnott and Bailey

The herbaceous habit may be attained in two ways, firstly and more simply by the reduction of cambial activity, so that the stem is furnished with a continuous, but thin, ring of wood which soon ceass to expand. Secondly by the reduction of cambial activity in the sectors between the leaf traces, the wood there formed being first lessened and finally abolished, leaving broad parenchyma tracts between the separate wedges of wood to which the vascular ring is thus reduced. Detailed anatomical evidence can be produced for all these processes, and it seems possible that each of them may have been operative in different cases. The anatomical features which we have discussed above have more relation to the Dicotyledons than to Monocotyledons, in which no process of timber building occurs.

Some Monocotyledons, however, such as the Palms, form apparently woody trunks by a process which is quite special to them, and which may be called diffuse thickening. In the early stages of their development, the apical meristem broadens year by year, with very little growth in length. On this broad meristem the number of leaf primordia increases progressively, with a consequent increase in the number of trace bundles in the stem. The shape of the stem is that of an inverted cone, usually partly buried in the soil.

When growth in length begins the stem shoots up into a cylinder, no further growth in girth occurring. No cambium is formed in the bundles, which remain permanently distinct. Each bundle is surrounded or partly surrounded, from an early stage, by a belt of sclerenchyma, and the only alteration of the tissues with age is the increased thickening of the cell walls, which may in time involve the whole of the ground tissue.

A very solid stem may thus be formed, even though it contains no mass of wood, and although the parenchyma cells of the ground tissue may remain alive and are often filled with stored starch grains. A concentration of smaller trace bundles and associated sclerenchyma strands at the periphery of the stem forms a mechanically strong ring, which in stems takes the place of periderm. The inner tissues are relatively unimportant mechanically.

A small group of genera in the Agarales also produce woody stems, notably *Yucca*, *Aloe*, *Dracaena* and *Cordyline*, by a different and peculiar process. We have already mentioned the existence of abunle cambium in many Monocotyledons, which produces a very limited amount of secondary xylem, though no continuous cambium is formed. In the genera cited above a broad meristematic zone arises in the inner cortex, closely surrounding the central group of primary bundles, that is to say, it occupies differentiated in the apical meristematic region, while in others it appears only towards the base of the main stem, where all the tissues are fully differentiated. It is composed of squarish, thin walled parenchyma, arranged in radial rows. It resembles a cambium and is usually so called, but it differes both in origin, structure and behaviour from typical cambia.

The radial rows are not the products of division of a single initial cell, but are built up by the successive divisions of three, four or more cortical cells in radial files. The individual cells are not cambiform and do not segregate xylem inwards and phloem outwards, but produce entire, closed, fibro-vascular bundles, within the zone, by the sub-division of single or of grouped meristematic cells. These new bundles arise close together in centrifugal succession, and as they mature the parenchyma between them becomes lignified, so that a remarkably solid band of secondary tissue may be built up in the course of years. Old stems of *Aloe* and some other genera have even been observed to show annual rings of growth.

10

ANATOMY OF LEAF

The flat blade or lamina of a dicotyledonous leaf is typically a dorsiventral structure, the upper of adaxial face differing from the lower or abaxial face. Among Monocotyledons on the other hand, dorsiventrality is confined to types with horizontal leaves, those which are borne vertically being often bifacial, i.e, with both sides alike, or they may be round, triangular or square in section, without distinction of upper and lower sides, when they are called *unifacial*. There is an upper and a lower epidermis, both composed of somewhat flattened cells and covered with a cuticle, which varies greatly in thickness, and may be itself covered by a waxy incrustation. The epidermis of both surfaces frequently bears trichomes in the form of hairs or secretory glands.

In the epidermis are the *stomata*, which are normally confined to the lower surface, but which may be found on both surfaces in a majority of Dicotyledons and very frequently in Monocotyledons. Each stoma consits of a pair of more or less curved guard cells, whose upper and lower walls are thickened. The guard cells enclose between them a slit-like opening, called the stomatal pore, which communicates with the intercellular spaces of the leaf. The guard cells are distinguished from the other epidermal cells by possessing chloroplasts. In certain families there are also subsidiary or auxiliary cells, which differ from the other epidermal cells by their smaller size and sometimes by their shape. There may be only two of these, one adjacent to each guard cell, or there may be a group of four or more surrounding the stomata. The guard cells show characteristic variations of turgor, often with a day and night rhythm. Increased turgor increases their curvature and this widens the pore between them, and vice versa. When they become flaccid the pore is closed, which usually occurs at night. The variations of turgor may be osmotic, due to increased sugar concentration in their sap, and in this connection it should be remembered that the presence of chloroplasts implies the power of photosynthesizing sugars.

There is evidence, however, for another view, that the swelling of guard cell is due to an increased imbibitional power of their contents, conditioned by a rise of acidity in the sap. The physiology of the movement is not yet fully understood. The stomata are exceedingly numerous, 100 to 200 per sq. mm. being not uncommon, but the number is very variable not only between plants growing under different conditions, but between younger and older leaves of the some plant, and even in different part of the same leaf. The latter variation is probably due to differences in the growth rate of particular areas of the leaf during its development, which we have described in an earlier part of this chapter.

In Dicotyledons the stomata are scattered irregularly over the surface, but in Monocotyledons they usually form longitudinal rows. Between the upper and lower epidermis lies the tissue called collectively the *mesophyll*. It is not normally homogenous but is differentiated into two layers. The upper layer is formed of cylindrical cells lying closely together, with their long axes perpendicular to the leaf surface. This is called the *palisade layer*, from its fancied resemblance to a row of fence slata. Its cells are rich in chloroplasts which form a practically continuous layer on the long walls. Between the cells are narrow intercellular passages. The lower portion of the mesophyll consists of very irregularly shaped cells which are only in contact at their protuberances and among which, therefore,there is a continuous system of large intercellular spaces. This is given the apt name of the lacunar or *spongy mesophyll*. Its cells also contain chloroplasts, but not usually so many as in the palisade layer.

The distinction of these two layers is clearest in Dicotyledons and is often lost in Monocotyledons, where the whole mesophyll may be spongy. At all level where the two mesophylls are in contact lie the vascular bundles or veins of the leaf, varying in size between the massive midrib, which may show some secondary thickening, down to single tracheids at the extremities of the smallest veins. They are normally orientated with their xylem towards the upper surface by the leaf. In Dicotyledons they are often surrounded by a layer of non-chlorophyllous cells, rich in sugars, called the bundle sheath. The leaves of Monocotyledons differ from those of Dicotyledons in having, in general, a lower degree of differentiation in the mesophyll. The palisade layer is seldom well marked and the whole of the mesophyll tissue may be spongy or parenchymatous in chapter. The veins are parallel and each vein is frequently flanked, above and below, by masses of sclerenchyma, thus forming a rib across the mesophyll, and dividing it into a series of isolated strips between the veins. As the leaves are usually held vertically the two sides of the leaf tend to be similar in structure, i.e., bifacial, with stomata equally numerous on both faces.

Elementary Anatomy of Leaf

Epidermis

The epidermal cells of leaves are very frequently *tabular*, that is to say, they are flattened in the anticlinal direction and broader than they are deep. A characteristic feature is likewise their wavy or undulatory outline, in surface view, especially in the Dicotyledons. In soft-leaved Monocotyledons the epidermal cells are generally elongated, with practically straight walls, but in a few families with hard leaves, such as Gramineae and Cyperaceae, the waviness of the outline reaches an extremity of complexity, resembling the suturing between the bony plates of the skull. Waviness is usually more pronounced in sun leaves than in shade leaves, even in plants of the same species, and it has been attributed to the progressive hardening of the cuticle from the middle of the cell surface outwards to the margin, its consistency becoming uniform only at maturity, when stretching stops. The increase

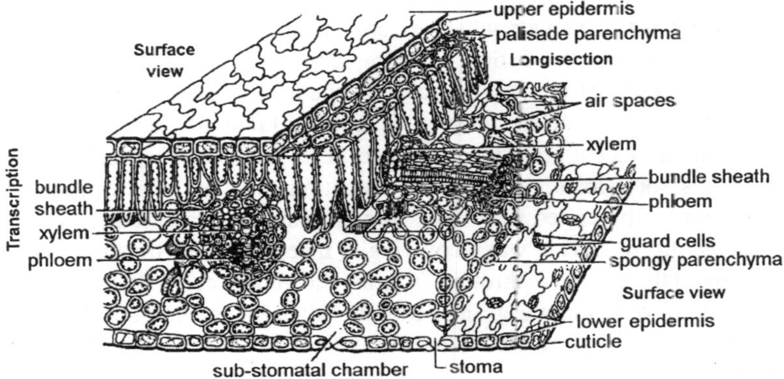

Fig. 10.1. Three-dimensional sectional view of a dorsiventral leaf.

of the cell-diameter by the stretching of the radial walls will naturally stop first at the points where the hardening first reaches the margins, while other parts of the wall will continue to expand. The view is supported by the fact that waviness affects only the outer contours of the cells, that is to say, on the cuticularized surface.

It is not shown by the inner edges of the radial walls, where they meet the inner tangential walls. The degree of waviness is related to ecological conditions, being usually less in plants of dry situations and vice versa. This is probably related to the extent of leaf growth, which is greater in moist conditions. It is quite exceptional to find chloroplasts or, as a natural consequence, starch grains in the epidermis. Instances to the contrary may, however, be found among plants which habitually grow in deep shade and among submerged aquatics, where the epidermis has largely lost its protective functions. The foliar epidermis resembles that of the stem in being covered with a continous layer of cuticle of varying thickness. This is usually sharply marked off from the individual cell walls below it. The latter are composed of an inner cellulose layer, and between this and the cuticle there is interposed a variable thickness of cutinized wall, consisting of a mixture of cutin and cellulose. This layer may sometimes be absent, even in highly thickened cells. The thickness of outer wall is not, therefore, a reliable index of its impermeability to water, since the intensity as well as the amount of cutinization may vary considerably.

Nevertheless it is generally true that more exposed surfaces have a thicker cutinization than those less exposed, and it follows that the upper surfaces of dorsiventral leaves have usually a thicker coating than the lower surfaces, sometimes twice as thick. Compared with the amount of water vapour which escapes through the stomata, the loss of water through the cuticle is extremely small. Haberlandt showed by experiment that the loss of water attributable to evaporation through the cuticle itself varied between 0.23 and 0.006 of the evaporation from an equivalent free water surface. The protective capacity of the cuticle is sometimes enhanced by a coating of wax, which produces the greyish film called popularly "bloom" and known botanically as a *glaucous surface*. This waxy covering is independent of the cuticle and is apparently excreted through it from the living cells. It commonly consists of very minute grains, loosely adherent to the cuticle but in some Grasses and related plants it takes the form of slender rod attached to the surface by their ends. A third type of waxy covering, in a continuous sheet, occurs especially in Palms.

The Wax Palm, *Ceroxylon andicoluni*, produces a layer nearly 5 mm. thick, which is scraped off and used to make candles, a single tree frequently producing as much as 25 lb. of somewhat resinous wax. These waxy coverings are of considerable value as a protection against evaporation, not only as an addition to the cuticular covering but also by building up protective ridges round the stomata. In many cases the covering, if rubed off, can be regenerated. The outer surface of epidermal cells are usually more or less convex, and in many plants they are markedly domed or even conical. The latter type of extension of the cell, when well developed, produces a velvety surface, which is of common occurrence among tropical jungle plants. An ecological advantage has been attributed to this type of surface, in that water falling on the leaf tends to spread by capillarity and hence to dry more quickly, thus offering less encouragement to the growth of epiphyllous Lichens and Bryophyta. The idea, though often quoted as an example of adaptations, rests only on supposition and should not be accepted uncritically. The epidermis of the leaf bears trichomes of the same types as those borne by the epidermis of the stem, though often in greater abundance, especially on the lower surface. When only a few scattered hairs are present the leaf surface may be described as *hispid*, whereas a closer covering is spoken of as *tomentose* or *lanate*.

Such densely wooly coats as are implied by these latter terms may be influential in reducing evaporation from the stomatal openings and they are, in fact, commonest in species which habitually grow in exposed places and dry soils, but it must not be inferred that they are confined to such plants or that all plants with a woolly covering stand in need of such additional protection. Only direct experimental evidence can justify such a conclusion. What experience in cultivation does bear out is that woolly plants are usually intolerant of prolonged damp, especially in winter. The thick hairy covering holds moisture tenaciously and probably facilitates fungal and bacterial attack on the tissues. A number of plants, such as *Platanus orientalis*, *Tussilago farfara*, and *Verbascum olympicum*, have only temporary coverings of hairs on the young leaves, the hairs later disarticulating at their bases, sometimes at a point marked by a specially thin zone of the cell wall, and fall off or are rubed off, leaving the mature leaf surface bare. The mechanical strength of the epidermis may be very servieable in preventing the tearing of the lamina, and this is generally due to the cuticle, which not only forms a continuous sheet binding the individual cells together, but may often be additionally attached to the epidermis by wedge-like processes extending inwards between the cells.

The thickening of the epidermal walls is always most prominent at the leaf margins, which are obviously the parts most vulnerable to tearing, and here the epidermis may also be supported by collenchyma or sclerenchyma. Many species have, however, outer epidermal walls which are calcified or silicified, more commonly the later. Silicification of epidermis is particularly strong in Gramineae and in Cyperaceae. Haberlandt drew attention to the probable importance in the perception by the leaf of the direction of the incident light, and hence in the adjustment of the position of the leaf with regard to it. He showed that parallel rays of light falling on the upper epidermis of the leaf, are, in many species, condensed to a bright spot on the inner cell surface, and that in at leat a few cases it was exactly focussed on this surface, so that an image of surrounding objects is formed in each cell, in a manner recalling the multiple eyes of insects. He very justly points out that there is no likelihood of any perception of such images by the protoplasmic lining of the cell, and that the operation of the mechanisms depends solely on the difference in brightness between the focused spot of light and the rest of the cell surface.

When the leaf surface is in the stable diaphototrophic position, i.e., normal to the direction of the sun is rays, this spot is central, but it shifts laterally with any movement of the leaf or of the sun, thus disturbing the equilibrium and providing a stimulus which is supposed to result in phototropic adjustment of the leaf. This hypothesis is supported by observations that is phototropic leaves are plunged in water or if their upper surface are painted with liquid paraffin thus altering the conditions of refraction at the leaf surface, the phototropic reaction is apparently abolished. It is possible, however, that this may be due to disturbance of some other condition in the leaf, and the hypothesis, though very interesting, cannot be regarded as proven. In a few cases, specially enlarged cells of the epidermis with papillate outer walls, are seemingly the organs of this light perception. These have been called *ocelli*, a name borrowed from Zoology.

Particularly good examples occur in *Lithops* pseudotruncatella, a small South African succulent, on the leaves of which the ocelli appear as minute dark spots, due to the greater size and depth of the ocellar cells. A function of the epidermis which must not be overlooked is that of water storage. The amount stored is not normally large, and it would not suffice to supply for long the quantities which are lost by stomatal transpiration, but under conditions of wilting, when the stomata are closed, the epidermis is the first line of defence against cuticular transpiration, and it can be observed that in wilting leaves the epidermal cells shrink and collapse from water loss before the mesophyll cells are affected. As the radial walls of the epidermal cells are usually thin and are in any case covered with numerous pit areas, it is possible for water to move laterally in the epidermis without much difficulty, and water may in this way be supplied to specially exposed spots on the leaf without diminishing the water content of the photosynthetic cells. A case where this may be important is that of undergrowth plants in tropical forests, subjected to the passage of sun flecks which produce for short periods a marked local heating and drying of the air in contact with the leaf.

Stomata

Although the stomata are a part of the epidermal structure they deserve special description, not only for their intrinsic importance,but also because they are not physiologically a part of protective system which is formed by the rest of the epidermis. They are the portals of gaseous exchange between the inner tissues and the environment, and they seem to be fundamental to land life, since we find them, or structures analogous to them, in all land plants, from the Bryophyta upwards. Stomata in land plants are chiefly distributed on the lower leaf surface, a fact which has been interpreted as giving protection against the blocking of the stomatal opening by dust, rain, and dew. There are, however, many exceptions in which stomata occur on both surfaces, especially among those plants, principally Monocotyle-dons, which have leaves held vertically. Centric and bifacial leaves which do not possess the usual dorsiventral anatomy have stomata distributed more or less over the whole surface, but usually with more on the lower side.

Stomata confined to the upper surface only are very rare, except in Grasses like Ammophila, with grooved leaves, and in the floating leaves of water plants, where the advantage is obvious. It is not, however, as might be supposed, the direct result of the conditions, since certain water plants, notably Polygonum amphibium, grown in air under moist, warm conditions, will produce in succession leaves of the submerged type, then leaves of the floating type with numerous stomata only on the upper surface, and lastly leaves of the aerial type with stomata on both sides, but more numerous on the lower side. Here the difference is clearly not due to the direct action of the environmental conditions, but is the result of changes in the inner organization of the plant, the causation of which is unknown, but which fit in satisfactorily with the normal characteristics of the plant's habitat in water of medium depth. This is an important aspect of the question of adaptation. The numbers of stomata per unit are of leaf surface are very variable but are normally between 50 and 400 per sq. mm. Salisbury, comparing British plants of various habitats, found that the frequency tended to be greater in plants from dry, exposed situations than from humid places, though the individual stomata were usually smaller in the former case. This difference he showed to be correlated with differences in the growth of the leaf as a whole.

During the primary phase of development in the leaf primordium the number of stomata is determined and thier initial cells are formed, but the subsequent phase of expansion determines their spacing and hence their frequency. When leaves from plants of the same species, growing under different conditions of humidity, are compared, it is found that the ratio of the number of stomata to the number of epidermal cells per unit area is remarkably constant. Salisoury expressed this by the formula:

$$I = \frac{S}{E+S} \times 100$$

and called I the *stomatal index*, Here, S stands for the number of stomata and E for the number of epidermal cells per unit area, the latter being chosen to avoid the apex, the margins and the large veins. It implies that the proportion of superficial cells which become stomata is constant for the species, and the index is, indeed, independent of habitat and of the age and position of the leaf, and is so specifically constant that it may be used as a diagnostic character in the separation of closely related species.

These important observations of the relation of frequency to the amount of expansion during leaf development, serve as a clue to the variations of frequency not only between different plants, but between different parts of the same plant. The frequency, especially in herbaceous plants, increases with the height of a leaf above the ground, and in trees with its distance from the base of the branch, that is inversely to its age. Likewise, there is often an increase of frequency from the base to the tip of the lamina and from the midrib towards the margins in the individual leaf. Stomata in Dicotyledons are characteristicaly orientated at random, that is to say, their long axes do not lie in any particular direction. Careful analysis has shown, however, that this resynthesis of starch is favoured, resulting in a drop of osmotic potential and hence of turgor. This account of the reactions, though it is consistent with observations and with experimental results, so far as they go, must be regarded as still hypothetical, since it has not been definitely proved that photosynthesis occurs in stomatal plastids, nor has the presence of amylase in the guard cells been demonstrated. Scarth considers that the opening and closing movements are too rapid to be accounted for by enzymatic reactions, and considers that the changes of acidity, which do take place, affect the imbibitional swelling of the cell colloids and that the changes of turgidity are, at least in part, due to changes of water content due to collodial imbibition.

As the water drawn in to the guard cells must come from the leaf tissues, it follows that the possibility of movement will be affected by the water content of the leaf as a whole. A decrease of mesophyll water content, due to wilting, will increase the osmotic potential of the mesophyll cells and when this has passed a certain value it will no longer be possible for the guard cells to withdraw water from them and the stomata will remain permanently shut, thus minimizing the further loss of water from the leaf. Finally it may be said that recent observations have detected a rhythmical pulsation in guard cells, with a fifteen-minute period, which is even maintained for twelve hours in continuous light. To what this is due to not known, though a short-period opening and closing can be produced, experimentally, by variations of tissue water content, under extreme environmental conditions. The classic researches of Loftfield have shown that stomata do not all react similarly to light and that they generally conform to one of three types.

Type 1. Typical of Grasses, including the cereals. During the day the stomata may open and close rapidly, but they are only open for an hour or two altogether and only under the most favourable conditions. There is no opening at night.

Type 2. Includes most thin leaved mesophytes.* Usually the stomata open all day and close all night. Under less favourable conditions they may close at midday, and under very unfavourable conditions, especially heat and drought, this may extend to the whole day. Night opening is the observe of day opening, so that under very dry conditions stomata may be closed all day and open all night, depending on the water content of the tissues.

Type 3. The "Potato Type". Under favourable conditions the stomata are open more or less all day and all night, closure being caused by increased evaporation rather than by darkness. With diminished water content they become more responsive to light and then they tend to comform to Type 2.

The stomata of many water plants have no power of movement and remain open even when the leaves are wilted. Immobile and enlarged stomata may also function as water stomata through which water is excreted under very moist conditions. They occur principally at the margins of leaves, as in Fuchsia, Tropaeolum and Primula, or at the leaf tips, as in the primary leaves of many Grasses. These water stomata form one of the types of water-secreting structures, or hydathodes.

Mesophyll

Typically the leaf mesophyll consists of one or more layers of prismatic cells arranged anticlinally, with narrow spaces between them, constituting the palisade layer, while below this lies the lacunar or spongy tissue, made up of irregularly shaped cells, between which is a connected system of large intercellular spaces, which almost completely surround the cells and contain an internal atmosphere which is kept in communication with the exterior air through the stomata. The elongation of the palisade cells has been compared with the elongation of epidermal cells into hairs, and it has been suggested that both may be due to the operation of similar factors, such as strong light or rapid transpiration associated with exposure, which check the growth of the leaf in area and promote the lateral extension of the cells.

There may be some truth in this, but it is obviously not a complete explanation and further information is needed. The fundamental structrue of the mesophyll is found in all normal dorsiventral leaves, both dicotyledonous and monocotyledonous, though in the latter the palisade layer is usually less developed than in the former. Leaves of the grass type, however, are exceptional. They have parallel veins which lie close together, with ribs of sclerenchyma above and below each vascular bundle. This cuts up the mesophyll into a number of independent strips in each of which there are, at the top and the bottom respectively, two belts of tissue containing chlorophyll, while the central portion is sometimes lacunar in structure, but more often composed of thin-walled, colourless cells which store water. Dicotyledons may also have a palisade on both sides of the mesophyll. This is formed, for example, in plants with a normally vertical leaf position and in others which grow under such conditions that the under sides of the leaves receive strong lighting. Such leaves are called *equifacial*, or sometimes irregularity is only apparent and is not related to the irregularity of the epidermal cells, but is, in fact, closely correlated with the direction of the underlying veins in the mesophyll.

Here again differential growth rates are probably responsible. More rapid stretching of the leaf tissue along the veins is shown by the more elongated epidermal cells usually formed above them and this may account for the elongation of stomata in the same general direction. The linear leaves of Monocotyledons, in which growth is continuous from the leaf base, usually have the epidermal cells and the stomata in longitudinal rows, the latter also having their long axes regularly in the same direction. Monocotyledons, such as *Arum*, with net venation, have randomized stomata like those of Dicotyledons.

Transversely orientated stomata are uncommon and are nearly always associated with reduced leaves and xeromorphic structure. The differentiation of stomatal mother-cells from epidermal cells begins at a very early stage of leaf development. It involves that rare occurrence, the division of a cell into two markedly unequal portions, a large cell which becomes epidermal and a smaller cell, always on the side towards the apex, which becomes a stomatal initial cell. The first differentiation of stomata is usually coincident with the beginning of expansion in the leaf rudiment. It is not simultaneous all over the surface, but begins near the leaf apex, the differentiation in other parts being concurrent with the appearance of air spaces in the mesophyll.

Stomata are formed successively for some time, the earliest to appear showing an orientation related to the direction of growth tensions, while those which appear later are seemingly irregular in alignment. The stomatal initial cell is in most cases also the mother cell of the stomata, but in certain families the initial cell may divide once or several times, producing a stomatal mother cell surrounded by a variable number of subsidiary cells. The majority of families, however, produce subsidiary cells, if at all, by sub-division of the neighbouring epidermal cells and the initial cell divides only once to form the two guard cells of the stoma. Between these the pore appears, owing to the splitting of the wall which separates them. There is much variation in the details of stoma formation, many of which are peculiar to certain families and can be used as systematic characters. When first formed the guard cells equal the epidermal cells in depth, and lie in the same general plane.

The sunken position of many stomata is due to later growth of the epidermis and sometimes to its periclinal division into two or more layers. The shape of the guard cells is remarkably uniform, though they vary considerably in size. They are curved in two directions. Firstly, there is the obvious concave-convex curvature which separates them from each other, except at their ends. This curvature is variable and affords the means of controlling the stomatal opening. Secondly, there is a curvature less easily seen, perpendicular to the surface, the extremities of the cells being turned somewhat downwards. This curvature does not vary as the first does. The upper and lower walls are highly thickened, while the lateral walls are relatively thin. The Gramineae are remarkable for guard cells of a peculiar pattern. They are straight, not curved, and shaped like a pair of dumb-bells, side by side.

The middle portions are highly thickened, while the expanded ends are thin walled. It is the expansion and contraction of these thin walled portions which varies the distance between the cells and hence the width of the stomatal openings. The mechanism of stomatal movement is not yet fully understood. The turgor pressure in the guard cells varies and with increasing turgor they become more curved and the opening between them widens, while the reverse occurs with decreasing turgor. These changes are associated with corresponding changes in their osmotic potential. The guard cells are unique among the superficial cells in containing green plastids. They are present even in plants grown in the dark, and are also present on the colourless portions of variegated leaves.

Starch accumulates in these plastids, even in plants which do not form it in the mesophyll cells, and this starch is subject to reversible hydrolysis into sugars, presumably by enzyme action, without, however, any change in the total carbohydrate content of the cells. The carbohydrate is apparently not manufactured by the guard cell, but seems to be contained from the mesophyll and it is even doubtful whether their plastids are photosynthetically functional. The starch content is greatest in the early morning hours and decreases in daylight, with a concomitant rise in the osmotic potential of the cell sap and increased opening of the stomata. The reverse changes occur during the evening and no further change appears to occur during the night.

It is supposed that light affects the balance by causing absorption of carbon dioxide by the plastids thus increasing the alkalinity of the cell sap, and thereby activating the amylase enzyme to hydrolyse the starch. With decreasing light, the carbon dioxide produced by respiration begins to accumulate, the cell sap becomes more acid and the

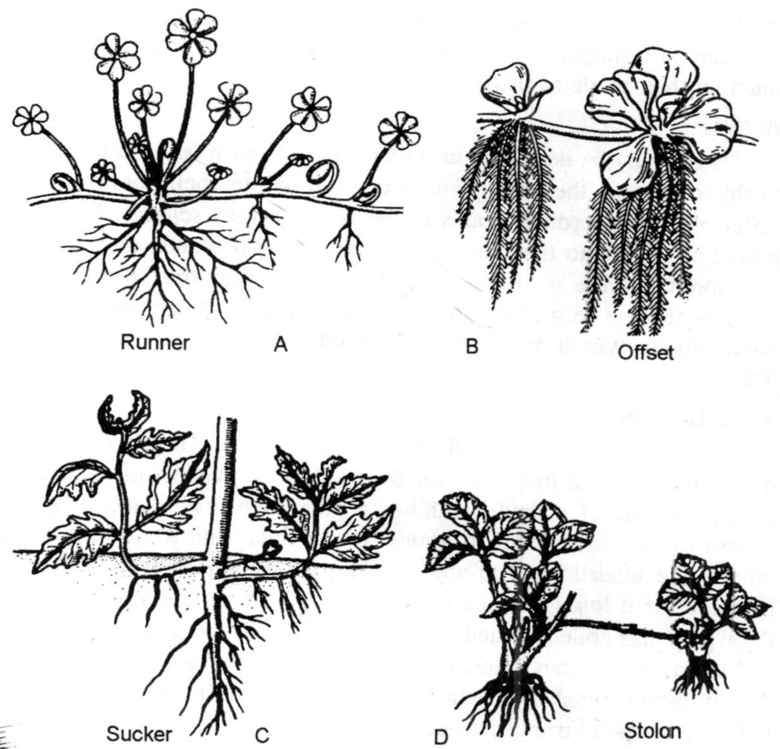

Fig. 8.3. Sub-aerial modification of stem: A, runner of Oxalis; B, offset of Pistia; C, sucker of Chrysanthemum; D, stolon of Fragaria.

reduced leaf at the base of this rosette a new runner emerges, and so on. The whole runner is therefore sympodial. Each rosette eventually produces adventitious roots and becomes independent, for the internodes die off in winter. Runners are commonest in herbaceous plants, but some shrubs also produce, them, such as *Lonicera japonica* and *Rhus toxicodendron*.

Offset and Sockers

The offset resembles the runner in origin, but is shorter and stouter. It is a short runner which turns up at the end to form a new plant, and occurs in *Agave*, water cabbage (Pistia stratiotes) and house-leek. The sucker is an *underground* runner or branch which grows upwards, and develops roots and aerial shoots. The sucker may be a branch arising from an axillary bud on an underground stem, e.g., *Chrysanthemum*, mint and dead-nettle. It may also arise from an adventitious bud on a root, as in plum and rose. Underground suckers have a root-like appearance but they can be distinguished as stem by the possession of scale-leaves, and often by their mode of origin.

Bulbils

These may be described as axillary buds, which become large and fleshy owing to the storage of food-material in their leaves. They differ also from ordinary buds in the fact that they separate from the parent plant, fall to the ground, and produce new plants, thus serving for reproduction (e.g., Remusatia some lilies). They may also take the place of flowers (e.g., in anion, Globba Agave, some grasses etc.). In those plants which produce them seed-formation is usually very uncertain.

Stem-Tendrils

These are highly specialised climbing organs. They are slender, often branched and may bear small scale-leaves. When still elongating the apical part of a tendril exhibits regular movement, and if in the course of this movement it comes into contact with a slender, solid object, the tendril coils around it. The young tendril is sensitive to contact, but it loses this sensitivity with age. When the apical part of the tendril has coiled around the support, that part between the support and the plant becomes spirally coiled, but forms two spirals in opposite directions separated by a point of reversal. Thus the plant is fixed to the support by a strong but elastic connexion.

The tendril may represent an axillary shoot, e.g., *Passiflora*. The tendrils of *Vitis* and *Ampelopsis*, are variously interpreted, e.g., as the modified growing points of the successive axes of a sympodium, or as derived from leaf primordia of a monopodium.

Modification of Underground Shoots

Rhizomes

These are plagiotropic underground stems of the most varied kind. They are found in a vast number of plants. In many species they are thin, tough and rapid in growth, as for example, in *Ammophila arenaria*, *Carex arenaria*, *Convolvulus arvensis* and *Agropyron repens*. Others have relatively fleshy, though rapidly growing, rhizomes, as *Tussilago farfara*, *Convallaria majalis*, *Mercur alis perennis*, Aegopodium podagraria and Mentha spp. A third group of rhizomes are fleshy and stumpy and serve more for storage than for spreading; such are the rhizomes of *Iris germanica*, Polygonatum multiflorum and Symphytum tuberosum. Of this latter class a number are dorsiventral, producing

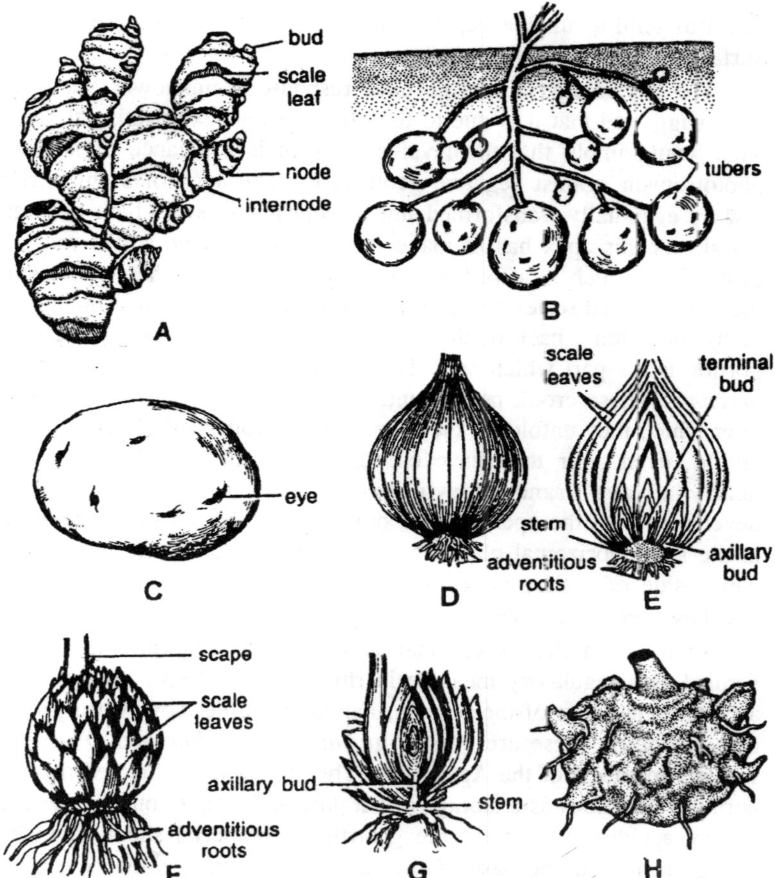

Fig. 8.4. Underground modification of stem: A, rhizome of ginger; B-C, tubers of potato; D, E, tunicated bulbs of onion (D, entire; E, longitudinally cut); F, G, scaly bulbs of garlic (F, entire; G, longitudinally cut); H, corn of elephant's foot.

leaves on their upper sides and roots only on their lower sides, besides being often flattened in section. In *Nuphar lutea* and *Nymphaea alba* the dorsiventrality only appears in old rhizomes and is conditioned by light, as they grow on the bottom of ponds. If they are buried in the mud they grow out into radial shoots which turn upwards until they reach the light, when dorsiventral growth is resumed. Many rhizomes display an extraordinary capacity for regulating the depth at which they grow. For example, if the dorsiventral rhizome of *Polygonatum multiflorum* is planted too deep, the growth of new branches is directed upwards until a suitable depth is reached. If it is planted too near the surface the reverse takes place.

The physiology of this peculiar response is unknown, but it has been suggested that it is due to the effect of light penetrating the soil. This would imply that the level of growth is a balance of negative phototropism against negative geotropism. The apex of a rhizome is always eventually transformed into an upright shoot, which becomes aerial. As this apex has to make its way through the soil it requires protection, which is afforded in one of two ways; either by smooth, hard and pointed scales which provide a boring point, as in Ammophila, or by the folding back of the apex into a crook, of which the curved portion is the part which actually penetrates the soil, as in *Menurialis perennis*. These crook persist until the apex grows up into the light, after which they unfold. If they are covered from the light they do not unfold, even after they have emerged from the soil. Rhizomes are nearly always branched sympodially, owing to the orthotropic development of the apex, lateral branches spread out plagiotropically in the same horizontal plane as the original rhizome, no matter from which side of it they arise. The plagiotropic growth of rhizomes is not, however, invariable.

Apart from the exceptional cases of upwardly directed shoots formed as a regulatory measure in rhizomes too deeply buried, there are some instances of the reverse, namely of direct downward growth of a rhizome. These are chiefly in *Yucca*, *Cordyline* and *Dracaena*, shrubby members of the Agavaceae. The downwardly directed rhizome normally functions as a storage organ pure and simple, but if flowering or any accident terminates the growth of the aerial shoot, branches spring from near the base of the rhizome, which become new aerial shoots. The apex of the rhizome does not itself develop thus and will only grow into an aerial shoot if it is turned upside down. Although most rhizomes are well provided with adventitious roots, there are a few cases in the Orchids, e.g., *Neottia* where no roots occur and the closely branched rhizome system itself functions as an absorbing organ.

Most rhizomes are perennial structures, but many examples occur among herbaceous plants in which a crop of short rhizomes is formed round the base of each aerial shoot. They grow only for one season and are next season replaced by a fresh crop, springing from each new shoot, while the old rhizomes rot away. Such plants occupy a limited area of ground with extreme density. The term stolon is often to be found in older books applied indiscriminately to various kinds of sclerenchyma (hypodermis) to the centre. It is not differentiated into cortex. It is not differentiated into cortex, endodermis, pericycle and pith.

Vascular System

It is composed of many collateral and closed vascular bundles scattered in the ground tissue. The vascular bundles lie toward periphery in greater number than the centre. Comparatively the peripheral bundles are smaller in size than the central ones. Each bundle is more or less surrounded by a sheath, which is more conspicuous towards upper and lower sides of the bundle. The bundle consists of two parts, i.e., xylem and phloem.

Usually the xylem is Y-shaped and consists of pitted and bigger vessel of metaxylem and smaller vessels (annular and spital) of protoxylem. In between metaxylem vessels, small pitted tracheids are also found. Around the lysigenous or water cavity wood parenchyma is present. The lysigenous cavity is formed by the breaking down of the inner protoxylem vessel.

Phloem consists of sieve tubes and campanion cells. Phloem parenchyma is altogether absent in most of monocotyledonous stems. The outer phloem which is broken mass may be called as protophloem and the inner portion is metaphloem. Sieve tubes and companion cells are quite conspicuous.

Anatomy of the Stem of Asparagus

Epidermis

It is outermost uniseriate layer composed of approximately rounded cells with cuticularized outer walls.

Ground Tissue System

Just beneath the epidermis a few layers of parenchyma are found which contain chloroplasts in them. This may be called cortex. The innermost layer of the cortex consists of compact cells and called the *starch sheath*. Below the starch sheath a multilayered complete band of sclerenchyma occurs, which gives mechanical support to the stem. The rest of the portion is ground tissue which consists of thin walled parenchyma cells having well developed intercellular spaces among them. This vascular bundles remain scattered in the ground tissue.

Vascular System

The vascular bundles remain scattered in the ground tissue. The central bundles are comparatively larger than the peripheral ones. They replaced by primary cortical cells becoming suberised. But in a few instances (Dracaena, Yucca, Zingiber, etc.) such suberised cells may be associated with groups of periderm to form a *rhytidome*.

ANATOMOY OF THE STEM OF ZEA MAYS

Epidermis

The epidermis consisting of a single layer of compact cells having no intercellular spaces among them. It is covered with thick cuticle. The epidermal hairs are altogether absent.

Hypodermis

Below the epidermis, usually two or three layers of sclerenchyma cells represent hypodermis.

Ground Tissue System

It consists of thin walled parenchyma cells having well-defined intercellular spaces among them. This tissue extends from below the

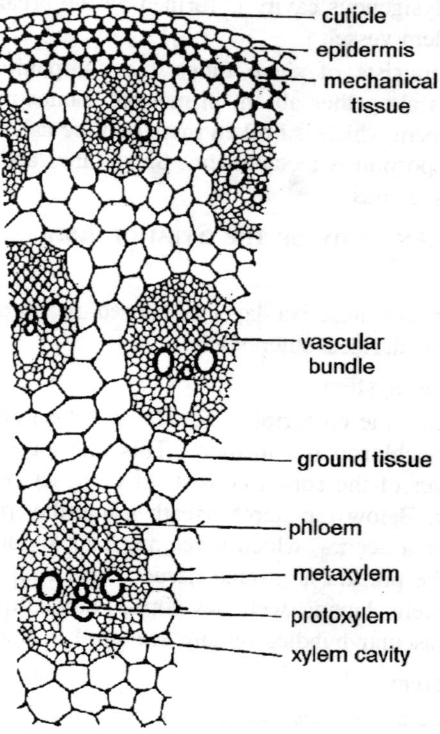

Fig. 9.4. T.S. of young Zea mays stem.

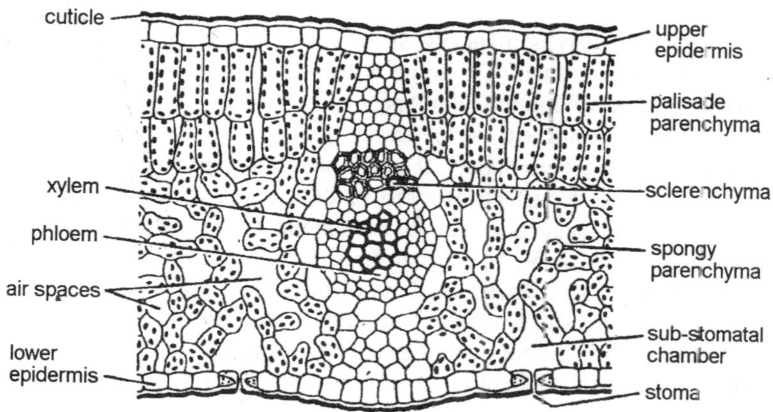

Fig. 10.2. Transection of a dorsiventral leaf.

isobilateral. A further modification of of leaf structure is that called *centric*. The extreme type of centric leaf is cylindrical, with all the mesophyll tissue arranged radially round one or more centrally placed veins. A less extreme type is found in some species, especially of Grasses and Sedges, in which the leaf as a whole is bilateral, but the veins are surrounded by radiating palisade cells, examples being *Papyrus* and *Portulaca*.

The truly centric leaf is really a special case of equifacial development, but there is another closely similar type of centric leaf which is derived in quite a different way, by the suppression of the upper surface during development, and the overgrowth of the lower surface, which thus surrounds the whole leaf. They are really extreme cases of the rolled or folded leaf. Such leaves are called *unifacial*. They may be cylindrical, as in *Juncus* and *Scirpus*, or they may be flattened in the vertical plane, as in *Rochea falcata*, but they are always recognizable by the fact that in cross section they show a line of vascular bundles, usually curved into a horse-shoe form or into an almost complete ring, instead of the centrally placed bundles of the truly centric leaf. The lacunar structure of the mesophyll, though most marked in the spongy tissue, extends to the palisade also between all the cells of which run lefts connecting with the spaces in the spongy layer. The cells are normally rounded in cross section and a view perpendicular to the leaf epidermis shows clearly the extent of the intercellular space system in the palisade, which is not so apparent when viewed in the usual transverse leaf section. The size and closeness of the palisade cells varies with external conditions.

In leaves growing in sunshine they are long, large, and close together, and in leaves in heavy shade they tend to become short, thin, and loosely arrange. The same is true of the spongy tissue, and the total volume of intercellular space may be twice as great in a shade leaf as in a sun leaf of the same species. The average volume of the internal space is about 20 per cent of the leaf volume. The internal area of cell surface exposed in the mesophyll is difficult to estimate was any accuracy, but investigations by Turrill established that the ration (R) between internal and external surface was lowest in shade leaves, with reduced numbers of mesophyll cells, (R = 6.8 – 9.9) and highest in sun leaves of xerophytic type (R = 17.2 - 39.3) the average ratio in mesophytic leaves being between 11.6 and 19.2. Contrary to expectation he found that the palisade tissue exposes from 1.6–3.5 times as much free cell surface as the spongy tissue. This is due to the form of the cells which lends itself to close ranking. In spite of variations in the size of the palisade cells, the palisade ratio, which is the ratio of the number of palisade cells to epidermal cells in a unit area, is remarkably constant and is used as a diagnostic character in pharmacognosy. This constancy is due to the mode of growth in the young leaf. Cell division ceases first in the epidermis, but the epidermal cells continue to expand, as marginal growth of the leaf continues, after the mesophyll cells have ceased to grow.

The latter are, therefore, drawn apart and the intercellular space are thus created. Obviously, therefore, wider spacing of the palisade cells will be correlated with larger epidermal cells, and the constancy of the palisade ratio is analogous to the constancy of the stomatal index. Not only does the palisade layer possess the larger cell surface, but it also contains a much greater number of chloroplasts than the spongy tissue. It also occupies the better lighted side of the leaf, and for these reasons it must be regarded as the principal photosynthetic tissue. At first sight the cylindrical form of the cells might not seem to be the best fitted for the exposure of the maximal area of chlorophyll to the sun's rays, towards which they present their narrow ends. Vertical light, passing through the epidermal cells is, however, refracted as we have seen above, and passes obliquely into the cell below, thus ensuring the illumination of the longitudinal walls of the palisade cells, and of the chloroplasts which line them.

The efficiency of illumination is thus high, and it is increased by internal reflection from the cell walls, which carries the light still further down in deep palisades. There is a general correlation between the depth of the palisade layer and the intensity of the incident light. Leaves growing under intense sunlight may have three or even more superimposed palisade layers. Leaves in the shade have only one layer, and that of slight depth. Indeed in deep shade the palisade may consist of funnel-shaped cells with curved walls and with the apex downwards, the chloroplasts being concentrated at the apex where the light is also concentrated by internal reflection from the cell walls, thus applying the principle of the searchlight in reverse. We have described in *Pinus* the flanged mesophyll cells which are characteristic of that genus, and have suggested that they are due to the construction of the cells during growth by cuticularized rings on the walls. Similar flanged or lobed palisade cells are not uncommon in Angiosperms. The II-shaped type is perhaps the commonest, and occurs widely among the Ranunculaceae. Whether it originates in the same way as in Pinus is uncertain, but physiologically the effect of the flanges in increasing the internal surface of the cell is identical, and hence the surface available for the exposure of chloroplasts is also increased. Such cells are not confined to the family mentioned above and they occur sporadically in many families both of Dicotyledons and Monocotyledons. The lacunar tissue plays a subsidiary part in photosynthesis, and it is uncertain whether its cell walls are sufficiently permeable to allow for much gaseous interchange or for the evaporation of water.

Lewis has recently shown that the walls of the spongy tissue cells are highly hydrophobic and unwettable by water. They show no tendency to absorb water which is injected into the leaf, though it is taken up by the cells of the veins. These observations have not been extended to the palisade cells, and it may prove that there is a differentiation in this respect between the two layers and that the palisade is the chief seat of transpiration as well as photosynthesis. The external cuticle of the leaf is often continued inwards through the stomatal pore and extends as a relatively delicate film over some part of the spongy mesophyll.

In some cases there is evidence that it may extend over the whole surface of that tissue, which may account for its unwettable characters. Injury to the mesophyll may induce the growth of tyloses from the spongy cells into the intercellular spaces, blocking them in the same way that the cavities of old vessels become blocked. Mesophyll cells exposed by injury may also form a phellogen and the wounded surface is thereby covered and protected by the development of *wound cork*. The lacunar tissue clearly serves at least two functions; the intercellular

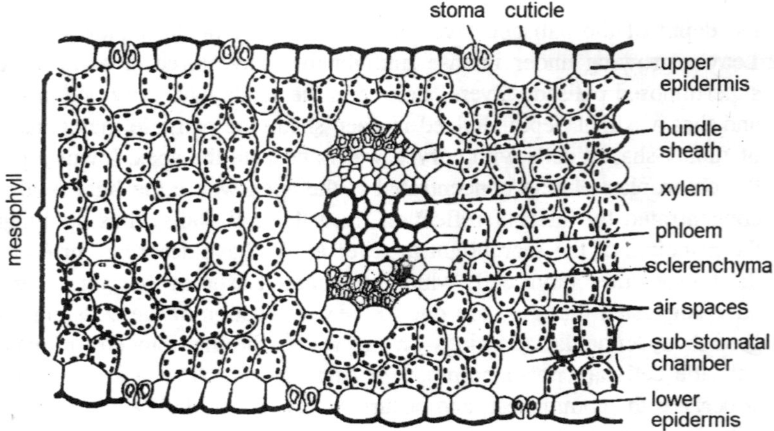

Fig. 10.3. Transection of an isobilateral leaf.

spaces provide a reservoir of air and water vapour which may be of great benefit to the palisade cells as a buffer between them and external conditions, while its cells serve as channels of transport for materials between the palisade cells and the veins.

Anatomically this illustrated by the system of what are called collecting cells, that is, mushroom-shaped cells the flattened heads of which each make contact with the bases of a group of palisade cells. The relative isolation of palisade cells from each other implies that there can be little, if any, lateral transport between them, and that carbohydrate translocation from the palisade layer must be through the base of each cell individually. The rest of the spongy parenchyma forms an anastomosing system of pipe lines connecting with the veins, which are both the source of the water supply and the channels of translocation of organic materials from the leaf to other parts of the plant. The veins are always differentiated at the boundary between the palisade and the spongy layer, and are hence practically median. Each vein is surrounded by a sheath of elongated parenchyma, normally devoid of chlorophyll, which is one cell thick in the finer veins, but increases in thickness as the leaf base is approached, where the sheath cells pass into the massive parenchymatous coat of the midrib which in turn is continuous with the ground tissue of the petiole. These appears to be a concentration of sugars in these sheath cells, and they are sometimes called the glucose sheath. They are probably important in translocation, especially at the vein extremities; where the amount of vascular tissue in each vein is very small. The question of an homology between these vein sheaths and the endodermis cannot be definitely answered.

A typical endodermis, with Casparian bands on the cell walls, does occur in the leaves of several families, notably the Primulaceae, the Plant-aginaceae and the Rosaceae. At the leaf base it say even be secondarily suberized, but among the smaller veins it is usually irregular and incomplete, and the characteristic band on the cell walls is absent from the smallest veins. In the majority of families on endodermis is altogether absent from the leaf. The free end of each vein is formed of a single tracheid, which is often enlarged. The phloem stops short of the end, and the last phloem cells are peculiar, and may be regarded either as enlarged companion cells or as sieve-tube mother cells which have not divided into distinct sieve-tube and companion cell units. The vein sheath extends like a cap over the end of each vein and is thus continuous. The vascular tissues of the veins are characteristically arranged with the xylem uppermost. This is necessary consequene of the position of these tissues in the stem bundles, for as a leaf trace bundle turns outwards into the petiole, its xylem becomes uppermost and this position is usually maintained, without torsion, through the petiole into the lamina.

Many accessory structures may occur in the mesophyll. Crystal sacs, containing Calcium oxalate crystals are common, while cystoliths in specially enlarged cells are characteristic of the Urticaceae. Glandular sacs and canals occur in certain families and contain either mucilage or aromatic terpenes. Where the sacs are large they form colourless points in the tissue, as may be seen with the naked eye in such species as *Hypercium perforatum*, which gets it specific name from this circumstance. Sclerid cells, often of complex shape also occur, particularly in those hard-leaved plants such as *Camellia* and *Olea*, which are called *sclerophyllous*. The orientation of these sclereids is usually definite in each case, lying either transversely, like pillars across the mesophyll, or longitudinally, in the spongy mesophyll or between the upper epidermis and the palisade layer.

Development of Vascular Tissue in the Leaf

The primary vascular tissues of the leaf blade and petiole form a system continuous with the leaf trace with which they are joined. All parts of this system differentiate from procambium, though the time of maturity differs in the different parts of the leaf. Ordinarily, the vascular tissue of the leaf to mature is the part of the leaf trace near the junction of leaf and stem. Here the tissue often differentiates soon after the formation of the leaf primordium near the growing tip. Recent evidence indicates that the phloem tissues, particularly the sieve tubes, differentiate acropetally in the procambium strands from the functioning phloem below. No xylem is formed in the developing leaf trace until the progressive differentiation of sieve tube elements reaches the base of the leaf primordium. Xylem differentiation begins at this point, and extends both acropetally and basipetally.

The order of maturating of the sieve tube elements may be acropetal from the mature vascular tissue below or may be discontinuous in the protophloem strand. As the leaf primordium elongates and expands, the vascular bundles are extended distally from the trace. Within the developing leaf blade, the intricate network of veins has its origin in procambium strands that arise from the division of the meristematic cells in the central mesophyll of the young leaf. Such division may take place close to marginal meristem or elsewhere in the leaf where cell division is going on. Differentiation of vascular elements may begin in more or less isolated strands and develop in both directions to join with other parts of the vascular system. In the formation of the conducting strands of the leaf, the same type of cells are developed, and the other is usually the same as in the vascular strands of the stems of the same plant. There is commonly, however, a larger proportion of the extensible protoxylem elements, especially near the bundle ends. In leaves with secondary growth, this growth takes place soon after other parts of the leaf attain nearly full size.

Orientation of Vascular Tissue in the Leaf

In the leaf traces of angiosperms and gymnosperms, before they leave the stele, the phloem is toward the outside of the stem. As the traces pass out of the stele and enter the petiole and blade, the xylem and phloem maintain their relative position, so that commonly in the petiole and usually in the blade, the phloem is on the lower or dorsal side of the leaf and the xylem on the upper or ventral side. Ventral and dorsal are used in the sense that the ventral side is the side next the axis and the dorsal side that awəy from the axis. Although the orientation of the xylem and phloem is fairly constant for typical leaf blades, many variations occur in the petiole because of the different methods of fusion, division, or twisting of the leaf traces in their course through the petiole.

In many plants, traces that enter the petiole separately remain distinct and pass to the blade without change of structure or orientation. In others, the traces fuse in the petiole to form a single strand of various cross-sectional shape, a hollow cylinder, or a group of more or less complete stele-like cylinders; in still others, the traces divide into several or many ways. Many of the bundles so formed may be amphicribral within the petiole. On passing out of the petiole into the blade, the vascular bundles usually again assume collateral structure with dorsal phloem; sometimes, however, the petiolar arrangement is maintained in the larger veins. In many ferns the traces to a single leaf make up a considerable portion of the stele below the point of attachment. These large traces are in most ferns a part of an amphiphloic siphonostele, and as soon as freed from the stele become amphicribral bundles by the union of the external and internal phloem at the sides of the trace. The external and the internal endodermis also unite about the traces. These bundles pass up the petiole and eventually divide to form the smaller bundles of the compound leaves. Dichotomous branching of the veins is common in the smaller divisions of the leaves of many general. Ultimately, the minute divisions come into close contact with the mesophyll, much as in angiosperm leaves.

In small-leaved pteriodophytes, for example, Lycopodium and Equisetum, the vascular supply consists of a single bundle which passes unbranched to the leaf tip. The vascular supply of stipules is derived from the lateral leaf traces after the leaf traces have left the stele. Where the stipules are not attached to the petiole and where they sheath the stem, the stipular traces depart from the leaf traces in the cortex. Where the stipules are attached to the petiole and are an integral part of the leaf, the stipular traces may not depart from the leaf traces until after the latter have entered the petiole. Except in plants with foliaceous, persistent stipules, the stipular traces are usually smaller vascular bundles; the xylem consists of relatively few conducting element and the phloem chiefly of parenchyma.

Elements of Xylem and Phloem in Leaves

With the exception of some monocotyledons both primary and secondary vascular tissue of the petioles and larger veins of leaves resemble the stelar tissues of the same plant in the type of elements present in both xylem and phloem, for example, if scalariform vessels are characteristic of the stelar tissues, they occur in the leaves also. The vessels, sieve tubes, and parenchyma cells of the petioles and large veins of dicotyledonous leaves are usually smaller than those of the stem and the bundles of these parts have relatively less secondary tissue as compared with primary tissue. The amount of secondary tissue in leaves varies with the species. As the vascular bundles branch and become progressively smaller, the secondary tissues are reduced in amount until in veins or medium and smaller size, there is none.

At the same time the size of the vascular elements is greatly reduced. In the leaf of a woody dicotyledon like the apple, the smaller veins consits of several close-spiral or reticulate elements accompanied by about the same number of small sieve tubes, companion cells, and parenchyma cells enclosed in the bundle sheath. With further reduction the xylem may consist of only a single spiral element and the phloem of a single elongate parenchyma cell.

In the transition from phloem consisting of sieve tubes, companion cells, and parenchyma cells to phloem made up of a single parenchyma cell, the sieve tube before its disappearance is reduced in diameter to that or less than that of the companion cell. In the smallest veins, the phloem mother cell fails to divide. The resulting parenchyma cell, which resembles a companion cell in form, has been called a *transition cell*.

Bundle Ends

The ultimate divisions of the vascular bundles in leaves terminate in what are known as "bundle ends". In leaves that have more or less definite vein islets, the bundle ends bend into the mesophyll of the islet and end abruptly near its centre. Some bundle ends may be somewhat enlarged at the tip or branched in various ways. In leaves with parallel veins for example, the leaves of grasses, or in leaves without definite vein islets, the bundle ends may be merely short spurs from the veins extending into the mesophyll.

The bundle ends and the small veins forming the vein islets supply the mesophyll with water and nutrients and absorb and remove the products of photosynthesis. Because of the distribution of the vein islets and bundle ends the distance through which water and materials in solution travel through the mesophyll is always about the same. In the apple leaf, for example, the distance between bundle ends or between bundle ends and veins bordering the vein islets is about 88 micra. The structure of bundle ends varies somewhat with the species. Probably in mesophytic dicotyledons the bundle end commonly consists of a single spiral or reticulate xylem element accompanied by a single specialized parenchyma cell, the two elements surrounded by the parenchymatous bundle sheath. The bundles ends of hydathodes and some glands, especially the glands of insectivorous plants which function in digestion, have more elaborate structure than typical bundle ends.

Bundle Sheath

With but few exceptions, notably in some aquatic plants and delicate ferns, the vascular bundles of leaves are enclosed in a structure known as the bundle sheath. The elements forming this sheath are parenchyma cells elongated parallel with the vein axis and joined laterally in such a way as to surround the vascular tissue with a tight sheath resembling an endodermis but lacking Casparian strips. In many species, for example, Malus pumila, these cells contain chloroplasts and resemble the adjacent mesophyll cell except that they are more regular and elongated in shape and contain relatively fewer plastids than the mesophyll cells. In other species, for example, Carya Pecan, the bundle sheath cells are more sharply differentiated from the mesophyll in that they take a different stain and do not contain chloroplasts. The cells of the bundle sheath are in intimate contact on their inner faces with the conducting elements of the vascular bundle or adjacent parenchyma and on the outer face with the palisade and spongy mesophyll cells which commonly are joined to form filaments ending at the bundle sheath. In some veins, flanges or spurs from the bundle sheaths extend upward and downward to the upper and lower epidermis. Physiologically considered, the bundle sheath form a layer of living cells surrounding the vascular bundles through which water and solutes must pass to reach the tracheids, vessels, and sieve tubes. The fact that often the sheath is connected with the epidermis suggests that it may conduct materials to the epidermis also. In bundles with well-developed fibres above and below the vascular tissue, the bundle sheath has contact with the mesophyll on the sides of the bundle only.

Sclerenchyma in the Grass Leaf

Selerenchyma in grass leaves is usually associated with the vascular bundles. In the larger bundles there are often two strands of fibres, one above and one below the vascular tissues. These join with the xylem and phloem to form a plate of hand tissue extending through the leaf. In smaller veins the strand of fibres may be on the dorsal side of the leaf only and joined either with the vascular tissue or free from it. In some species, strands of fibres occur beneath the epidermis on both sides of the bundle but free from it. Small bundles may lack sclerenchyma altogether. There is thus great variation in the amount and position of the fibrous strands. In addition to the sclerenchyma associated with the bundles, nearly all grasses have a strand of fibres along each edge of the leaf. In leaves of some xerophytes, hard tissues may extend over the entire dorsal side.

Bundle Sheaths in the Grass Leaf

Among the characteristic features of grass leaves are the specialized tissues immediately surrounding the vascular bundles. These occur commonly as two layers of cells, each one cell thick, completely surrounding the bundle next to the vascular tissues. The inner sheath is more or less thick-walled and lignified, resembling an endodermis in general appearance. Whether this is an endodermis in the strictly morphological sense is doubtful. The thickening of the cell walls of this layer is frequently heavier on the inner tangential and radial than upon the outer tangential walls, although in some plants it extends entirely around the cells. The inner walls are pitted and presumably allow transfusion between the vascular tissue and this layer. Various functions have been ascribed to this layer, but protection against crushing, especially of the softer phloem cells, seems most probable.

In bundles with well-developed sclerenchyma, thus adding rigidity to the whole structure by completing the so-called "girder structure." In some species, the inner sheath is absent or present in the larger bundles only. The outer sheath, often known as the *mestome sheath*, is made up of thin-walled parenchyma cells, which appear mostly isodiamteric in transverse sections of the leaf, but much elongated in longitudinal section. These cells form about the bundle a parenchymatous girdle, to which has been ascribed the function of conducting the soluble food products from the photosynthetic tissues to the conducting tissues. In the larger bundles of some species the cells of this girdle lack chlorophyll, and are hence readily recognized. In the smaller bundles the cells may contain chlorophyll but usually in less amount than other photosynthetic cells. In some species the cells of this sheath extend to the epidermis on one or both sides of the bundle.

Mesophyll of the Grass Leaf

In the leaves of grasses, as in other leaves, the photosynthetic tissue fills the space between the vascular bundles more or less solidly. Usually there is no well-developed palisade layer of elongate cells such as is characteristic of dicotyledonous leaves generally. Sometimes, however, there is a weakly developed palisade layer next to the epidermis on one or both sides of the leaf. The cells of such a palisade layer are nearly isodiametric and differ from the spongy mesophyll cells chiefly in their more compact arrangement. The cells in the spongy mesophyll are also frequently irregular in shape, and arranged in branching conduction systems, as in analogous tissues of other types of leaves. In a few xerophytic grasses and some species of Cyperus the green tissues themselves are limited to a sheathing girdle about the vascular bundle.

Epidermis of the Grass Leaf

Although variations in the structure of the epidermis of the grasses often occur, especially in plants adapted to growth in extremes in environmental conditions, the general features are fairly constant. Great variation occurs in the extent of cultivilization. In surface view the cells do not show marked irregularity of outline as does the epidermis of many dicotyledons. A type of cell common to the leaves of nearly all grasses is the so-called *motor cell*, or *bulliform cell*, which, apparently, constitutes the mechanisms that functions in the rolling or grass leaves in dry weather. These cells have much more depth than the ordinary epidermal cells and are arranged in rows extending throughout the length of the leaf upon its upper surface, frequently lying at the bottom of well-defined grooves. The cells are thin-walled and lack chlorophyll, and with decrease in their turgor the leaf rolls upward and inward. Some species possess only one or two rows of this type of cells, whereas others have many. Few species lack them. At the edges of the groups of motor cells, as seen in cross section of the leaf, there are cell types transitional between typical motor cells and the ordinary epidermal cells.

Arrangement of the Stomata in the Grass Leaf

The type of stoma found in the leaves of the grasses is surprisingly uniform considering the diversity of habitat in which grasses are found. The stomata are elongate with their long axes parallel with that of the leaf and the epidermal cells surrounding then. They are usually arranged in rows alternating with rows of epidermal cells. Frequently several rows of stomata are spaced close together between wider bands of epidermal cells. Although in a considerable number of species stoamta occur upon both surfaces of the leaf in approximately equal numbers, in the majority of species these openings are more numerous in the upper epidermis and in some species, chiefly xerophytic may be wanting upon the lower side. Various types of hairs are found in the epidermis of many species. Short, stiff projections, which give the surface of the leaf a harsh texture, are particularly common. In some forms these teeth give the leaf margin an effective cutting edge, as for example, in speices of *Leersia*.

Persistence of Leaves

A large proportion of the gymnosperms, many tropical and some broad-leaved, temperate-zone angiosperms, retain their leaves for more than one season. In evergreen plants possessing secondary growth, as in the gymnosperms and some dicotyledons, the persistence of leaves for more than one year involves the lengthening of the leaf trace as successive annual layers of xylem are added by the cambium. This lengthening is accomplished by the activity of a special mersitematic layer in the trace itself. Ordinarily, the needle leaves of the gymnosperms and the broad leaves of angiosperms do not persist more than three to five years, after which time they are cut off by abscission layers in the same manner as are deciduous leaves. In a few general, evergreen leaves may persist for many years as, for example, on the trunks of Araucaria. In most ferns, some of the plants, and other monoco-tyledons the leaves persist for a considerable number of years, the older leaves gradually ceasing to function.

In many of these plants, particularly some palms, the leaves are not cut off by abscission layers but cling to the stem until they disintegrate by weathering. The leaves of many annual and perennial herbs are also "withering persistent." In the majority of temperate-zone plants, both woody and herbaceous the leaves function for only a single season at the end of which they wither or are abscised. The morphological nature of the bundle sheath is uncertain. The nature of the cells and their contents suggest that it is a part of the mesophyll which it resembles.

Mechanical Support in the Leaf

Considered mechanically, the dicotyledonous leaf is a complex network of strands which supports a relatively large blade or photosynthetic surface at the end of a slender petiole. In this support, collenchyma, sclerenchyma, and the woody xylem play an important part. In early stages of development collenchyma is of major importance in supporting the developing petiole and leaf blade. Collenchyma is most abundant in the outer cortical layers of the petiole and the larper veins. It is also commonly associated with veins of intermediate size where it lies above the vascular tissue. The turgor of the parenchyma itself is also important in support during the developmental stages of the leaf. In docotyledonous leaves fibres are usually associated with the vascular tissues of the petiole, where they occur as bundle caps adjacent to the phloem. Groups of fibres are found also on both the ventral and dorsal sides of many of the larger veins.

In this position, often in combination with collenchyma, these fibres above and below the vascular bundles build up vertical plates of supporting tissue extending through the leaf and connecting the upper

length and, except for occasional small connecting strands, are separate from one another. Frequently, the leaf blade has a large median bundle associated with a pronounced midrib projecting on the dorsal side. The other bundles may be of two or three sizes differing chiefly only in the amount of mechanical and conducting tissue present. The smaller bundles are spaced alternately between the larger. The entire leaf consists of a sheathing base and a linear blade; the former surrounds the culm for some distance and merges in a more or less prominent joint with the leaf blade, which is set at an angle with the culm. In many species the sheathing base extends above the joint in a short membranous structure called the ligule.

11

Structural Modification of Leaf

When we turn to the consideration of the leaf we are confronted by the same difficulty of definition that we found in the case of the stem but in an accentuated degree, in as much as the range of morphological variation is much greater among leaves than among stems. Whatever formal definition of a leaf we attempt to frame, it will be possible to find some cases which appear to contradict it. The science of comparative morphology, which arose from Goethe's famous "Essay on Metamorphosis," was based upon the concept of organ categories, that is to say, on the principle that the plant body is built up of a limited number of kinds of categories of organ, primarily distinct from each other, but related, in the architecture of the plant, in a fundamentally uniform way, no matter how variable their external appearance may be.

According to the principle of metamorphosis which Goethe formulated, any one kind of organ may vary, according to its position and function, so far from the normal that it may become difficult to recognise. Nevertheless it should not, so the argument runs, take over wholly the character of an organ of different category, and it was considered to be business of the morphologist to trace out and determine its true nature. So rooted did this idealistic theory become that controversies were carried on, sometimes without any acceptable conclusion, regarding the placing of such structures as the ovute in their proper category of leaf or axis. The most familiar type of leaf, namely the green foliage leaf, was tacitly assumed to be the ideal form, and all other "leaf" organs were supposedly metamorphosed foliage leaves. With the growth of our knowledge of plant evolution such formal ideas of morphology have become obsolete. Goebel, in his great "Organography of Plants," was the first to break away from them by directing attention first and foremost to the functional aspect of organs and their relationships to the plant's conditions of life. He was careful, however to avoid the naive adaptationism which confuses *post hoc* with *propter hoc* and seeks to relate every observed structure to some supposed function in relation to the environment.

The complexity of plant structure is not necessarily a direct reflection of the complexity of the environment. Even with our limited knowledge of the life conditions of plants it is quite evident that similar requirements have been met by plants in an infinity of different ways, which are not physiologically predetermined. We must combine both the morphological and physiological view points if we are to attain a natural understanding of plant structures. Are we then to abandon all categories and treat every plant structure as an independent entity? Not entirely so. A survey of the types of lower plants, such as we have made in former chapters of this work, shows us that the categories of stem, root and leaf in the Higher Plants are not primitively distinct but are differentiated from one primary structure, the axis, which, as we have previously suggested, may be derived from the primordial cell-filament. They have been differentiated moreover in very many ways and in very various degrees, so that fine-drawn distinctions between them must be artificial.

Nevertheless, if we compare many types of Higher Plants we find certain regularities in the relationship of organs, associated with the greatest diversity of individual form, which justifies us in using the old categories as general descriptive terms, although without their old rigidity, and thus simplifying and clarifying our ideas, while leaving room for the recognition of cases which cannot be fitted into the framework of any formal definition. We have already applied this principle of relationship when we described the stem as the organ which bears the leaves. What then are we to include under the term "leaf"? We may agree to apply the term to those structures which arise as lateral outgrowths from the superficial tissues of the stem without any subtending organ, and are arranged in a definite geometrical other.

Further, we may add as normal, though not invariable characters, that leaves are generally of bilateral symmetry and of limited growth, that they subtend buds in their axils and that they donot directly bear other leaves though they may produce buds. Thus, if we seek to apply the name to a particular organ we must be guided by the generality of its characters, recognizing that Nature has no regard for our subjective limitations, and that a given organ may contravene one or even several of the above characters and yet be most appropriately classified as a leaf. Much was made at one time of the restriction that a leaf could never be terminal on the axis. There are a number of instances, both normal and abnormal, in which leaves do assume this position by displacing or replacing the true stem apex. Apart from such disputable structures as carpels or stamens, which are frequently terminal, there are terminal leaves in *Polyganatum*, which owe their position to the abortion of the apical bud. Among the Gymnosperms, *Pinus monophylla* also appears to have its single leaves terminal on the spurs. Yet in this case also, traces of an abortive apex may be found, so that the terminal position is only secondary.

In *Funcus*, "terminal" leaves may cover and include the stem apex in their bases, but they do not replace it. Instances which have also been cited in *Coridalis* and in a Bamboo appear to be similar to the foregoing, i.e., the apparently terminal leaves have sheathing. There are many examples of both rootless and stemless plants, but if we consider the formative importance of the leaf bases enclosing the stem apex, which is moreover abortive. In fact no case has been produced in which the stem apex is directly transformed into a terminal leaf and until this can be shown we may continue to treat leaves as characteristically lateral organs. When we consider the phylogeny of the angiosprmic leaf we find that it shares with the pteridophytic leaf a character which is of wide significance, namely that its vascular supply leaves a gap in the vascular system of the stem, which stamps it as belonging to the same megaphyllous cycle of affinity. This structural condition is called *phyllosiphonic*, in contrast to the *cladosiphonic* condition in which only the branch-traces leave gaps in the stem stele, a condition which is characteristic of the Lycopodiales and other microphyllous groups.

The megaphyllous leaf may have been evolved from a flattened branch system bearing microphyllous leaves, whose small separate laminae became merged into a larger general lamina. This evolutionary scheme, which is associated with the names of Lignier, Tansley and others, harmonizes with the apparent derivation of both microphyllous and megaphyllous stocks from primitive, leafless types like the

Psilophytales, with dichotomous axes, and it has much to commend it. Among the Psilophytales we find types which are covered with minute leaves, developed apparently like superficial emergences, which invite comparison with the microphylls of both Mosses and Lycopods. If we accept the homology of these leaf types we must regard the microphyllous state as the more primitive of the two and the megaphyllous state as derived from it. From the pteridophytic leaf, through the gymnospermis to the angiospermic leaf, the fundamental type has not altered, though specialized growth patterns have brought about many changes of form. One rather marked difference may be observed in the vein systems, or *venation*.

In Ferns an "open" dichotomous venation is characteristic, in which the branching of the veins is open towards the leaf margins, while in Angiosperms the lamina is usually prevaded by a vein-network with closed meshes. In the architecture of the stem, it is scarcely surprising that the existence of a truly leafless plant is doubtful. There are of course, very many plants in which the leaves have been reduced until they are no more than scales, yet these cannot be considered as other than metamorphosed leaves. Scales may be reduced almost to invisibility, as on the rhizome of the saprophytic Orchid *Carallorhiza innata*, or they may have lost even a vascular supply, as in *Cuscuta*, and consist only of undifferentiated parenchyma, but they are, nevertheless, so clearly linked by intermediate forms with larger and better developed leaves, that we must regard them as leaves in spite of their extreme reduction. If we exclude floral leaves and speak only of vegetative structures, the nearest approaches to a truly leafless condition are to be found in the following:

1. The early stages of the parasite *Orobanche*, which has no vestige of leaves or of an apical bud, until it has become well established on its host plant;
2. The cylindrical green stems of *Heleocharis* and *Scripus lacustris*, which consist of a single, greatly extended internode, bearing only two small scales at the apex;
3. The cladodes of *Asparagus*, if so extreme a case may be admitted, can also be cited as leafless stems, though of a very reduced kind.

Morphology of the Leaf

The morphology of the leaf varies with different major plant groups. In some of the more primitive groups it is apparently a lateral expansion of the axis in which epidermal, cortical, and stelar tissues take part.

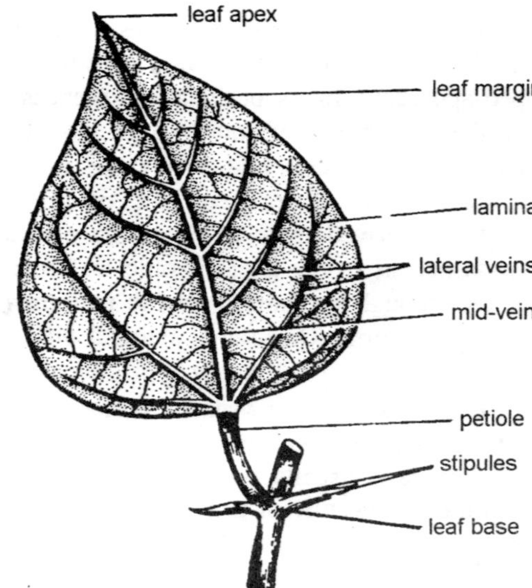

Fig. 11.1. Parts of leaf.

The stelar tissues—xylem and phloem, together with accompanying sclerenchyma—form the skeleton upon which the green tissues, which are cortical, are supported. The epidermis is continuous with the epidermis of the stem. The leaf in other groups—ferns and seed plants—is probably a reduced branch system with its members fused in various degrees. In many monocotyledons and some dicotyledons, the leaf is undoubtedly a phyllode, that is, flattened, expanded petiole.

In some other plants, for example *Phyllocladus*, the function of the leaf is carried on by fused branchlets, leaf-like in appearance, but in which the position and orientation of the vascular tissues within the flattened part indicate its branch nature. Leaves consist typically of three parts: the expanded portion, or blade, where most of the green tissue is located; the petiole, which supports the blade on the axis and functions also in conduction; the stipules, small, paired lobes at the base. Stipules in many plants are wanting or are soon lost by abscission, but in some plants are persistent structures which may form an appreciable part of the photosynthetic system, as in *Pisum*.

Anatomically, stipules are appendages at the base of the leaf, with a vascular supply derived exclusively from the foliar traces. Leaves are classified as parallel veined when the main bundles traverse the leaf without anastomosis, and *net veined* when the main branches of the vascular system form a network. Piunate and palmate, closed and open venation, and other familiar morphological classifications are also based upon the arrangement of the vascular bundles. Whatever the arrangement of the larger vascular bundles, the ultimate divisions of the conducting strands of angiosperm leaves completely or partly encircle minute areas of the photosynthetic tissue with which they are in close contact. These divisions of green tissue have been called *vein islets*, and in a way represent more or less well-defined photosynthetic units.

The vascular bundles surrounding the vein islets are in intimate contact physiologically with the photosynthetic tissue since they lack the sclerenchyma normally present about the larger bundles. The size and shape of the vein islets vary with different types of venation and with different species, and in some plants, especially in the ferns and grasses, definite islets do not exist. As the leaf matures, the islets increase in size, their growth being a part of the general growth throughout the leaf. In full-grown leaves of any species, the size of vein islets is apparently fairly constant, regardless of the size of the leaf or the age of the individual plant.

Ontogeny of the Leaf

Leaves originate in the promeristem of the growing point of the stem. The leaf primordium is first evident externally as a round or wedge-like protuberance on the side of the promeristem. This is initiated by anticlinal and periclinal divisions in the outer layers of the apical meristem just below the apex. At the tip of the protuberance is a group of meristematic cells that constitute the apical meristem of the leaf. The behaviour of this meristem varies with the form and structure of the developing leaf. In the formation of a simple dicotyledonous leaf this terminal meristem builds up a short, finger-like structure flattened on the adaxial side. If the leaf is stipulate, stipular meristems arise early as swellings at the base of this structure. Lateral ridges next appear except at the base. These are the marginal meristems that build the leaf blade. As development of the blade begins, the base of the finger-like structure remains unexpanded and forms an intercalary meristem that builds the petiole.

The body of the fingerlike meristematic structure becomes the midrib of the leaf. In pinnately compound leaves the first-formed apical meristem lays down the central axis of rachis of the leaf and the lateral leaflets originate from primordia which arise laterally on this structure and function as terminal meristems in forming the axis of each leaflet. Upon these leaflet axes, marginal meristems form the blades of the leaflets, as in simple leaves. The leaf blade develops according to a more or less definite pattern which varies only in minor details.

The epidermis is formed from the superficial layer of the marginal meristem by anticlinal divisions. Sub-epidermal cells on both sides of the leaf are formed by anticlinal divisions of subepidermal cells in the marginal meristem. These subepidermal layers remain fairly distinct in the early stages of development, keeping pace with leaf expansion by continuing anticlinal divisions. In later development, the adaxial layer gives rise to the upper layer of palisade cells and abaxial layer produces the lower layer of spongy parenchyma and even some of the adjacent, more deep-seated layers of mesophyll by periolinal division. The central tissues of the mesophyll are usually formed by the periclinal division of subepidermal initials of the marginal meristem. In the inner daughter cells formed in this way, division in all planes gives rise to the central mesophyll, consisting of the inner part of the spongy parenchyma, the vascular bundles, and sometimes to inner layers of palisade tissue. Periclinal divisions of the central-mesophyll area take place rapidly near the marginal meristem so that close to this meristem the leaf is quickly built up to its maximum number of layers in thickness.

Ontogeny of the Petiole and Stipules

The development of the petiole takes place simultaneously with the expansion of the leaf blade. Cell division occurs throughout the petiolar meristem followed by rapid elongation. The ontogeny of stipules varies with the relation of these structures to the leaf base. In the majority of dicotyledons, the stipules are attached to the leaf base. In such plants, the young leaf in early stages is three-pronged, because the stipules develop rapidly from apical and marginal meristems. In dormant buds and near the terminal meristem of young shoots the stipules may be as large as or larger than the leaf petiole as seen in transverse section. Most temperate-zone trees shed the stipules before or soon after the leaves are mature. Sheathing stipules, stipular thorns, stipules attached laterally to the petiole have a somewhat different ontogeny.

Leaf Differentiation

The development of leaf rudiments from their meristematic primordia at the growing point of the stem is by no means simple and are still far from a full understanding of the process. Yet the importance of the leaf rudiments, not only morphologically but physiologically, and their dominant part in the development of the shoot can understood

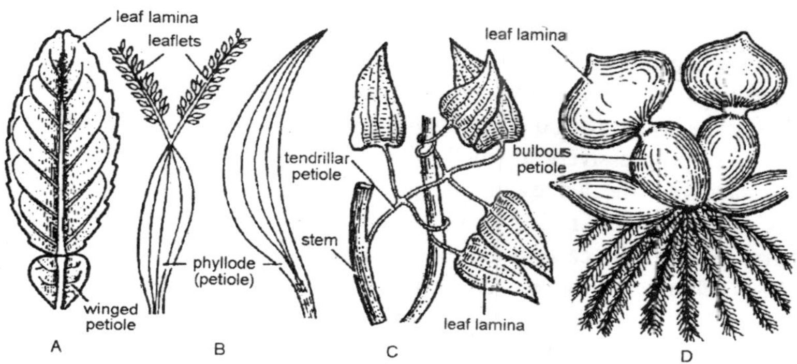

Fig. 11.2. Petioles: A, winged; B, phyllode; C, tendrillar; D, bulbous floating.

when we realize that these minute meristematic rudiments produce the auxins which both stimulate and inhibit growth in all parts of the shoot. We know that leaves are invariably lateral outgrowths; to this we may now add that they are invariably outgrowths from meristem tissue and are never adventitious, though, in some exceptional cases, the meristem from which they come may not be that of the stem apex.

In some monocotyledonous seedlings, as in the sporelings of some Ferns, the formation of the first few leaves precedes the organization of an apical growing point, while the cotyledons are never in any case, the product of a stem apex. Indeed the cotyledon of Monocotyledons such as *Iris* is truly terminal. The anomalous cases of leaf origin in *Lemna* and in the Podostemaceae have already been mentioned. The extent of the apical meristem which is involved in the formation of a leaf primordium appears to be very variable. Among the known cases the majority seem to originate from the tunica, especially where this is two or more layers deep, but in other cases the corpus certainly takes part as well.

The earliest part of the rudiment to appear is that which later forms only the apex of the mature leaf, and in some plants, especially some tropical climbers, this first portion of the rudiment grows precociously to form an elongated point or forerunner in which tissue differentiation, including venation, may precede the appearance of the rest of the lamina, of which, however, it may later become an indistinguishable part. The earliest growth of the rudiment is strictly apical, but this lasts, with few exceptions, only for a short period. In most Monocotyledons apical growth ceases when the rudiment is less than 0.5 mms. long, and the greatest part of leaf growth is therefore intercalary.

In Dicotyledons the end of apical growth is usually somewhat later, when the leaf is several millimetres in length. Certain striking exceptions are known in which the apex retains its embryonic character during the whole period of the development of the leaf, which therefore increases progressively in length, like a Fern leaf, and is indeed usually circinate like the leaves of that group, the growing apex being thus protected by being rolled up within a covering of older more resistant parts. Such is the case in the long leaves of *Drosophyllum*, which is a close relative of the insectivorous *Drosera*, and it is also true of leaves of some species of Drosera and of Utricularia. Leaves are normally organs of limited growth, but in this latter genus the place and functions of roots, which are absent, are taken over by long thread-like leaves, the rhizophylls, which have a downwardly directed and indefinite apical growth. Only their origin and the occasional occurrence of imperfect insect traps on the rhizophylls remain to show that they have, originated from modified leaves and are not, in fact, as in appearance, true roots. In some species of Utricularia the persistence of apical growth in the leaves has led to a situation where all distinction of stem, root, and leaf appear to be lost.

The vegetative plant consists of creeping axes which may develop either from or into foliage leaves or into rhizophylls with complete promiscuity. In this connection one may recall that the leaves of *Welwitschia* are also of unlimited growth, though in their case the growth is intercalary. During the brief period of normal apical growth in the leaf rudiment, the basal portion expands laterally, in accordance with the transverse expansion of the meristem of the stem apex, so that it occupies an increased are of the circumference of the apical dome.

The leaf rudiment in Monocotyledons expands laterally so much that it rapidly encircles the whole growing point like a collar, and the close succession of these collars, each growing actively, quickly pushes the leaf bases of older rudiments centrifugally away from the growing point. This produces the sharp flexure of the leaf trace strands at the point of their entry into the leaf base, which we have commented upon in the previous chapter, as characteristic of most monocotyledonous stems. In describing differentiation at the apex of the stem, we pointed out that the procambium usually makes its first appearance in the ring of residual meristem, at separate points in the ring, each point lying, at the base of a leaf primordium. These vertical strands of the earliest procambium extend themselves continuously both downwards in the tissue of the internode below and also in an upward and outward direction into the leaf primordium itself. This latter extension is due to the appearance in the primordium, at a very early stage, of a median band of meristem, connecting the end of the procambium advancing from the stem with the apical meristem of the primordium, which is still active in this early phase of development.

The rapid elongation of these meristem cells to form procambiun in the leaf is probably the reason for the relatively rapid growth in length which occurs in the young rudiment. This growth in length is accompanied by radial growth in thickness, so that the primordium assumes the form of a slender cone, somewhat flattened on the axial side. This constitutes the petiolar-midrib region of the leaf, which is differentiated before the lamina makes its appearance. The petiole does not, therefore, arise as a structure intercalated between the lamina and the leaf base, as older accounts maintained, but is one of the primary structure of the leaf. The lamina begins to differentiate from the upper or adaxial portion of this petiolar-midrib structure in the form of two thin marginal ridges of meristem which appear before the rudiment is as much as 1 mm. long.

Very little is known about the early development of compound leaves, but it has been observed that the pinnae arise on the side of the rudimentary rachis as hemispherical cushions, replacing the continuous marginal ridges of meristem which are characteristic of the simple leaf blade. These separate rudiments usually arise in basipetal succession, but cases occur of the opposite succession, and also of double succession from the middle in both directions simultaneously. The rudimentary lamina consists of five to eight layers of cells and it grows by a marginal meristem. The outer cells of this margin seem to divide chiefly in the anticlinal direction, producing the upper and lower epidermis respectively. By periclinal division, however, they produce a line of submarginal initials and it is from these latter that the internal tissues of the leaf are produced. The course of differentiation may follow various sequences, which differ in detail, but in most cases there seem to be three primitive cell-layers formed. The upper or adaxial layer produces the palisade, the middle layer forms the veins and part of the spongy layer, while the lower or abaxial layer develops into the lower part of the spongy parenchyma. Marginal growth ceases as soon as the main outline of the mature leaf has been formed and is succeeded by a protracted period of generalized surface growth, characterized by the predominantly anticlinal division of the cells. The result of this is that the number of layers laid down by the marginal meristem remains practically constant throughout the later development, the only important exception being the formation, by periclinal divisions in the middle layer, of procambium strands from which the veins are developed.

The simultaneous longitudinal extension of the veins and superficial growth of the other parts of the lamina, lead to adjustments of position, which are responsible for the characteristic ptyxis or folding of the young leaf within the bud. The final shape of the mature leaf is determined by local variation of growth rates in different parts of the lamina, which may be due to variation in the distribution of auxins in the tissues. The mature leaf consists of three portions, the *lamina*, the *petiole*, and the leaf base or *phyllopodium*. In most Dicotyledons the leaf base is relatively undeveloped, though it may in some cases give rise to the *pulvinus*, a short cylindrical cushion at the base of the petiole, which acts as a joint between leaf and stem, and by means of which the position of the leaf may be changed. In other cases the phyllopodium gives rise to the stipules.

Many Monocotyledons, however, have greatly enlarged leaf based, forming sheaths surrounding the stem, which may, as in some Grasses and Sedges, be the longest part of the leaf. In Palms with palmate leaves, such as *Chamaerops*, moreover, there is ground for the belief that the petiole also belongs to the phyllopodium, since the ligule, which in the Grasses and Sedges occurs at the top of the sheath, is, in these Palms, at the top of the petiole, immediately below the lamina. These peculiarities are associated with the persistence of intercalary growth in monocotyledonous leaves, which in many instances continue to elongate throughout their life-period by the activity of a basal zone of meristematic cells. An analogous persistence of growth at the base of the lamina in other cases leads to the development of cordate or sagittate leaves with outstanding basal lobes, or in extreme cases to peltate leaves in which there is an extension of the lamina below the point of junction with the petiole, which then appears to be inserted in the middle of the lamina. The suppression of laminar growth at certain points, which may also occur, gives rise to the irregularities of outline which we have previously mentioned as characteristic of certain species of *Artocarpus* or *Broussonetia*.

In some of the Araceae it may affect, not only the margins but the central parts of the lamina, giving rise to characteristic perforations which are seen in their most striking development in the large leaves

of the tropical climbing Aroid, *Monstera deliciosa*. Patches of meristematic tissue in the leaf rudiments of this plant die and drop out at a very early stage of development and the margins of the lacunae are covered by a secondary epidermis formed from interior cells of the young leaf. The early-arrest of growth in the midrib, associated with continued growth of the lamina on each side of it may produce either an emarginate leaf such as that of *Liriodendron*, or in extreme cases, an apparent dichotomy of the leaf, which is characteristic of certain genera, such as *Bauhinia*, a leguminous tropical climber. Leaves of this type may indeed have originated from compound leaves, the arrested growth of the rachis leading to the coalescence of opposite pinnae at their bases, but the principle of development is the same in either case, namely arrest of the principal axis, associated with extended lateral growth. This is not to be confused with true dichotomy of the leaf blade which may sometimes occur as an abnormality, for example in *Salix*. Palmatifid and pinnatifid leaves may, on the other hand, originate from the opposite type of growth distribution, namely, the greater rate or longer continuance of growth in and close to the main veins, with the result that promontories of leaf tissue are formed, the shape and arrangeent of which will depend on the venation.

The cases mentioned above are only a few conspicuous examples of variations in the leaf from which are readily traceable to local variations of growth during the later phases of leaf development, and it is probable that nearly all the normal as well as the abnormal forms of leaves owe their characteristics to similar causes, though direct information on the matter is lacking. Such variations are, however, not entirely due to internal or heritable factors and the environment undoubtedly has a modifying influence. Sinnott and Bailey have pointed out the ecological relationship of certain leaf types to climate, in that certain forms of leaf are predominantly characteristic of certain habitats. For example, leaves or leaflets with entire margins are characteristic of low-lying tropical regions, while leaves or leaflets with non-entire margins are predominant in mesophytic, cold temperate areas, though dry temperate regions usually produce entire margins.

The effect of submersion in causing the development of filamentous leaves in aquatics is also well known. The first appearance of primordia at the stem apex is frequently far in advance of the time for their full development, and, after reaching a certain size, a period of dormancy ensues. This is conspicuously the case in the winter buds of trees, which are often formed quite early in the preceding summer. Inside these buds are development of leaf primordia for the next season goes on until the dormant period begins in October or November. When spring comes, the development of this performed foliage is often surprisingly rapid. In a few cases, such as the Horse Chestnut (*Aesculus*), the whole of the foliage for a given season is prepared beforehand. In others the young shoot apex continues to form new leaves for a short time in the early part of the year, but this does not continue much beyond the end of May, after which the apex is again transformed into a winter bud in which the later formed primordia are retained till the following spring.

Structure of Leaf

A typical leaf has three parts, the blade or the expanded portion, a stalk or the petiole and the basal portion of the leaf which connects the leaf with the stem. All leaves do not possess all these three parts. For instance, we have leaves such as those of Radish, Lactuca, Sonchus and Argemone without stalks. The leaf blade, on the other hand, is very well developed in most of the flowering plants. Similarly the basal portion of the leaf must be expected to be present in all plants. Young leaves which are enfolded in the bud do not possess stalks, or if they happen to have them they are very short and almost invisible. However, the stalk develops subsequently, and elongates after the folding of the leaves. This is so, because in the enfolded leaves the basal portion and the lamina become developed very much earlier than the stalk or the petiole.

Stipule

From the basal portion of the leaves outgrowths arise in many plants. Generally two outgrowths, one on each side, are seen and they are called stipules.

In very many plants the stipules are small and not prominent. There are also plants in which the stipules are very conspicuous and large. The Peepul (*Ficus religiosa*), the Banyan (*F. bengalensis*), the Jack tree (*Artocarpus integrifolia*) and *Cassia auriculata* have large stipules. In the first three the stipules serve to protect the young leaves in the bud; they are large and semilunar in *Cassia auriculata*. Sometimes they are tubular, as in *Polygonum gladerum*.

Blade

When leaves are young they are generally folded in different ways. In many plants the two halves of the blade are folded along the midrib, in such a way that the inner surface and the edges of the blade meet,

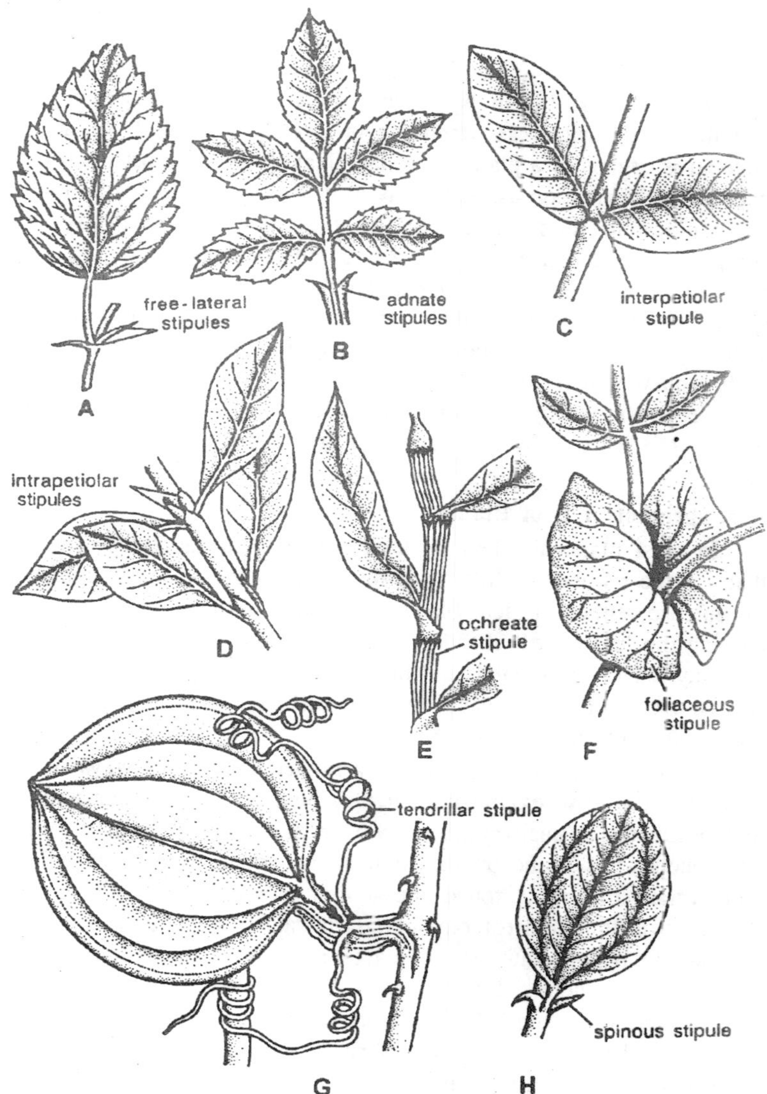

Fig. 11.3. Stipules: A, free-lateral; B, adnate; C, interpetiolar; D, intrapetiolar; E, ochreate; F, foliaceous; G, tendrillar; H, spinous.

as in Anona, Abutilon, Thespesia, Morinda and Tephrosia. This kind of folding is called *conduplicate*. Sometimes the leaf is rolled on itself in such a manner that one margin being rolled towards the midrib remains inside and the other over it outside, as in Musa and Canna, and the folding is termed convolute.

It is not unusual to find both the margins of a leaf rolled inwards towards the midrib as in Nymphara, Nelumbium, Ottelia and Viola and this is described as *incolute*. As already remarked, of all the vegetative organs of a plant the foliage leaves are the most important, because the formation as well as the transformation of organic matter takes place in them. The work that leaves have to do is very complicated and varied. The green colour enables them to do the work of forming organic matter under the influence of sunlight. This special work of the leaf cannot be carried out unless the leaves are exposed to sunlight. So we expect leaves to disposed on the axis in such a manner that all the leaves may get as much light as possible. And yet too much light is injurious to leaves. The adjustment of the position of the leaves with reference to light is therefore a very delicate one.

Arrangement of Leaf Blades

In most plants the leaf blades maintain a horizontal position so that a large number of light rays may fall on them. If the light gets too intense, either the leaf blade will change its position or the twig bearing the leaves will shift its position. Every green plant produces as many leaves as possible and their position with regard to the light will be such that all the leaves may get sufficient light without shading one another to any large extent.

Leaves generally appear to be arranged in various ways on the stem, but on close examination it will be found that either a single leaf originates at each node, or two or more spring from it. When only one leaf is borne by the stem at the nodes, the leaves will be found arranged spirally around 'the stem or alternately. This arrangement of leaves on the stem (or phyllotaxis) is said to be *alternate*. Leaves are said to be opposite if two leaves arise *opposite* to one another from the same node. If there are more than two leaves at a node, they are said to be whorled. Leaves are opposite in Morinda, Calotropis, and Vinca. They are whorled in Nerium, Alstonia and Clerodendron. As examples for the alternate phyllotaxis we may mention Thespesia, Abutilon, Pavonia and Melia. In stems with alternate leaves no two successive leaves will be found to be one exactly above the other and yet the leaves will be in vertical rows. Between two successive leaves on the same vertical line, a number of leaves will be found, but at different heights and in different positions.

To make out the arrangement of the leaves in such cases, a close examination of the nodes of a normally growing shoot is necessary. As an example a branch of Thespesia, or Abutlion may be taken. In these branches any leaf chosen will be exactly above or below the sixth leaf and the five leaves spirally arranged round the stem. The spiral will consist of two turns around the stem and the first leaf will be separated from the second by a space equal to two-fifths of the circumference. If the circumference were one inch the second, and the second the same distance from the third and so on. In a sprial of five leaves we get five vertical rows arranged round the stem at equal distances. We also meet with spirals having two, three, or four leaves or more. When the leaves are opposite the successive pairs will be at right angles to one another. In other words the leaves at any one node will be across the leaves of the nodes immediately above and below it. This arrangement of the leaves is called *decussate* arrangement. When the leaves are in a whorl, the leaves of the alternate nodes are exactly one above the other, and the leaves of the successive nodes will be found on different vertical lines side by side. The phyllotaxis will become clear, only if we remember that the leaves require the play of sunlight on their blades.

Usually the most advantageous position for the leaf blade is to place its surface across the direction of the rays of light. But if the light becomes very intense, this position is certain to injure the leaf blade. So, under such circumstances, the leaf should be able to shift its position. When the sun's rays pour straight down, the leaf will shift its position, so that the blade may be pyrallel to the rays instead of being at right angles. Almost all leaves possess the power of changing their position according to the nature and the intensity of the light of the sun. As all the leaves of a plant need light, it is obvious that they should grow in such a manner, as not to shade each other. One of the means by which the leaves are helped in this matter is the phyllotaxis. There are also other means, besides the phyllotaxis, by which the leaves avoid the shading. Choose some erect branches from a plant, say *Acalypha indica* or an Amaranth, and look down upon them from above. Then the leaves will be found arranged in rows round the stem, in such a manner that the space round the stem is utilised to the greatest possible extent.

Further, the leaves below are not shaded by those above, because the petioles of the former are longer than those of the latter. Again the size of the leaf blade also has some bearing on the disposition of the leaves around the stem. If the leaves have broad laminas, the rows round the stem will be three or four; there will be five or six rows, when they are moderately broad; the rows will be many in cases where the blade is very narrow. Whatever the number of rows around the stem, all the leaves get their share of light, because the rows are not likely to shade one another. But, in the same row leaves above are likely to throw those below into shade. As a matter of fact this does not happen so as to interfere with one another. By the adaptation of the length of the internode and the direction and length of the leaf-blade, shading is avoided. Thus it is seen that the adjustment of the leaves for the sake of light is very varied and complicated. In some cases the leaf surfaces present beautiful mosaics and rosettes. For instance n many Solanaceous plants, such as *Physalis minima*, the leaves are not all uniform in size and so the leaves arrange themselves in a mosaic fashion. Many plants of the order Composite may be cited as examples for the rosette habit of the leaves. The leaves in *Lactuca Heyneana*, Sonchus and some Blumeas are mostly confined to the base of plant. In other words, the leaves are radical.

In all these plants the leaves are narrow at least in the lower part and so they are disposed round the stem, so as to form a rosette. It is easy to give more examples; for instance Trapa, Elytraria and *Elephantopus scaber*. The same principle, that the leaves should spread themselves so as to enable all the leaves to be lighted and, at the same time, avoid shading one another is very well brought home by the heads of trees. For instance, the heads of trees such as those of the Banyan are generally found covered by leaves so as to form a framework or covering with as few gaps as possible. To enable all the leaves to obtain sufficient light, this is the best possible arrangement. If we look at the head of the same tree from inside and from near its trunk towards the branches, we find large spaces between boughs and twigs, and no leaves are found distributed in these gaps.

Shape of Leaves

Leaves, vary very much in their shape. For purposes of description names are given to the shapes that are striking. A leaf whose lamina is narrow with the sides parallel, is said to be *linear*. Leaves of grasses and those of the cereal plants are linear. When a leaf is somewhat broad at the middle or a little below and tapers towards the apex, as in Nerium, Polygonum and Polyalthia it is described as *lanceolate*. A leaf is oblong when the margins are almost straight and the blade uniformly broad. Guava, Calotropis and Banyan leaves are good examples of oblong leaves. When the leaf blade is broad and rounded at the base and also tapering to a point at the apex, it is described as ovate. Leaves of Hibiscus Rosa-sinensis, Acalypha indica, Solanum nigrum and Physalis minima are good examples.

When the blade is broader at the apex and narrowed towards the base, as in the leaflets of *Cassia obovata*, it is *obovate*. Leaves like those of Nelumbium are described as *orbicular* or *rotund*. Sometimes the blade is hollowed out at the base, and pointed at the apex so as to be roughly like the heart spot on a playing card, and such are called *cardate* or *heart-shaped*. The leaves of Thespesia and Aristolochia may be cited as examples. If the apex be rounded, instead of being pointed, the outline is said to be reinform or kidney-shaped. The leaves of *Hydrocotyle asiatica* are reniform. There are some leaves whose lamina are sagittate. Several species of the Natural order Aroideae have sagittate leaves. If the basal lobes of such a blade are straight and at right angles to the blade, the leaf is then *hastate*. If the basal lobes are rounded and prominent, the leaf is said to be *auricled*.

Margin of Leaves

The leaf margin is *entire*, when it is quite even without any indentations. It is *dentate*, if the margins is cut up into prominent teeth, as in *Hibiscus Rosasinensis*; *serrate* if the teeth are small and directed upwards; as in *Acalypha indica*; *crenate* when the teeth are rounded, as in Bryophyllum calycinum, Hydrocotyle asiatica and Stachytarpheta indica. Sometimes the margins of leaves become deeply indented and then the leaf is said to be lobed, if the cut does not go more than half way to the centre, as in the cotton leaf, and it is deep reaching the middle of the leaf, as in Lactuca and Radish, then the leaf is said to be *cleft*. When a leaf is lobed. the lobes are arranged either on the sides of the midrib as in Lactuca and Radish, or they may all spread like the fingers of a hand, as in Jatropha and *Hibiscus ficulneus*.

The former is described as pinnate and the latter palmate or digitate. Sometimes the lobing becomes so deep as to cut the blade into distinct pieces so that a piece may be plucked off without in the least affecting the others. In such cases the leaf is said to be a compound leaf. The leaflets may either be disposed prenately as in *Cassia auriculata* and *Cassia siamea* or palmetely as in *Eriodendron anfractnosum* and *Gynandropsis pealaphylla*. The division of lamina may take place only once and then the leaf is simply compound, or the leaflets may become divided further, as in the case of *Moringa*,

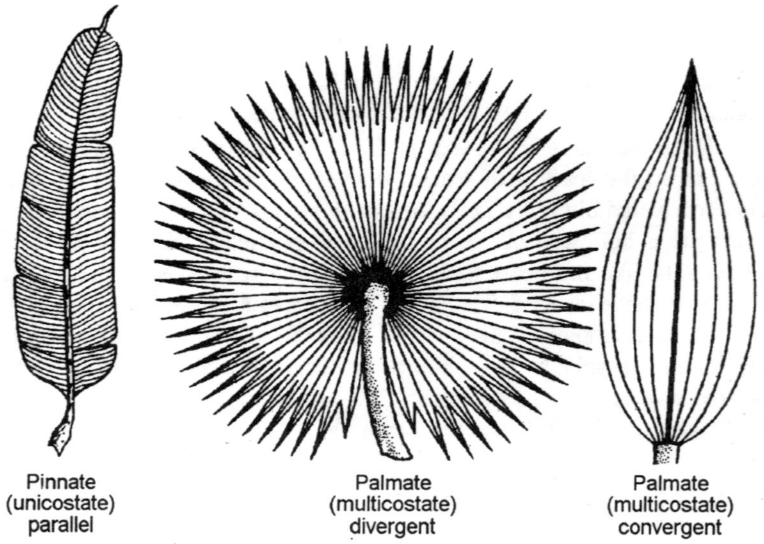

Fig. 11.4. Parallel venation.

Cadiospermum, *Acacia arabica*, *Melia Azedarach* and *Millingtonia hortensis*. Then the leaf is described as bi-or tri-pinnately compound according to the division.

Apex of Leaves

If the apex of a leaf tapers to a point gradually it is said to be acuminate, and if it is merely sharp pointed it is acute. The apex is obtuse. When it is rounded. Sometimes the apex will be straight as though cut off and then it is truncate, and it there is a notch it is said to be retuse, if the notch is shallow and emarginate if it is deep. In some cases we see a sharp point projecting from the apex and then it is said to be mucronate. We neet with leaves having a triangular piece at the apex and then it is said to be *cuspidale*.

Kinds of Leaves

The organ directely concerned in the work of manufacturing organic substances is the foliage leaf. This is the kind of leaf which is very general occurrence. Sometimes leaves are forced to do some kind of work, other than the preparation of organic substacnes. For instance, in seedlings, the cotyledons in most cases have very little to do, at any rate in the beginning, with the work of making organic stuff. The work that the cotyledons are mainly concerned in is to store food and make it available to the growing seedlings. Another kind of leaf

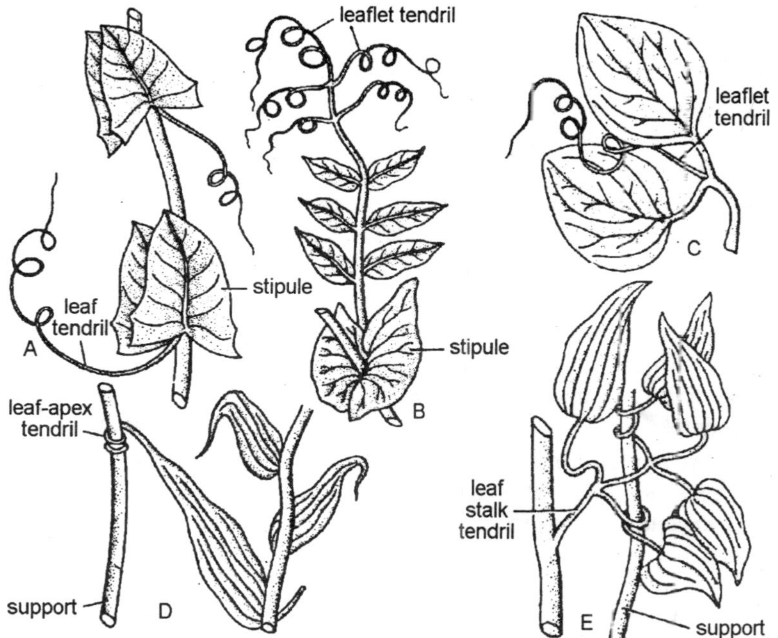

Fig. 11.5. Leaf tendrils: A, Lathyrus aphaca; B, Pisum sativum; C, Naravelia; D, Gloriosa superba; E, Clematis.

occasionally met with in some plants is what is known as the scale-leaf. These leaves are met with in connection with scaly buds.

In Mango trees when growth is at a standstill the terminal buds at the ends of twigs are covered by small scales. These scales are really leaves remaining undeveloped and small so as to afford protection to the growing point. And yet another kind of leaf is the bract found in connection with the flower. Bracts are generally small, but in some plants they are large. For instance, the terminal branches of the plant Gynandropis pentaphylla, the foliage leaves gradually pass into bracts and so, they are very conspicuous and leaf-like.

Organs of Reduced Foliar Type

Every Higher Plant produces various foliar organs which differ from the foliage leaves and are usually spoken of as "reduced," which does not necessarily imply their evolution from foliage leaves, but merely that in comparison with them they have a lower grade of organization. We shall not deal here with floral leaves, but among vegetative structures we may cite as examples:

1. *Cataphylls* reduced leaves on the lower part of a plant or a shoot, including the cotyledons, the scale leaves on rhizomes, and especially, the bud scales, which mark the base of futures shoots.
2. *Hypsophyllus* reduced leaves on the upper part of a plant or shoot, especially the bracts in the inflorescence region, and including the prophyllus or bracteoles of the individual flowers.
3. *Stipules* foliar appendages attached to the leaf base or the petiole, and including the *stipellae* which are attached to individual leaflets in a compound leaf.
4. *Ligules* and other scale-like appendages of certain leaves, such as the *intravaginal scales* in some Monocotyledons.

The morphology of these structures and their relation to foliage leaves is still, in many cases, obscure, but some have been investigated in detail.

Cataphyllus are in most cases homologous with leaves, but only in a minority of cases is a scale equivalent to a whole leaf or has a recognizable likeness to the fully developed leaves, though in some examples, e.g., bud scales of *Syringa vulgaris* (Lilac), a series of forms may be traced between the outer typical bud scales and juvenile leaves within the bud. Another condition is that in which the scale corresponds to the leaf base, and this is probably the commonest case. There is a difference in the rates of development in the upper and lower zones of the scale rudiment, the upper zone, corresponding to the petiole and leaf blade, failing to develop beyond the rudimentary state. In some buds, however, a succession of forms may be found, showing the gradual development of the lamina at the apex of the scale, the proportion of lamina to leaf base increasing towards the centre of the bud until the normal form of the leaf is reached. Such intermediate stages are well shown during the opening of the bud in such woody plants as *Ribes* and *Aesculus*. The divergence from typical leaf development begins at a very early stage of scale formation. In opposition to the usual idea of reduction is the fact that there is no diminution of growth rate in the development of a bud scale such as that of *Aeculus*.

Actually the development of the scale rudiment is faster than that of the leaf. There is more rapid vacuolization and maturation of the cells, but growth is limited to the margin and soon ceases. Scale formation and leaf formation are two distinct periodic phenomena, and are apparently determined by oragnismal factors, the fate of the rudiment being decided before its appearance, though the resultant structures may finally differ so greatly that it is not possible to homologise their parts. The cataphylls of the winter bud mark the base of the each year's shoot-generation during the growth of a perennial plant. Similarly the base of the first shoot in the seedling frequently bears cataphylls. Following the cotyledons a succession of reduced leaves may be produced, increasing in size at each node and leading eventually to the mature leaf form. Unlike the bud scales these cataphylls are separated by internodes, the length of which increases upwards on the stem. Cataphylls, like other leaves, usually subtend axillary buds, which are, however, of reduced size and are generally dormant. If they are stimulated to growth as, for example, by severe pruning, they form dwarf and slow growing shoots, or spurs, on which flowers are borne.

A third type of scale formation is that from the stipules of leaves. Many buds have scales of this type, for example, *Alnus*, where the bud is covered by three scales two being the stipules of the outermost leaf and the third being one stipule of the second leaf. Other conspicuous examples are *Magnolia* and *Liriodendron*, in which the stipules fall off when the buds open and their protective function is finished. In the buds of the Oak and Beech there may be a considerable number of such stipular bud scales, each of which is a double structure formed of a pair of stipules, the lamina being reduced to a microscopic point. Only the two outermost scales are simple.

Hypsophylls originally included all the floral leaves, petals and sepals as well as bracts, but the term is now generally limited to the latter. Bracts may be green organs with the same assimilatory capacities as foliage leaves, or they may act as protective coverings to the flower or inflorescence buds and in some cases they may be specially coloured and serve the biological end of increasing the conspicuousness of flowers. If a hypsophyll subtends an axillary structure, this is usually, though not invatiably, a flower or a flowering branch and its peculiarities of form seem to be linked with this fact, since in cases where an inflorescence produces fresh vegetative shoots, the normal foliage generally reappears on these.

The term *bract* is usually applied to a hypsophyll subtending an inflorescence, and the term *bracteole* to one which subtends a single flower, but the distinction is not universally applied, and the term hypsophyll itself, which merely serves to unite the two categories of organ, is not, therefore, of much general utility. Like the cataphylls, the hypsophylls are often connected with the normal leaves by intermediate forms which make a sharp line of division impossible, and again like the cataphylls, they may arise in three different ways. Firstly, by a reduction in development of the entire leaf; secondly, and more commonly, by a reduction or abortion of the lamina and petiole, and a correspondingly enhanced development of the leaf base. This may often be perceived by a comparison of venation with that of the normal leaf. The leaf base usually has a distinctive venation, different from that of the lamina, even when it is coalescent with it, and this type of venation will be found to correspond to that in the hypsophylls. Thirdly, they may be formed by the modification of stipules in some plants belonging to families such as the Rosaceae, where these latter organs are present.

The term *stipules* has been loosely applied to outgrowth at the base of a leaf of very various forms and natures. Some of these are, however, not of true stipular character and the name should be limited to paired lateral outgrowths of the leaf base, corresponding to the lateral lobes or leaflets which may be produced from the laminar portion of the leaf. Simple though such a definition may appear, it is by no means easy to apply it in all cases. Fusions and displacements during the process of development may obscure the lateral or the paired nature of stipules, so that opinions differ regarding the interpretation of many structures, such as for example, the ochrea in the Polygonaceae.

We shall here relegate the ligules and intravaginal scales of the Monocotyledons to a separate category, while admitting that this is a disputable opinion. There is a widespread belief that true lateral stipules are altogether wanting in the Monocotyledons, but Gluck has shown that genera in several families, notably *Hydrocharis*, *Potamogeton*, and *Dioscorea* posse free lateral out-growths of the leaf sheath which must be accepted as stipules.

In Dicotyledons examples of *pseudo-stipules* are not uncommon, especially in general with compound leaves, such as*Cobaea* and *Vicia*, in which the lower pair of leaflets may develop so close to the leaf base that they stmulate stipules and may even envelop the stem. *Pisum* is apparently the reverse case, where true stipules become enlarged until they appear to be the lowermost pair of leaflets. Pseudo-stipules may arise in a variety of ways other than by modification of basal leaflets. They may, as in *Cestrum* for example, be the paired basal leaves of an axillary shoot. In the simple leaves of *Viburnum opulus*, the leaf base carries a number of paired projections, which are usually tipped with glands and might be taken for true stipules if gradations did not occur between them and the nectary glands further up the petiole. Both sets of structures are modified leaf teeth and belong to the laminar portion of the leaf which runs down the petiole to the base in the form of two parallel ridges. Similar examples may be found in other species.

Stipules and pseudo-stipules may both exist together in the same plant, as in Lotus, where the foliolar pseudo-stipules cover and hide the glandular stipules. Plants with opposite leaves often have stipules which are placed between the leaf bases and ensheath the young leaves of the next node above. Such intepetiolar stipules are particularly striking in Galium, where only two of the apparent leaves at each node are true foliage leaves, the rest, two to eight in number, being stipules which closely resemble the leaves but do not receive trace bundles directly from the stem stele as do the true leaves. True stipules may often be found united together. When the union is along their inner margins they may form a single organ bridging across the leaf axil. If, on the other hand, they fuse by their outer margins, the united structure may be antidromous, that is, apparently opposite to the leaf. Stipules belonging to opposite leaves may also unite to form partial or complete sheaths around the note, as in the Rubiaceae.

The *ochrea*, mentioned before, is a membranous sheath, arising from the leaf base and surrounding the axillary bud and the stem for a short distance above the node. It is a family character in the Polygonaceae, and also occurs in some Ranunculaceae, such as *Caltha*. It has been interpreted as derived from the axillary fusion of two stipules, but it develops as a single sheathing organ, and seems to be better regarded as a tubular upgrowth of the leaf base and not of stipular nature. Whatever be their morphological origin, stipules and pseudostipules definitely function as protections for the leaf rudiment, which they often precede in development, and for the axillary buds.

In many cases they are caducous, that is, they are shed when the leaf approaches full development, thus showing that their chief importance is then past, but in others they are retained, and, by carrying on photosynthesis, they contribute something, though in most cases probably not much, to the nutrition of the plant. Three types of leaf structure remain to the mentioned. The first are called intravaginal scales because they appear in the axil within the leaf sheath or vagina of certain Monocotyledons, especially members of the Helobiae. They take the form of small tooth-like scales, generally linear or lanceolate, and they actively secrete mucilage. They are not stipules, since they may occur in association with stipules or even, as in *Hydrocharis*, in their axils, and they are generally regarded as trichomes. They may be single or serial, and according to Arber they arise either from the surface of the axis or from the dorsal tissues of the base of the leaf immediately above them. Secondly, the leaf base may be enlarged into, a wide sheath, or vagina, more or less encasing the stem. This is commonest in Monocotyledons, but is also characteristic of certain dicotyledonous families, notably the Umbelliferae.

In many Monocotyledons the vegetative stem remains very short, and a large part of what appears to the leafy axis is really a cylinder formed by the tightly wrapped leaf sheaths of successful leaves, up through the middle of which eventually grows the flowering axis. This structure is familiar in many Grasses and Sedges but reaches its greatest development in the Banana, where it attains tree-like proportions. Eichler distinguished leaf sheaths which were composed of stipules more or less adnate to the leaf base from those in which the sheath was a simple expansion of the petiole or leaf base itself. Most leaf sheaths in Dicotyledons are of the former nature; only in a few families, such as the Compositae, where no stipules exist, is the leaf sheath indubitably petiolar. The Monocotyledons rarely possess free stipules, but in them the leaf sheath, which is almost universal, is probably also the product of fused stipules, united to the leaf base.

In both groups of Angiosperms the tips of fused stipules may remain free, appearing as a scale at the top of the sheath, known as the *ligule*, which often shows, by its division into two lobes, an indication of its double nature. Indeed, it has been pointed out that ligule and sheath bear an inverse relationship to each other and that the shorter the sheath is, the larger and more stipule-like is the ligule (e.g., in *Trifolium*). Thirdly, when the laminar portion of a leaf rudiment is suppressed during its development, and the resulting mature structure assumes the appearance and assimilatory function of a leaf, although actually incomplete, it is called a *phyllode*. True phyllodes are not as numerous as they may appear to be on a superficial survey of Angiosperms. The name has been freely applied to cases where the developmental history is unknown and some of these at least have proved, on close investigation, to be true laminate leaves.

A well-known case in point is the leaf of certain American species of the umbelliferous genus Eryngium (e.g., *E. agavifolium*), which are linear and have parallel venation. The suggestion that here the leaf consists of a base and a broadened petiole only is not borne out by the ontogeny, which shows that the lamina becomes long and narrow, and that its pinnae are reduced to long teeth on the leaf margin. There is good reason to believe that the same is true of a number of other apparently phyllodic leaves. Genuine phyllodes, in which the biological function and the dorsiventral structure of the lamina have been taken over by the petiole or by the leaf base, or a combination of the two, are best known in speices of *Acacia*. Many species of this genus have the multipinnate, compound leaves which are so frequent among the Leguminosae, but certain species, such as *A. longifolia* and *A. glaucescens*, bear only phyllodes which are simple and mostly somewhat leathery in texture. Their venation is reticulate, like that of a typical dicotyledonous lamina, but one peculiarity marks all these phyllodes, namely, the presence of a strong marginal vein, which runs all round the phyllode. In most phyllodic Acacias, and in the similar structures in species of *Oxalis*, such as *O. ruscifolia*, the seedlings show a striking "recapitulation", that is to say, a series of developmental stages connecting the pinnate leaf type with the phyllodic type.

The first leaves are normally pinnate, with narrow peticles, but in subsequent leaves the petiole is broadened and the pinnate lamina is reduced until a stage is reached at which the petiole has assumed the mature phyllodic structure and the lamina has disappeared or survives only as a small pointed tip to the phyllode. The phyllodes of Oxalis differ from those of Acacia only in that the former are horizontally flattened while the latter are usually flattened vertically. Phyllodes, like most leaves, arise almost entirely in the tunica of the apical stem meristem.

At an early stage in the development of the rudiment a superficial meristem appears over the adaxial surface, which is responsible for the broadening of the structure. At the top of the phyllode is a minute point which represents a rudimentary terminal leaflet, while at its base two stipules are formed. The developmental history, therefore supports the view that the phyllode is simply a broadened petiole. A. P. de Candolle in 1827 raised the question whether the linear outline and parallel venation which distinguishes so markedly the leaves of Monocotyledons might not be due to their phyllodic nature. This view has been strongly upheld by Arber, who bases much of her argument on anatomical grounds, especially the widespread occurrence of vascular bundles with inverted orientation on the adaxial face of the "leaf". The broadened lamina which occurs in many Monocotyledons (e.g., *Alisma*, *Arum*, *Tamus*, etc.), she interprets as due to a secondary expansion of the apex of the petiolar phyllode, and she further comments that the prevalence among such cases of simple lanceolate or cordate shapes indicates an origin distinct from that of the multiform laminae of Dicotyledons. Even among the Monocotyledons which have leaves with a lamina and a petiole, e.g., the Araceae, the phyllodic tendency shows itself in the frequent lateral expansion of the petiole, associated with the diminution of the lamina, so that the blade is formed of two portions, one above the other, which may be either separate or more or less confluent.

There can be little doubt that the leaves of many Monocotyledons are rightly to be classed as phyllode, but it is perhaps unwise to extend the theory too far, still more to elevate it into a morphological dogma. A comparison of phyllodes with cataphylls is obvious and it would be difficult to draw a clear line between them. Cataphylls generally consist to a leaf base only, and in phyllodes the petiole is usually, though by no meant always involved. The term phyllode is, however, by usage, restricted the organs which carry out photosynthetic functions or replace normal leave in other ways, for instance, as spines; while cataphylls are, as we have pointed out above, periodic structures limited to certain growth phases of the plant. The presence of flower buds on true leaves, as distinct from cladodes, is, at first sight, a puzzling phenomenon from the morphological standpoint. The best known example is in *Helwingia ruscifolia*, a plant often cultivated in greenshouses. A cyme of small flower appears from the midrib about half-way up on the adaxial leaf surface.

A study of the development shows, however, that the flowers originate from the lower of two axillary buds, the upper one of which remains as a normal axillary bud, while the lower one, the flowering bud, coalesces with the young leaf and is carried outwards by it during its development. In one very peculiar case, *Erythrochitor* (Rutaceae), it is the upper bud which is the flowering one and and it coalesces with the leaf rudiment above it, so that the flowers appear on the lower side of the leaf above the node to which they properly belong. This coalescence of axillary shoots with their subtending leaves is no uncommon occurrence and in the solavacae example, if often gives rise to striking partive from the normal relationships of leaf of shoots.

General Descriptive Terms

If the petiole is present, the leaf is petiolate or stalked; if absent, sessile, If in a leaf the membrane runs vertically down the stem for some distance, the leaf is decurrent. In grasses a ligule is developed on the base of the lamina, and the leaf is said to be ligulate. A leaf is stipulate or exstipulate according as stipules are present or absent. Stipules vary much in position, colour, size, and form. In the shoe-flower (Hibiscus) they are lateral and free. In species of *Pisum* and *Lathyrus* they are foliaceous; in this case they help in the work of photosynthesis. When the stipules are dry, small, pale, and membranous, they are usually functionless.

In some buds (e.g., in the banyan) as already stated, they form the outer protective scales which fall off as the leaves expand. Occasionally the stipules are modified into spines, as in *Zizyphus jujuba* (ber or bor) and Acacia. If they are adnate to the petiole, they are called *petiolar*. Where there is only one leaf at the node, if they run round to the other side of the stem and fuse there, an opposite stipule is formed (banyan); if there inner margins cohere between the leaf and the stem, an axillary stipule is formed; if they cohere in both ways, a tubular sheath called an ochrea is formed round the base of the internode.

Insertion of the Leaf

The point at which a leaf-base joins the stem is called insertion of the leaf. Leaves are described as *cauline* or *ramal* according as they are developed on the main stem or on the branches. Leaves developed on very short "reduced" stems so that they *appear* to come off from the root are called radical leaves.

Texture and Duration of Leaves

Shade-and moisture-loving plants usually have thin leaves with poorly-developed cuticle. In plants exposed to intense insolation on the other hand the leaf is usually thicker and more resistant, and possesses a well-developed cuticle. This condition is very marked in the leaves of many tropical plants and in evergreen plants in temperate regions. Leaves which are thin and membranous are described as herbaceous. Some are succulent and fleshy. The leaves are *caducous* if they fall off very early; *deciduous*, if they fall at the end of each season; *persistent*, if they remain on the plant for more than one season. Plants with persistent foliage leaves are evergreens.

Duration of Leaf Development

Apical and marginal growth in leaves is of comparatively short duration in most plants but in fern leaves the apical growing point persists for sometime, the tip continuing to develop after the base is mature. In some genera, apical growth continues even for more than a year and very long leaves are formed. In plants other than ferns, apical and marginal growth ceases early and general growth throughout the young leaf continues. The outline and fundamental structure of most leaves re-developed while the leaf is yet minute. In the winter buds of many temperate-zone trees, for example, *Liriodendron*, small leaves are present, in shape resembling full-grown, leaves, with the main vascular structure outlined, and a considerable part of the cells already present. Later growth consists in large measure of rapid increase in cell size and maturation of the masophyll cells and vascular tissues, which takes place during the period of leaf expansion. New cells may be formed generally throughout the leaf during this time. After the leaf attains full size, the larger vascular bundles usually increase in diameter by secondary growth. There is, naturally, great difference in the time taken for leaf development in different plants.

In most woody plants of the temperate zone, complete, expansion and maturity are attained rapidly, the actual time depending upon seasonal temperatures. In many tropical plants and in herbaceous ferns with very large leaves, development goes on over a longer period. Intercalary leaf meristems are common in leaves of the linear type occurring in the grasses generally, and in such genera as *Iris*, *Allium* and *Pirius*. However, such meristems also present for only a comparatively short time, with the exception perhaps of those of the anomalous. Iyonosperms, Welwitschia where they are apparently undeterminate in time of activity.

Leaf Duration

The duration of leaves, at least in perennial plants, is usually short compared with the life of the plant which bears them. Deciduous leaves life for only one vegetative season and are then shed in a senescent condition, although still alive. This only takes into account their life as adult leaves, but in many instances they are formed in the bud during the early part of the preceding season and remain enclosed in the bud in an immature state throughout the following winter before developing to maturity, so that they may properly be said to live for two seasons. Even the so-called evergreen leaves have a relatively short lifetime, indeed it may be said that the difference between a deciduous plant and an evergreen is only that the latter never sheds all its leaves together. In many evergreens among the Dicotyledons the leaves last no more than one year, while those which are longest lived endure for not more than five years. Among some of the slow growing Monocotyledons, such as *Agave*, this age is probably exceeded, but details are lacking.

Prefoliation

The form and arrangement of the young leaves in the bud is termed prefoliation. Prefoliation includes (a) Ptyxis, or the form of the young leaves in the bud, i.e., the way in which they are folded or rolled on themselves; (b) Vernation, or the relation between the different leaves in the bud, i.e., the manner in which they are arranged with regard to each other. These points may be determined either by removing the leaves of a bud one by one, better by taking cross-section of the bud. In flower-buds aestivation is usually used in the same sense as vernation in vegetative buds.

Ptyxis of the Leaf

It is plane if there is no folding or rolling at all; *conduplicate*, if folded lengthwise along the midrib, with the two halves face to face; plaited or plicate, if there are numerous longitudinal folds; *crumpled*, if folded in all directions; *convolute*, if rolled from one margin to the other; *involute*, if rolled from both margins to the middle of the upper surface; *revolute*, if rolled similarly to the middle of the lower surface.

Vernation

Vernation is *valvate* if the young leaves touch each other laterally, but do not overlap; imbricate, if some overlap others, but not regularly; twisted, or contorted, if one margin of each leaf is directed inwards, and is overlapped, while the other margin is directed outwards, and overlaps the margin of the adjacent leaf.

Modified Leaves

There are a number of plants feeding on insects. They do not thrive well, unless they obtain insects. In these plants there are adaptations of a special kind to enable them to catch and digest small insects. As examples of such plants we may mention *Drosera Utricularia* and the pitcher plant *Nepenthes*. In Drosera the leaves are provided with glandular hairs secreting muncilage in such profusion as to imprison the insects, when the hairs come in contact with them. The leaves of Utricularia are modified into bladders with trap doors to catch insects. In the pitcher plants the leaves are prolonged at the apex into long processes ending in cups. These cups or pitchers as they are called, contain a liquid. The insects are allured by these pitchers so that they may fall into them and get digested by the water contained therein. Leaves are able to produce organic matter, as they contain chlorophyll. Young parts of plants are also able to do this because they also contain the green colouring matter.

There are many plants without leaves, as in the case of the Pricklypear. In such plants the chlorophyll grains become imbedded in the cells of the cortex of the stem. Such stems look quite different from the ordinary stem and they are called *cladophylla*. In some leaves the petiole develops into a flat structure very much resembling the leaf and the blade either remains very insignificant, or it is not at all developed. Such petioles are called *phyllodes*. As example of cladophylla we may mention the stem of Boucerosia, Opuntia, Casuarina and some Euphorbias. Phyllodes are found in *Acacia auriculiformis* and in *Parkinsonia aculeata*. Petioles have sometimes wings on both sides as in the orange leaf and the compound leaves of some species of Vitex and *Filicium decipiens*.

Special Modifications of Leaves

Water storage leaves are found in many succulent plants. The leaf is usually very thick and only the peripheral regions ar green and carry on the normal leaf functions. The bulk of the central region of the leaf consists of special water-storage tissues. As would be expected, we find leaves of this type on some plants inhabiting dry situations, and they are enabled to with stand extreme water shortage as they can draw on their internal water reserves during periods of drought. The aloes and agaves furnish examples of this type of leaf.

Leaf Tendrils

Leaves or parts of leave frequently have the form of tendrils. The Leguminosae, Bignoniaceae and Cucurbitaceae furnish examples. In Lathyrus hirsutus the tendrils represent the upper leaflets of a compound imparipinnate leaf. Accompanying the reduction in leaf-surface that this entails we sometimes find, as in this vetch, "wings" developed on the stem so that the total surface available for photosynthesis is not reduced. In some species of Pea, e.g., *Lathyrus aphaca*, the whole leaf is elongated to form a tendril. In Smilax (*Liliaceae*) the tendrils have been referred to as stipules, though it is more probable that they represent lateral organs due to the sub-division of the lamina into three parts of which only the middle one functions as a leaf.

Leaf Spines

These may represent leaves or parts of leaves. The whole leaf may be modified into a spiny structure, as in barberry where it is 3-5 rayed. In the seedling of gorse, the leaves are terifoliate, but in the adult plant the leaves are reduced to slender spines with a firmer branch-spine in each axil. In Mexican poppy and the thistles the leaf margin forms a series of spines. In species of Acacia, of Zizyphus and of Euphorbia the stipulus are represented by spines, whilst in *Azima* spines in the axils of normal leaves represent the leaves of lateral shoots.

Phyllodes

In same Australian species of Acacia the lamina of the leaf is absent, but the petiole is so flattened as to appear leaf-like. These flattened petioles are known as phyllodes and they are so developed as to place their surfaces, in the vertical plane. Inspection shows the phyllode to be a leaf structure, because there is a bud in its axil, but does not reveal that it is not a lamina turned edgewise. In young seedlings, however, normal compound leaves occur and the transition from normal leaf to phyllode can be traced, leaving no doubt as to the petiolar nature of the phyllode.

Pitchers

In various insectivorous plants the leaves develop into pitcher-like structures, and show other peculiarities of form connected with their ability to trap and digest insects. Succulent plants often have fleshy leaves which store water. The normal mesophyll is replaced in such leaves by two zones of cells, one towards the upper surface, which is relatively shallow and is composed of small parenchyma cells containing chloroplasts, and another much deeper zone, making up the bulk of leaf, which consists of very large, thin-walled cells with no chloroplasts, in which the water is held.

Ascidial Leaves

These are specialized leaves in which the lamina is not flat but encloses a sac-like space. They are usually the result of overgrowth of one side of a dorsiventral leaf and may, therefore, be regarded as extreme cases of rolled leaves, or even as related to unifacial centric leaves. Ascidial leaves may occur as abnormalities (e.g., in Rosa), but where they are normal developments in a species, they are almost always associated with some specialized life conditions, such as insect capture water storage, myrmecophily, etc., and examples of these will be dealt. They are perhaps commoner among floral leaves than among foliage leaves, and in the flower they are often used as nectar containers. The "spurs" and "honey sacs" of many flowers belong to this category, and also the floral pouches of Cypripedium and Calceolaria.

Storage Leaves

The best known examples of these organs are the scales of bulbs. The latter very greatly in the number of such scales they contain, from one only, in some species of Allium, to the bulbs of *Lilium auratum* with upwards of a hundred scales. The scales likewise vary in their nature. In the bulbs of Tulipa Saxifraga granulata, Fritillaria, Gagea and Allium cepa, the common Onion, the scales are cataphylls, that is, modified leaves; but in Narcissus, Galanthus, Ornithogalum and many other genera, the scales are formed from the persistent basis of foliage leaves, which become secondarily fleshy after the lamina has withered.

In Lilium the outer scales are persistent leaf bases, but the inner scales, which in large bulbs may be very numerous, are cataphylls. Thickened storage leaves are not necessarily only formed underground. Axillary bulbils with storage cataphylls are not uncommon on aerial shoots, as in *Dentaria bulbifera* and Lilium bubliferum, in both of which they form a means of vegetative propagation. Neither are storage leaves confined to bulbous structures. In many species of Oxalis, for example, fleshy storage leaf bases are formed in clusters at intervals on the rhizomes, that is, on longated shoots. Similar elongated groups of storage leaves are formed on the underground stems of Gesneraceae, while in the will-known root parasite, *Lathrea*, the entire shoot bears only fleshy storage leaves. Lastly, we may mention the specialized storage leaves which make up the detachable winter buds of some aquatics, such as *Hydrocharis*, which are used for vegetative propagation.

Buds on Leaves

These are exceptional occurrences and almost always associated with vegetative propagation. The buds are rarely of the bulbil type and mostly take the form of small leafy rosettes which are usually detachable. We may classify them into two biological categories:

Accessory buds

Which are formed on the normally growing leaf blade or the petiole, either on the ventral surface as in Cardamine pratensis, Tellima, Drosera, Lycopersicum or Nymphaea zanzibarensis, or on the margins as in Bryophyllum tenuifolium.

Reparative buds

Which develop after separation or injury of the leaf, as in Begonia, Ficus elastica and Bryophyllum calycinum.

HETEROPHYLLY

It is no uncommon thing to find leaves of different types on one and the same plant, a phenomenon known by the general name of heterophylly, although the features of the condition are by no means uniform. In the widest sense of the term, and including as leaves all the various categories of bracts, cataphylls, etc., one might say that heterophylly is the universal rule, but confining ourselves only to modifications of the foliage leaves, we may distinguish at least three main types of heterophylly. The first type we may call environmental heterophylly, because the difference is conditioned by environmental factors. The clearest examples of this are to be seen among water plants, where the submerged and the floating or aerial leaves may be very distinct. Among dicotyledonous aquatics the submerged leaves are often so much subdivided that their segments are reduced to mere filaments while the aerial leaves have a fully developed lamina. Monocotyledonous aquatics on the other hand frequently have linear submerged leaves with no lamina. Certain plants, especially shrubs growing in dry elimates, may also show a variation of leaf form according to the season, the normal leaves only appearing during the rainy season and being substituted by scale-leave during the dry part of the year.

Closely associated with the above instances are those in which certain leaves are specially modified from the normal type in connection with some special function, such as water or food storage, climbing or the capture of insects. Some adaptive modifications of this type we shall deal with later in this chapter. The second type or habitual heterophylly involves the formation on the main shoots, of foliage leaves of different, sizes and to a lesser extent, of different shapes, apparently without any special functional significance. Among the commonest cases of such pure heterophylly are those of plants with opposite leaves in which one leaf of a pair is smaller than the others. This is quite common on plagiotropic* shoots, where it is normally the upper leaves which are reduced in size. This is most pronounced in those species where the leaves on plagiotropic shoots are twisted laterally, so as to form flattened shoots on which there are two lateral rows of large leaves, and two dorsal rows of small leaves, representing unequally developed leaf-pairs. This lateral heterophylly may perhaps confer an advantage by avoiding the overshadowing of leaves by other leaves but it is not inseparable from this condition, because habitual heterophylly is also found on the orthotropic shoots of many plants where the factor of overshadowing does not come in.

The reduction of one member of each pair of leaves often goes so far as its complete disappearance, the originally opposite leaf arrangement being sometimes traceable by the presence of supernumerary stipules at each node and sometimes, as in the Elm, traceable only during the development of the young plant, in which a primitively isophyllous, opposite arrangement of the leaves gives place, at an early stage, to an alternate arrangement with leaves in two ranks only, through a short intermediate zone of true heterophylly. These latter cases are of particular interest in that they suggest the possibility that alternate leaf arrangements in which no trace of the change now remains, may have been derived from the opposite arrangement. A very marked, but exceptional, type of habitual heterophylly is exhibited by certain plants which bear leaves of variable form, distributed at random. Three well known examples of this are: Broussonetia papyrifera, Artocarpus integrifolia and Lonicera japonica, the last named producing irregularities especially under starved conditions.

In all these cases the normal leaf shape is entire, but only a minority of the leaves retain this form, the majority developing in most irregular fashion, various parts of the lamina being suppressed, so that hardly any two leaves are alike. Before leaving this type of heterophylly we may also mention cases where, as in the common Ivy, the leaves of flowering shoots, although retaining the full size and character of foliage leaves, differ in outline from those of vegetative shoots. Such differences probably merge into the widespread condition of the production of specialized bracts associated with the flowers. The third type of *developmental heterophylly* is that associated with the change from juvenile to mature foliage in the development of the individual. This touches upon the vexed question of evolutionary recapitulation during ontogeny, but with a few exceptions the changes involved are not such as to throw much light upon the evolution of the species concerned certainly not, at any rate, in regard to their remote ancestry, though they may indicate comparatively recent changes of foliage type in the history of the race.

The case of the Ivy, mentioned before, may be considered as falling into this class, in as much as the palmate leaves, arranged in two ranks, which are produced on the climbing vegetative shoots, are naturally characteristic of the earlier stages of development, while the entire ovate leaves of the flowering shoots are only produced when the flowering stage has been reached. It is preferable, however, to regard the change in this case as correlated with bract development, which is in every case associated with the flowering stage of the plant rather than as purely a phenomenon of maturity, for if flowers are not produced, no change of foliage occurs. Juvenile foliage which gives place to mature foliage in the ordinary course of vegetative development is well known in *Pinus* and some other conifers and occurs also in many Angiosperms, especially among woody climbers.

The genus *Pothos* provides a very striking example. Starting with sessile, entire leaves in the young plants, there is first a change to petiolate leaves and later to divided leaves, which are only irregularly incised to begin with, but finally become regularly multipinnate. So great is the difference that young specimens in cultivation have been regarded as types of a new genus until their mature foliage appeared, in the same way that juvenile forms of some Conifers are vegetatively propagated and cultivated under the names of *Prumnopitys* and *Retinospora*, which are applied only to these juvenile states. The genus *Eucalyptus* consists of large, rapidly growing tree species, which show a very marked difference between the juvenile and the mature foliage. The later leaves hang vertically; they are always narrowly lanceolate, alternate and petiolate. The juvenile leaves, on the other hand, are variously shaped; frequently obtuse, sessile, or even connate in pairs, and standing out stiffly from the branches.

The difference is so marked that one could easily believe that the mature shoots belong to some other species grafted on the juvenile tree. Almost equally marked, though less spectacular changes of leaf shape may be seen in many herbaceous plants such as *Campanula rotnudifolia*, *Malva moschata*, and in several species of *Scabiosa*, in which the juvenile leaves are entire and the mature leaves are deeply divided. In many plants, indeed perhaps in most, there is some degree of change in leaf form, from node to node. The rate of this change, which is genetically controlled, has been used as an index of the physiological maturity of the shoot. Certain species of Acacia develop only phyllodes in the mature state but produce the typical multipinnate leaves of the genus on their seedlings. The change over takes place through intermediate stages, marked by progressive broadening of the petioles and the reduction and final disappearance of pinnae.

Here, at any rate, we seem to see a case of recapitulation of an evolutionary change, since the pinnate leaf is almost certainly the primitive type for the genus, which has been replaced by the specialized phyllode formation. The same replacement of laminar leaves by phyllodes can also be seen in some species of *Oxalis*, where the primary leaves are ternate. Juvenility of foliage is not entirely confined to seedlings, for in many trees similar changes of leaf type may be seen on the vigorous stool shoots that spring from the base of felied individuals. Arising as they do in most cases from dormant buds formed when the tree was young, they frequently display juvenile characters and recapitulate changes observable in the seedling development. Along with heterophylly there may be mentioned the asymmetry of leaves which prevails in certain plants, in which the two halves of each leaf, or the two sides of a pinnately compound leaf, are unequally developed. This may occur, as in the Elm, even on orthotropic shoots, but it is much more frequent on plagiotropic shoots. In the latter case it is often associated with dorsiventrality of the plagiotropic axis itself.

The genus *Begonia* provides examples of this in nearly every species, but the relationship to the axis varies, sometimes the larger and sometimes the smaller half of the leaf being upwards. In the latter case, however, the larger half is usually turned upwards by secondary twisting of the leaf-stalks. Experiment shows that asymmetry can easily be induced in leaves by severing a vein on one side of the young leaf and thus reducing its nutrition, and in the naturally occurring examples an analogous difference of nutrition seems to be operative, the larger leaf-half being formed towards the better developed side of the dorsoventral axis, from which its trace bundles come. When stipules are present in such cases they are also often asymmetrical, the larger of the two being on the larger side of the leaf. Asymmetry has been interpreted as an adaptation to minimise the overshadowing of each other by closely set leaves. It may have such a value of certain cases, but in orthotropic shoots with widely spaced leaves this need does not arise, nor can it in any case, be regarded as the cause of the asymmetry.

Nevertheless the avoidance of overshadowing has a definite biological value, especially those with plagiotropic shots, whether creeping, climbing or woody, do in fact place their leaves, by the twisting of the petioles and by differential growth of the petioles in length, in the position of minimum overlap. This fitting of the leaves together has been given the name of the *leaf mosaic* and some ecological importance has been attributed to it. Apart from plagiotropic shoots, the basal rosette of leaves in biennials, the so-called "radical leaves," often show very well-marked mosaics, the petioles of the lower leaves in the rosette being so much elongated that their laminae are carried outwards beyond the laminae of the upper leaves. The leaves produced on the elongated flowering shoots produced later by the same plants may not, however, show any mosaic arrangement, unless they happen to be unusually large or closely placed on the stem. In treating of adjustment of leaf position mention should be made of the peculiar change called *resupination*, or turning of the leaf upside down, which is most familiar in *Alstromeria*. No definite advantage can be indicated for this peculiar change, which is habitual in the genus, but in other cases the twisting of the leaf may have an ecological significance.

Such are, for example, the "compass plants" of the American prairies, tall Composites which place their leaves pointing north and south with the edges vertical. In this way they receive equal lighting on both sides of the leaf, which has, correspondingly, an equifacial structure, both sides being alike. Similar twisting movements may also occur temporarily in plants exposed to very strong sunlight, while in *Eucalyptus*, although the twisting into the vertical position is permanent it occurs only in the elongate mature leaves and not in the rounded juvenile foliage.

Leaf Folding

The young leaves which are still enclosed in a bud often show remarkable foldings due to the tight packing imposed by limitations of space.This is not usually a mere crumpling of the young famina, such as one might except to arise from the independent growth of each lamina within the limits of the bud, but is evidently the result of co-ordinated growth, the manner of which is not yet understood.The particular type of folding is remarkably constant in each species and is known as the *ptyxis*. The variety of such arrangements is very great. In Monocotyledons the conditions are usually relatively simple, the leaves being either tightly rolled together into a single bundle or else folded together oppositely. The large leaves of the Palms are an exception, for they are often complexly pleated like a folded fan. In Dicotyledons individual rolling or folding or pleating or combinations of these, are all known. The mode of unfolding, when the bud begins to grow, depends naturally on the particular ptyxis, and the appearance and relationships of the young leaves at this time are called the *vernation*.

PHYLLOTAXIS

The placing of the leaves on the shoot, known as phyllotaxis, is ultimately due to the sequence and the spacing of the leaf rudiments, as they appear on the surface of the meristematic apex of the stem, and it involves some of the subtlest and most obscure of the space-time relationships which we have already indicated as fundamental in apical development. Unfortunately the lack of factual knowledge and the facility with which the subject lends itself to geometrical and mathematical deductions, have combined to leaf attention away from the living plant and towards a labrinth of theory, which has repelled some of the ablest investigators, to say nothing of the great majority of students. Before entering into any details, there are three main considerations which should be borne in mind, which may be of help in getting a true perspective. *Firstly*, the symmetry relationships involved are not peculiar to the arrangement of leaves upon a Higher Plant.

Leaves are, as we have tried to emphasize, only a special type of shoot of determinate growth, and the problems involved in their location go far back in evolution, far beyond the evolution of the Higher Plants, indeed beyond the evolution of leaves themselves, and are inherrent in every form of branching which depends upon apical growth. It is only in the simplest Thallophyta, with no apical growth became an established habit, regularity and rhythm in branch formation became the rule. The axillary position of the branch initials in Higher plants tends to obscure this truth, for it seems to make the position of the branches dependent on the position of the leaves. Only in the few examples of non-axillary production of shoots, e.g., in the bractless inflorescences of some Compositae, do we see, and see in such cases with peculiar clarity and beauty, that the same rules of phyllotaxis apply to the arrangement of shoots which apply to the arrangement of foliage leaves.

The conclusion that leaf arrangement is only one expression of a fundamental organic symmetry which rules throughout the plant is strenghtened, not only by comparison with the branching of Thallophyta, already referred to,but by the more immediate comparison with the arrangement of vestigial scales, for example on rhizomes, or with the spines of Cacti, which show the same symmetrical order of formation, though remote from foliage leaves in form and function. *Secondly*, that the position of a leaf primordium on the stem apex is determined before even the smallest external indication of its growth is visible. It follows that the determining factors, whatever they may be, are internal and protoplasmic and are those factors which control the distribution of growth-potential in the meristem.

External factors, therefore, such as contact-pressures against neighbouring primordia, which only come into play after the rudiment has started to grow, cannot be determining factors with regard to its position, however important they become later in influencing its development. There are many reasons for believing that a leaf is closely identified physiologically with the portion of the axis immediately around and below its point of insertion and that a similar demarcation into leaf fields or primordium fields exists at the apex. Each leaf primordium, appears in the centre of such a field, and its position is predetermined by the mosaic of fields covering the surface of the growing point.

Thirdly, that the geometrical relationships which are exhibited by developing primordia when the dome-shaped surface of the apex is projected on to a plane surface, while beautiful and interesting in themselve, are irrelevant to the question of the origin of the primordia. They do not control the initial spacing of the primordia but follow as necessary consequences of that spacing. That this is so, is illustrated by the frequent irregularities or changes of phyllotaxis which may be observed on one plant or even sometimes on one shoot. If the established symmetry had a predetermining effect on the appearance of new rudiments at the apex such changes should not occur, but in fact they do, and they may even be to some extent experimentally controlled, which serves to show that the origination of new primordia is independent of the pattern which develops as an effect of their arrangement. A homely parallel may be drawn with the action of a gardener who plants his successive rows of seedlings equidistantly and alternates the positions of plants in successive rows. He might be interested or amused if it were pointed out to him that by so doing he was creating a quincuncial pattern on the ground, but that has no determining effect upon his choice of spacing, which is dictated solely by the wish to give his plants the maximum of light and air. The causation is, so to speak, physiological, the inevitable consequence is a morphological pattern.

A similar confusion of cause and effect is illustrated by the contention that leaf arrangement is dictated by the need to avoid the overshadowing of lower leaves by those higher up; a good example of teleological argument. Its weakness is easily perceived when we consider that some common types of leaf arrangement, e.g., the decussate type are not elicient in this respect; or again that the same arrangement may be shown by the scales on a rhizome and the leaves on the aerial shoots; or lastly that means exist for subsequent adjustment of the positions of growing leaves, whatever their phyllotaxis, to ensure a leaf mosaic. In other words the avoidance of overcrowding among leaves is an effect, not a cause. That it is a desirable effect in most circumstances in undeniable, and it may be well be that natural selection has favoured certain types of phyllotaxis in consequences, but that does not invalidate the argument against it as a true cause of phyllotaxis.

The study of phyllotaxis has suffered from the historical accident that it was approached at first from the wrong end, that is to say, from the examination of mature shoots with elongated internodes, while the really critical condition at the apex were not examined until recently. On mature shoots there are three fundamental types of leaf arrangement; i.e., in opposite pairs, or in circles called *whorls*, or singly at the nodes. A survey of the families of Angiosperms shows that in certain families one or other of these types may predominate almost exclusively, e.g., opposite pairs of leaves in Labiatae, but it is not possible to say, on the basis of their systematic occurrrence, that any one of them appears to be definitely more primitive than the others. Nor is this surprising if we recollect that the establishment of these types must have long antedated the evolution of the Angiosperms.

There is some evidence that, in certain families of genera, the arrangement in opposite pairs may have given place to the simple arrangement of leaves singly. We have touched upon this previously in speaking of heterophylly, and there is, additionally, the evidence from seedlings that not only the cotyledons but the first one or two pairs of epicotyledonous leaves are frequently opposite, while the later leaves are formed singly at the nodes. Such conclusions are, however, of strictly limited application. One of the most general rules of construction in radial axes is that of the equidistance of parts, which holds good not only for outer members but also for internal structures such as vascular bundles.This rule is followed both in the opposite and in the whorled phyllotaxis.

Opposite pairs of leaves, moreover, usually alternate at right angles with those above and below them, thus forming for ranks, the *decussate* arrangement. Paired leaves in two ranks only, the *distichous* arrangement, are relatively rare, though single leaves alternating in two ranks are common in Monocotyledons and are almost universal in the great families of the Iridaceae and Orchidacae. The opposite and the whorled phyllotaxis are closely similar in that they imply symmetrical, radial growth at the apex with equal spacing of the primordia, and the paired condition may be looked upon as simply a whorl of two membres. As both, except for the bilateral distichous cases, imply radial symmetry of growth at the apex, they are commoner in orthotropic than in plagiotropic shoots. A tendency towards the development of bilateral or even dorsiventral symmtery may, however, often be observed in the growth of the axillary buds of such leaves, when the leaves themselves do not show it. For example, all the axillary shoots on one side may be stronger than those on the other side of the axis, or may become flowers while the buds standing opposite develop vegetatively. This produces a secondary dorsiventrality even in an orthotropic shoot and is usually associated with the presence of truly dorsiventral shoots in members of the same family.

In other cases, strong and weak shoots may alternate on each side of the main axis, thus showing a tendency towards spiral organization.The close relationships of opposite and whorled arrangements is shown by their tendency to pass one into the other. Individuals can often be found, of species which normally bear opposite paired leaves, with three or even four leaves at each node, or sometimes only at certain nodes. The essential similarity of the two conditions lies in the simultaneous formation of several primordia at the apex, but the number actually formed seems to bear some relation to the vigour, if not the size, the apex, since it tends to increase as the plant matures. It should be noted that in whorls of several leaves, as in those with only two, the members of each whorl alternate in position with those of the whorls above and below. The rule of equal spacing applies also in the cases of leaves inserted singly, which are by far the more numerous. If we examine a shoot of this kind we shall see, that if we start from any leaf and trace a line from this to the next leaf above it and so on until we come to a leaf which is exactly above the first, in a vertical line, further, that the number of leaves intervening between any two in the same vertical line is constant, and so is also the number of turns of the spiral round the stem in passing from one to other. This means that successive leaves must be equally spaced around the stem, or in other words, that every leaf is separated from the next by a fixed proportion of the circumference of the stem. This is called the *leaf divergence*.

By comparing a number of different plants we find that the characteristic divergences are not all the same, but that the number of different divergences is nevertheless limited and that certain figures occur again and again. Expressed as fractions of the circumference, the divergences found experimentally can be arranged in order this: 1/2, 1/3, 2/5, 3/8, 5/13 etc., a numerical series called, from its discoverer, the Fibonacci series. This series has the property that the numerator and denominator of each fraction are equal to the respective sums of those of the two preceding fractions. It has the further property that the first term, 1/2, expresses the maximum divergence,and the second term, 1/3 expressed the minimum divergence in the series. The maximum divergence, 1/2, is not very common. It represents leaves which are arranged singly, in two rows, on opposite sides of the stem and is specially distinguished as the alternate or distichous arrangement. We have already mentioned that it is commoner among Monocotyledons than among Dicotyledons, and is especially characteristic of Gramineae, Orchidaceae, and Iridaceae. The second or minimum divergence is likewise restricted in occurrence, but it is noteworthy that it predominates in the Cyperaceae, a family closely related to the Gramineae. It is also frequent in the flowers of Monocotyledons but is rare among Dicotyledons.

A closer approximation than 1/3 is very rare among higher plants, but cases of 1/4 or 1/5 divergence are known, and they form the first term of another Fibonacci series. The divergences of 2/5 and 3/8 are the commonest among Dicotyledons Higher fractions are characteristic of shoots with either very narrow or very closely set leaves or branches, like some inflorescences. As the higher fractions of the series are approached, it becomes increasingly difficult to say which leaves are directly above each other, as the number of intervening turns of the spiral increases, approaching the limiting case in which no two leaves are directly superposed, which occurs in some very condensed shoots such as gymnospermic cones in which the fractional denominator approaches infinity. In describing these spiral arrangements it is customary to refer to the vertical rows of superposed leaves as *orthostichies*. The number of orthostichies present will obviously equal the denominator of the divergence fraction. The mature stem has normally only one spiral of leaves, traced conventionally in the anticlockwise direction round the stem, which represents the sequence in which the initials were produced at the apex and is therefore referred to as the *genetic spiral*. The existence of this spiral does not, however, imply and twisting of the axis during growth, nor does it imply any spirality or "spiral tendency" as it used to be called, at the apex. It is simply a geometrical figure which is "described" as the mathematicians would say, by the equiangular spacing of the primordia on an elongating axis.

As the genetic spiral approaches the apex it becomes more and more condensed towards a flat helix, and the young leaves or primordia approach closer and closer together. The number of orthostichies represents, in each case, the number of primordia which exist simultaneously at any given level of the apex, itself. These primordia

are not, of course, except in whorled phyllotaxis all formed simultaneously. For example in stems with decussate phyllotaxis, although there are four orthostichies, only the primordia of one pair of leaves are formed at the same time. They leave only a small flat apex unoccupied between them and it is not until after an interval or *plastochrone*, during which the apex again grows forward into a dome, that another pair is formed. As the primordia approaches each other towards the apex, secondary spirals become evident in their arrangement. These are called *parastichies*, and they are purely a geometrical consequence of the packing of the rudiments together. They can be traced both clockwise and anticlockwise. In whorled constructions the number of parastichies in each direction is equal. Spiral phyllotaxis always produces an unequal number.

The mathematical deductions which may be derived from these arrangements, however interesting, leave untouched the primary biological question of why a given rudiment arises at a given point on the apex, since its inception must be determined before it appears. The apex is a dome, the surface of which is undergoing uniform expansion outwards from its apex as centre. On this growth-field secondary centres are established which become primordia of leaves, and we have still no better account of the phenomenon to offer than that originally proposed by Hofmeister, namely, that each primordium tends to appear in the largest unoccupied space left by its predecessors. This seems to accord not only with observations, but with what we should expect from the conditions of competition for nutriment between primordia, such as must obtain at the apex.

It does not exclude the possibility that the whole apex may be physiologically mapped out into primordium fields, e.g., by intersecting gradients, even where it appears to be still undifferentiated; it simply points to the sequence in which these fields will tend to develop.

Leaf Structure by Clearing

Leaf structure may also be studied further by clearing a piece of the lamina with a reagent such as chloral hydrate or Eau de Javelle. If such preparations are examined under the microscope it is possible to see the size and shape of the epidermal cells, the distribution of the stomata and the characters of the hairs. Below the upper epidermis, the palisade cells appear circular and the intercellular spaces between them are recognisable. Cleared preparations are particularly useful for studying the structure of the smaller veins. Frequently these anastomose or fuse with others, enclosing "vein-islets" of mesophyll.

Sometimes the veinlets end bindly in the mesophyll; the sieve-tubes become smaller, the companion cells relatively more prominent, the larger xylem vessels give place to narrower and shorter spiral and reticulate tracheids. Finally there is no differentiation in the phloem and a short spiral tracheid terminates the xylem. Sometimes the vein-endings near the margins of the leaf are in close association with the small-celled glandular tissue, of hydathodes. The epidermis in contact with each hydathode has one or more water-pores.

These are usually larger than ordinary stomata, and their guard-cells have lost the power to control the size of the aperture, so that the pore remains permanently open. Water-pores are found at the tips of leaves of grasses, and on the margins of a number of dicotyledonous leaves associated with hydathodes. Chloral hydrate dissolves out the cell contents such as protoplasm and starch, but does not destroy calcium oxalate crystals. Hence the characters of the latter can be studied in carefully cleared preparations.

Isobilateral and Centric Leaves

In the isobilateral leaf of Iris the mesophyll presents the same appearance towards both surfaces. In a transverse section of the lower, sheathing part of the leaf there are two series of bundles, one towards each side, the phloem portions of the bundles in each series being directed outwards towards the epidermis. In the narrower upper region the bundles lie more or less in the same plane, some with xylem directed one way, others with it pointing the opposite way. In centric leaves, e.g., onion there is a radial arrangement of tissue.

Development of the Leaf Structure

The development of the stem-apex, including the young leaves has already been described. At first all the cells of the developing leaf are meristematic, but later the meristematic tissue is restricted to the middle of base, and growth is therefore intercalary. Finally, when the full number of cells has been produced, meristematic activity ceases. At this stage the young leaf is still very small and folded up in the bud with the other leaves.The expansion and increase in size of the leaf, when the bud unfolds, is due to the extension-growth of the individual cells, not to the formation of the new cells.

The Internal Structure of Leaves

The leaf-stalk or the petiole does not differ much from the stem in internal structure. In the petiole the vascular bundles are arranged generally like a semicircle, especially the larger ones. There are also smaller bundles between the larger ones. The cambium does not persist in these bundles, because there is no need for secondary thickening. The xylem lies towards the upper surface and the phloem towards the lower.

The lamina of an ordinary leaf consists of a mass of parenchymatous cells filling up the meshes of the network of veins, which is prominent feature of the foliage leaves. This network of veins not only supports the leaf tissue, but also serves as a channel for carrying water to the cells in the meshes of the network. If also prevents the tearing of the leaf blade and keeps it quite flat. As striking examples of the network of veins, we may mention the leaves of *Ficus religiosa*, *Quisqualis*, *Dolichos* and *Antigonon*.

A leaf considered from an anatomical point of view consists of three parts. They are an epidermis covering both the surfaces of the blade, the mass of parenchymatous cells having chloroplastids imbedded in them and the network of veins. The epidermis consists of flattened cells all fitting closely together, and they do not contain chlorophyll grains. The lower epidermis differs from the upper epidermis in having the characteristic openings, the stomata. Each stomata consists of a pair of semilunar cells bounding the opening. These cells are called the guard cells and they contain chlorophyll grain. Further these guard cells are capable of changing their shape and so very the size of the opening. These stomata are generally confined to the lower epidermis of leaves but there are also leaves having stomata on both sides. The number of stomata is generally very large and it may be anything from 50,000 to 450,000 or more to the square inch.

The parenchymatous cells, forming the bulk of the leaf tissue in leaves with the upper and the lower surfaces well marked, are arranged in two distinct ways. In the upper portion of such a leaf one or more layers of cells are elongated vertically and lie side by side without much of intercellular space. This is called the *palisade parenchyma*. The parenchyma lying below the *palisade parenchyma* consists of cells arranged so as to have a large number of space, and hence this part of the mesophyll is termed spony parenchyma. This distinction of the mesophyll, into a compact palisade parenchyma and spongy parenchyma, is liable to variations according to the habit and the species of the plant. In the leaves of a plant growing in an open place exposed to the sun, the palisade parenchyma will be deep and will consist of several layers of cells, e .g ., leaves of Nerium, Calotropis and the Banyan.

Plants growing in a shady place have leaves with a single layer of palisade cells, e.g., Vigna and Dolichos. In the case of plants whose leaves are exposed to the light of the sun on both the sides, this separation of the mesophyll into two kinds of parenchyma dose not exist. For instance, the leaves of grasses do not show this separation but the whole of the mesophyll consists of compactly arranged parenchymatous cells. The most prominent structure in a leaf is its network of veins, and this part may very well be observed by allowing leaves of plants to root. The mesophyll and all the soft tissues decay, leaving only the hard vascular frame work. This network of veins is sometimes called the "skeleton" of the leaf. The ground-tissue is essentially parenchyamtous, but collenchyma and sclerenchyma are often present as in stems.

The Lamina Bifacial (Dorsiventral) Type

We have chosen the leaf of privet (Ligustrum vulgare) as a suitable representative of a dorsiventral type of leaf. shows a digrammatic representation of the distribution of the tissues as seen in a vertical section of the lamina in the region of the midrib, and also detailed drawings. The cells of the upper and lower epidermis appear somewhat rectangular in section. They are living cells whose outer surface is covered with a cuticle. Stomata are confined to the lower epidermis and are cut in various directions by the section. Glandular hairs are also present in the lower epidermis, sunk in depressions. The mesophyll has an upper region consisting of about three layers of cylindrical cells whose long axes are at right angles to the epidermis. This is the palisade parenchyma. Narrow intercellular spaces run between these cells, as shown by the imprisioned air when a section of a fresh leaf is cut and mounted in water. The lower part of the mesophyll consists of irregular cells, loosely arranged, thus leaving large intercellular spaces between them.

All the intercellular spaces communicate with each other and with the stomata.The cells of the mesophyll contain numerous chloroplasts embedded in the cytoplasm lining the cell-wall. The close packing of the palisade cells renders the upper surface of the leaf a darker green than the lower surface. The palisade tissue is concerned chiefly with photosynthesis, and although the spongy tissue also discharges this function, its importance lies rather in its intercellular-space system which permits of free diffusion of gases and water vapour between the plant and its environment through the stomata. Thus carbon dioxide present in the air enters the stomata and ultimately penetrates between the palisade cells, whilst oxygen resulting from photosynthesis is able to diffuse out of the leaf through the stomata.

Similarly the reverse process takes place during respiration, whilst water-vapour given off from the mesophyll cells into the intercellular spaces diffuses out through the stomata into the surrounding air. Between the palisade and spongy mesophyll run the vascular bundles. Where a vein is cut transversely, as in the midrib, it is readily seen that the xylem is towards the upper and phloem towards the lower surface of the leaf. In a section of a reticulately veined leaf, such as privet, some veins will be cut obliquely or longitudinally. Endodermis and pericycle, although present round the larger bundles, are not easily recognisable. The endodermal cells contain small starch grains which stain with iodine solution. Hence the name starch-sheath. The pericycle is absent in the smaller veins. The cells of the upper epidermis have straight walls and are polygonal in shape, those of the lower epidermis have curved walls and are irregular in shape. Stomata are numerous in the lower epidermis, and the discoid heads of glandular hairs are seen.

In many leaves, particularly leathery ones such as *Ficus elastica* and *Prunus* lauro-cerasus, groups of palisade cells are seen to converge at their inner ends towards single mesophyll cells. The latter have been called collecting-cells, it being supposed that they collect the carbohydrates synthesised in the palisade cells and pass them on to the tissue of the vascular bundle. The structure of different leaves varies considerably in detail.

The palisade tissue is well developed in the leaves of plants which grow exposed to bright sunlight; it is poorly developed in shade plants. Sometimes in the same plant "sun-leaves" may be distinguished from "shade-leavs" by their thicker texture and greater development of palisade tissue. Strenghtening bands of scleren-chyma may be developed, usually in relation to the vascular bundles, or at the leaf-margins. Cells containing crystals, or cavities containing oil may be present. Epidermal cells may contain mucilage as in Buchu and Senna, or be modified and serve for water-storage as in some xerophytes. In holly and india-rubber plant the epidermis may have a hypodermis of one or more layers of cells below it. The leaves of many Monocotyledons stand more or less erect. Here well-marked palisade tissue is absent and the mesophyll consists of small rounded cells containing chloroplasts. The vascular bundles, however, are all similarly oriented.

12

Structural Modification of Roots

Among the simplest of these special cases are the *storage roots*. Normal roots usually store food reserves, but in some cases they are enlarged into *root tubers*, large, rounded structures with massive parenchyma tissues, packed with reserve materials. Sometimes the whole root is involved, as among soil-living Orchids, such as *Orchis morio*; sometimes the base of the root thicknes, as in the tubers of *Dahlia variabilis*, sometimes small tubers are formed as branches on normal roots, as in *Asparagus*. Frequently, as in *Oenanthe crocata*, these tubers are the means whereby the plant outlasts the winter season. Plants which belong to the so-called "root crops" are in a special category. Nearly all of these, such as the Carrot, Turnip, Beetroot, etc., are biennials.

Storage of Root

In the first season the storage organ is formed, with only a very contracted stem which remains at soil level, bearing a group of large leaves. In the second season the stem elongates and bears flowers, drawing on the stored food below ground. The plant thus loses much of its food value to man, if it is allowed to flower or "bolt", as gardeners say. These storage organs are compound structures, the lower part being the true primary root, and the upper part being the hypocotyl of the seedling. The two parts are completely united into one organ. The nature of the storage tissue varies. In the Turnip and the Radish it is formed from parenchymatous secondary xylem. In Carrot and Parship it consists of parenchymatous tissue which chiefly represents secondary phloem and secondary cortex. In the Beetroot and Mangold

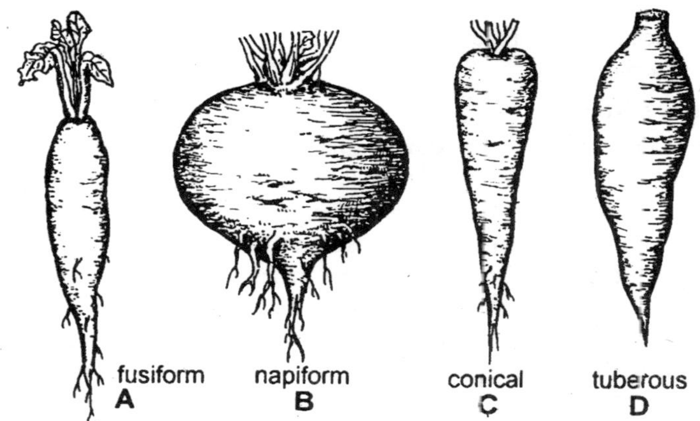

Fig. 12.1. Modification of tap root. A, fusiform root of radish; B, napiform root of turnip; C, conical root of carrot; D, tuberous root of four o'clock plant.

it is largely soft xylem tissue in concentric rings, alternating with parenchymatous phloem. The rings are formed by a succession of short-lived cambia which are produced by the pericycle. The same structure occurs in many storage roots.

Contractile Roots

A very interesting modification is shown by *contractile roots*, which are formed by many perennial herbaceous plants. The most striking examples are found on the bulbs and corms in Monocotyledons of the families Liliaceae and Iridaceae. Perennial tap roots such as those of the Dandelion are also often contractile though to a lesser extent. These specialized roots are long, straight, thick and fleshy. In young plants of *Gladiolus* they may be thicker than the corm itself. They grow straight downwards, without branches or root hairs. They only last for two or three months and they contain at first abundant stores of glucose, which are rapidly consumed.

As the reserves disappear, the cortex collapses in a series of transverse zones, which shortens the whole root by 30 to 40 per cent. This exercises a forceful pull upon the plant. In the case of the bulbs and corms, where these roots are produced only at the close of the growth season, the plant is drawn bodily downwards and the yearly repetition of the process carries the plant down to a certain level, after which the formation of contractile roots ceases. If the bulb is then dug up and replanted near the surface, however, the process begins again. Contractile roots are also formed in some rhizomatous species such as *Polygonatum* and *Asparagus*, where no drawing of the plant downwards results and their function seems to be only that of giving more secure anchorage in the soil against the effects of wind on the large aerial shoots.

Buttress Roots

Many tropical trees show remarkable *buttress roots* around the base of the trunk. They are formed by the bases of the main roots, in which secondary thickening is unsymmetrical, being chiefly on the upper side. This results in the formation of thick, woody walls, running obliquely between the trunk and the ground. They are sometimes so large that two men, standing one on each side of a buttress, would be hidden from each other. Some authors have suggested that this additional support of the trunk is rendered necessary by the great height of many tropical trees and perhaps by the prevalence of hurricane winds, but this is largely guesswork.

Horizontal Roots

A common habit among tropical trees is the production of roots which grow horizontally along the surface of the soil, spreading widely around the tree. Although they are a marked feature of tropical forests, it is not possible to say what advantage, if any, is gained by this method of growth.

Aerial Roots

Aerial roots are produced by most plants which grow as epiphytes. In many cases these roots may be of enormous length and of very rapid growth (4 to 6 in. per day). Roots of this kind are formed by many tropical Aroids and they hang down in large numbers, like cords, from the branches of the jungle trees, forming a remarkable feature of the forests. When they reach the ground they penetrate the soil and produce a normal, branching root system. Among epiphytic Orchids the aerial roots are, however, short and wholly aerial and hang loosely in the damp air of the jungle. They can absorb rain and condensed moisture directly from the air. Such supplies of water are naturally intermittent, and the ordinary equipment of delicate root hairs would be quite unsuitable for an environment which is liable to periods of drought. Instead of a piliferous layer these roots are clad with a special tissue called *velamen*, which may be many cells thick. The cells are dead and empty. Their walls are performed with round apertures and are striated with ribs of cuticle. In fact the tissue is functionally a sponge, like the cortex of *Spnagnum*. On its inner side the velamen is bounded by a layer of cells with thickened walls, corresponding to the exodermis. When, as frequently happens, this layer is highly thickened, there are unthickened "passage cells" produced at intervals, just as in the endodermis.

As the velamen is external to the exodermis it must be a derivative of the dermatogen and is therefore analogous to the multiple epidermis often found on shoots and leaves. The velamen is colourless and papery, so that from the outside the root looks dead, but the dead layer is also translucent and the cortex within is full of active chloroplasts to which light can penetrate. The obvious suitability of velamen to the conditions of epiphytic life has often caused it to be described as a choice example of "adaptation." Unfortunately for this idea, the tissue is not restricted to epiphytes. It is found also in many soil-rooting Orchids and in other Monocotyledons, e.g., *Asparagus*, where no epiphytism exists. Needless to say, this does not detract from the usefulness of the velamen in the epiphytes which possess it, but it does point to another and probably more correct view than the adaptational one namely, that the velamen is an anatomical peculiarity which has made possible the adoption of the epiphytic habit by plants which possessed it, though it has not obliged all of them to adopt epiphytism.

Pneumatophores

Analogous to the epiphytic roots which hang in the air are those which grow up into the air from beneath the slime of tropical swamps, especially in the tidal swamp forests called Mangrove. These are known as *pneumatophores* or "breathing roots". They arise from long horizontal roots, and as they are negatively geotropic they grow vertically upwards, projecting eventually several inches into the air. They are usually only a few inches apart, so that very large numbers may surround a single plant, strongly recalling in appearance a crop of young Asparagus shoots. There is a corky layer over each root, which covers even the apex, and the portion in the air bears numerous lenticels. Internally they possess an extremely spongy cortex, and in the cortex of some species there are also special cells, or idioblasts, which are empty and have thick cuticular ribs on their walls.

Roots hairs are not produced either by the pneumatophores themselves or by the horizontal roots from which they spring, but the pneumatophores form short absorbing branches from their lower portions and hairs are borne on these. Experiment has confirmed the old idea that pneumatophores serve as organs of aeration for the underground root system, buried as it is in tenacious mud with a low oxygen content. Diffusion into the horizontal root is, however, limited to short distances.

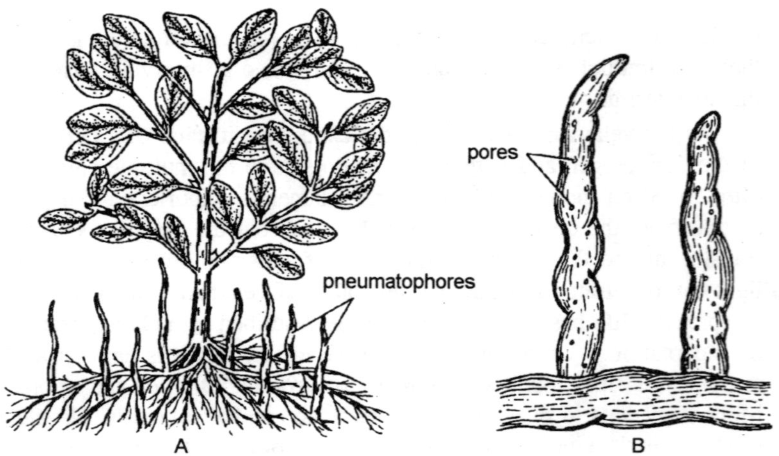

Fig. 12.2. Pneumatophores (respiratory roots) of a mangroove plant (Avicennia spp.): A, main plant; B, two pneumatophores enlarged.

They are also centres of active respiration, and their vertical growth may be necesssary to place the absorbing roots at the most favourable level. In sandy soils they are much fewer or may be absent. The submerged roots of some swamp plants, such as *fussieua* in warm countries, are surrounded by a greatly enlarged cortex called *aerenchyma*, which is formed of thin, dumb-bell shaped cells, with big intercellular spaces. Although this is submerged, its function is probably that of increasing the absorptive surface, not for water but for dissolved oxygen, which is in low supply in warm water.

Haustoria

Parasites among the Flowering Plants make use of modified roots as a means of penetrating the tissues of the host plant. In *Orobanche*, the Broomrape, it is the primary root of the seedling which performs this function. It has no root cap and is the only root formed by the parasite, except some short exogenous outgrowths, which serve only for fresh attachments and may be modified roots. In *Lathraea*, the Toothwort, and in the green semi-parasites on grass roots, of which *Euphrasia*, the Eyebright, is an example, it is the branches from an otherwise normal root system which attack and penetrate the roots of the hosts. The Dodder, *Cuscuta*, has only a temporary root, in the seedling stage. The embryo has no leaves, and when the root has withered the thread-like stem grows independently on the surface soil for a short while, but perishes if it does not quickly find a host plant. Thereafter it twines around the stem of the host, like its relative,

Convolvulus. Its nourishment is obtained by means of numerous penetrating suckers which spring from the surface of its stem nearest to the host.

The resemblance of these suckers to the adventitious hold-fast roots of the Ivy has led to the general belief that they also are adventitious roots, specially modified. If so they have been so far modified that it is difficult to find any root character which they possess. The stem of *Cuscuta* has four cortical layers. From the outer two there first develops a flat pad which attaches itself firmly to the epidermis of the host. In the centre of this pad, and involving all four cortical layers, there grows out a penetrating organ, the point of which is formed of elongated cells, which penetrates between the epidermal cells into the host cortex. Here the long cells of the point take on independent growth and spread through the tissues. A strand of xylem and phloem now differentiates in the sucker and makes connection between the vascular tissues of the host and the corresponding tissues in the *Cuscuta*. In all the above cases of parasitic attachments the vascular tissues of the host are the object of the attack, and substances are deflected from them into the parasite.

It is difficult to find a common morphological basis for all such parasitic attachments, but they are sufficiently uniform in function to be grouped together biologically, and they are given the name of haustoria. The foregoing modifications of the normal root in structure or function are for the most part shown by members of the primary root system. There are some further modifications which are characteristic of adventitious roots.

Climbing Roots

Many tropical climbers, especially of the family Araceae, make use of *clasping roots*, which are non-geotropic but respond negatively to light and are markedly sensitive to touch. These reactions cause them to cling closely to the bark of the supporting tree, which they envelop with a stout network.

Absorbing roots are produced separately in some cases. The latter grow directly downwards and eventually enter the soil. Both types are evidently modifications of one common root-form and differ from one another in little except in the larger vessels of the absorbing root and the thicker sclerenchyma of the clasping roots. A similar modification is familiar in the common Ivy (Hedera helix). Adventitious roots are formed in great numbers on the side of the stem next to the support. Normally these remain short and unbranched and become precociously corky. They attach themselves to the support by the formation of mucilage from the surface cells of the apex. The free-hanging shoots do not as a rule produce holdfast roots. Apparently some moisture is necessary for typical holdfasts to develop, but if there is an excess of moisture or if they are grown in water they become normal roots. They may be regarded therefore as normal roots impeded in development by their conditions of growth.

Stilt Roots

Adventitious roots sometimes form supporting stilts. A good example is shown by the Maize, which develops a cluster of roots from the first one or two nodes above ground level. They grow obliquely downwards into the soil and give added support to the stem, which, with its large leaves and heavy cobs, is very vulnerable to the effects of wind. Stilt roots are also well shown in *Pandanus*. Less familiar, but much more striking, are the stilt roots of of the Mangroves. Several species of *Rhizophora*, *Sonneratia* and *Avicennia* make up the tidal wood-lands in tropical bays and estuaries. All these trees are supported by scaffoldings of stilt roots, which may cover many square yards round each tree, making a formidable entanglement over the ground. Not frequently the base of the stem roots away, so that the tree is left perched on its stilts above the mud, like a giant vegetable crab.

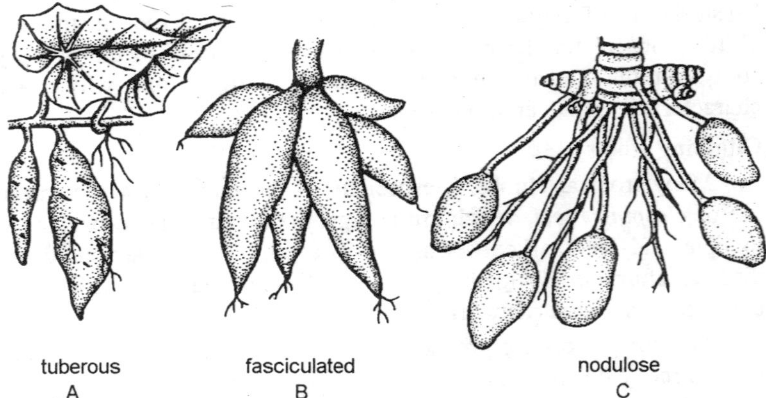

Fig. 12.3. Modification of adventitious root: A, tuberous roots of sweet potato; B, fasciculated roots of Dahlia; C, nodulose roots of turmeric.

Prop Roots

Prop roots are those formed from the main branches of a tree. They grow straight downwards to the soil, which they enter and form underground branch roots. As they become thickened they give support

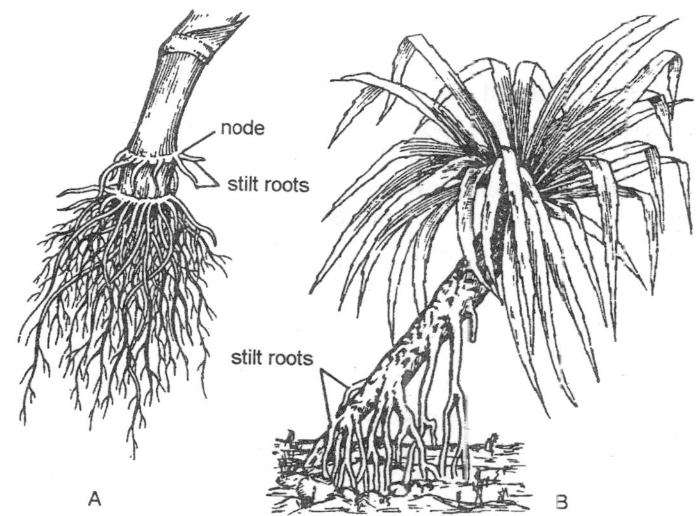

Fig. 12.4. Modification of adventitious root: stilt root of sugarcane (A) and screw pine (B).

to the branches, which are thus able to continue their horizontal growth, producing more prop roots at intervals. The best example of this is *Ficus benghalensis*, the Banyan Tree. Banyan is the vernacular name for a merchant, and it was given to this tree because it was the customary shelter for itinerant pedlars, who set up stalls in their shade. Some of the biggest specimens have been used as shelters for whole villages of native huts.

Thorn Roots

It only remain to mention two minor but curious modifications. The first is the formation of *thorn roots* of the base of the stem in several species of Palms. As in the case of stem thorns, these begin as normal structures but their tissues soon become densely sclerotic, including their apices, which then cease growth. In addition to root thorns, which are modifications of the roots themselves, some plants bear root spines which are outgrowths from the roots, *e.g.*, the Palm, *Iriartea exorhiza*, the big spiny stilt roots of which are used as greaters in Panama.

Cortical Roots

The second is the case of *internal* or *cortical roots* in several Monocotyledons, including especially *Tillandsia* and *Vellozia*. These roots grow directly downwards from their point of origin inthe pericycle of the stem, penetrating the cortex as they go. Eventually they emerge from near the base. In *Vellozia*, which is common on exposed rocky places in South Brazil, these roots are not indeed truly internal. It is more correct to say that they actually make up a pseudo-cortex, covered and protected by a sheath of persistent leaf bases. The intensely dry habitat would no doubt prevent the growth of exposed aerial roots. A comparison suggests, itself with the compact mantle of roots which encloses the stem in the Tree Ferms and in their fossil relative *Psaronius*.

Root Buds

One of the most fundamental characters of roots is that they bear no leaves, but they may and frequently do, in certain species, produce buds, which usually arise endogenously in the same manner as branch roots. Only in *Linaria* are the root buds exogenous. A few cases are known in which even the root apex may become transformed into a bud, namely in the Monocotyledons *Listera* and *Neottia* (Orehidaceae) and in *Anthurium* (Araceae). This formation of adventitious buds may be divided into three categories on biological grounds.

Reparative

Buds formed as a sequel to root injuries or to the destruction of the normal shoot system. This is common in woody plants and is often additional to the formation of buds of the next category. Among herbs there are fewer examples, and in most herbaceous plants where reparative buds are formed there are no others on the roots, e.g., *Crambe*, *Cochlearia*, *Anchusa* and *Morisia*. They form a valuable means of vegetative propagation.

Accessory

Buds formed as part of the normal development in addition to the usual stem buds. They may serve as a means of vegetative propagation and are often called "suckers". They are common among woody plants, e.g., *Populus*, *Rubus*, *Rhus*, *Prunus*, *Hydrangea*.They are rare among herbs. Well known examples, however, are *Rumex acetosella* and *Epilobium angustifolium* and certain other aggressive weeds.

Necessary

These are buds which are the plant's only means of survival, normal buds being abortive or lacking. The saprophytes Pyrola and Monotropa are in this category and here also may be classed *Lepidium latifolium* and *Armoracia vulgaris* (Horse Radish) since they scarcely ever set good seed.

Linaria vulgaris (Toadiflax) is a peculiar case. Not only are the root buds exogenous, but they appear in the seedling stage, and their growth replaces the main shoot, which aborts early. Some buds are also formed on the hypocotyl. This leads us to the most peculiar case of all, that of the family Podostemaceae, whose members grow only in tropical waterfalls and exhibit the most surprising variety of forms, most of which are thalloid. In their vegetative parts they are utterly unlike Flowering Plants. Their developmental histories are very various, but follow the general principle that the main shoot of the seedling is absortive and is replaced by the outgrowth of adventitious roots from the hypocotyl. These develop into the "thallus" and give rise to branch roots, to leafy shoots and to flowers. These roots are flattened and contain chlorophyll. They cling very closely to the submerged rock surfaces to which they are attached by root hairs and by short, exogenous outgrowths, called *haptera*, which resemble the holdfasts of the Ivy and serve the same purpose.

In extreme cases the outgrowth from the hypocotyl has lost even the form of a root and develops into a circular crust on the rock, almost exactly like a Lichen. The shoots which arise from these "roots" are endogenous, but they can scarcely be called adventitious since they form part of the regular course of development. The family presents some of the most interesting phenomena of adaptation to environment in the whole plant world.

The Evolution of Roots

The possession of true roots is generally regarded as a distinguishing character of vascular plants, but as a matter of fact some traces of root formation can be found among the Bryophyta. A few species of Hepaticae and Musci have leafless, branched, downward extensions of their axes which it would be arbitrary to exclude from the category of roots since they correspond both in their general morphology and their functions with those of higher plants.

There is indeed no reason why a gametophyte should not develop roots like a sporophyte. Among the lower Pteridophyta, in which the gametophyte and the young sporophyte are very much alike, it must have been an open question at one period of evolution which would become the rooted plant and the chief agent in the colonization of the land. In the majority of the Spermatophyta the embryonic plant normally begins life with a main root which appears to be a downward prolongation of the main axis, and in the embryos of many Seed Plants the axis is a unitary structure with growing points at both ends, while in the tissues between one cannot distinguish the root and stem portions until a late stage of germination. This condition in the embryo has suggested that the primary root is simply a downward extension of the primary axis. If we look, however, at those members of the Thallophyta which normally grow attached to a substratum we see that the base of the axis terminates, not in a root, but in an attaching organ or holdfast. In comparing these lowly plants with the Spermatophyta we would gain the impression that the essential step in evolution which has produced the root was the establishment of a basal growing point, whereby the axis became bi-polar and acquired the power of extending in both directions. But we would get a false idea of this evolution if we were to imagine it as taking place in a fully evolved axis such as that we are familiar with in the Seed Plants.

The evolution of the root took place place among the lower Pteridophyta, and some of the lowest of that group, notably the Psilotales, are rootless even at the present day. The evolution of the root did not therefore coincide with the development of the land habit of life but occurred among plants which were already land-living. Nevertheless its importance is due to the conditions of land life, which restrict water absorption to the underground parts. The lower end of the axis in the most primitive land plants –the "antiapex" as we may call it–did not form a root but a bulbous swelling, corresponding to the thallophytic holdfast, which was well developed in the fossil Psilophytales and perhaps survives among many of the Pteridophyta in the embryonic foot which attaches the young sporophyte to the gametophyte. This foot, not the primary root, is the true antiapex of the embryo in such types.

In the embryonic Lycopod the first root is normally a side growth, sometimes indeed arising close to or even actually at the stem apex. The anti-apical position of the primary root with which we are familiar in the Dicotyledons has only been reached in the higher plants in association with the enclosed embyo and the disappearance of the attaching foot. Bower has called the young stages in the development of the Bryophyte embryo the "primitive spindle," the anti-apex of which is also a foot, not a root.

The view here put forward is therefore in sharp contrast to that of Compbell, who traces the origin of the root to the meristematic base of the sporophyte in *Authoceros*, on the assumption that the primary root is itself the anti-apex. It seems much more probable that root and shoot have both alike been differentiated from the branches of a

Some parasitic plants attach themselves to the roots of their host plants, instead of the stem or branches. *Striga lutea* and the Sandal wood tree are well known parasites. Both these plants do not look like parasites, because the haustoria are attached to the roots of the host plants that are underground. A very remarkable root parasite thrives in the island of Sumatra. It consists of only a gigantic flower attached to the root of vines. When in full bloom the flower is said to be about a metre in diameter and many pounds in weight.

ROOTS AND FUNGI

The roots of a considerable number of plants are regularly associated with the filaments or hyphae of soil fungi. Thus the roots of certain moorland plants, orchids, and many forest trees have fungal threads growing inside the cortical cells, or closely and pressed to the root surface. The term *mycorrhiza* is applied to such an association of fungus and root. The view has frequently been taken that this association is mutually beneficial to the two organisms concerned, and that it can therefore be classed as an example of *symbiosis*. As will be indicated later, this interpretation has not been universally accepted. Mycorrhiza usually occurs on plants growing in soils which are rich in humus (that is, decaying organic matter) but poor in mineral salts, for example, the soils of woods and heaths.

The majority of such plants possess chlorophyll and so are able to manufacture their own carbohydrates, but, since the soil is deficient in nitrates, they need an alternative source for their nitrogen supplies. This appears to be provided by the fungus which, unlike the roots of the higher plant, is able to break down the protein constituents of the humus into simpler, soluble organic compounds. The fungus absorbs these compounds, and some of them may be transferred to the root with which they are associated.

It is possible that in return the fungus receives supplies of carbohydrates from the green plant. There are, however, some plants with mycorrhizal roots which possess little or no chlorophyll, such as the birds's-nest orchis (*Neottia nidus-avis*) mentioned below. Here the higher plant must derive not only its nitrogen but also its carbohydrates from the humus, presumbaly through the intermediary action of the fungus. Since *Neottia* manufactures no carbohydrates it is difficult to see what nutritional advantage the fungus gains, and it may be that the root merely provides the fungus with a suitable habitat.

Two main forms of mycorrhiza are recognized namely:

(a) The ectotrophic type in which the fungus grows only on the surface of the root and between its cortical cells and

(b) The endotrophic type, in which the fungus penetrates into and occupies the cells of the root. Intermediate conditions also occur.

(a) Ectotrophic Mycorrhiza

Pines (*Pinus* species), larch (*Larix decidua*) and spruce firs (*Picea* species) form symbiotic relationships with many toad-stool fungi. These fungi, such as species of *Boletus*, *Scleroderma*, *Amanita* and *Russula*, consists of many threads which grow between the soil particles, like those of the mushroom; the familiar aerial parts being the spore-bearing structures. These threads form an ectotrophic mycorrhiza around and in the young roots of such coniferous trees. Experiments have shown that the presence of the fungus is beneficial to the trees. For example, conifer seedlings grown in soil containing the mycorrhizal fungus, resulting in the fungus-root association being set up, grow more strongly than similar plants grown in soil lacking the fungus. The reason for this benefit is not known for certain.

One possibility is that, through the action of the fungus, the tree gains food materials derived from humus. Other theories have suggested that the absorption of water and mineral salts is promoted by the presence of the fungus. Beeches (*Fagus* species), oaks (*Quercus* species) and many other trees also possess ectotrophic mycorrhiza. The mycorrhizal roots of such trees are very different in external appearance from the normal roots. For example, in the pine, the roots remain short and fork repeatedly to give a coral-like mass of swollen rootlets. The yellow bird's nest (*Monotropa hypopithys*)—not to be confused with the bird's orchis—also possesses an ectotrophic mycorrhiza. The aerial part of this plant, which grows in fir and beech woods, consists of a flowering shoot on which however, there are no green leaves. Below ground there is a much branched system of roots, each covered with a felt of fungal threads. Since the plant is unable to carry out photosynthesis, all its nourishment must be obtained from the humus presumably with the aid of the fungus.

(b) Endotrophic Mycorrhiza

Endotrophic mycorrhiza often occurs in the roots of heather (*Calluna vulgaris*), a plant which usually grows on poor, peaty soils of an acid nature. The new roots produced in the spring are infected with the threads of a fungus which grow not only on the root surface but also penetrate and grow inside the cells of the narrow cortex. Later, the cells of the cortex appear to digest the fungal threads and may thus obtain organic food-stuffs, some of which may have been absorbed from the humus by the fungus.

There may be more than one type of fungus concerned but a species of *Phoma* is usually regarded as the important one. It has been stated that this fungus can build up nitrogenous compounds from atmospheric nitrogen (that is, carry out the process of nitrogen fixation) and that *Calluna* seedlings benefit from this additional supply of nitrogenous foodstuff. This experimental evidence is, however, not very sound. Many orchids possess endotrophic mycorrhiza, the fungus concerned being a species of *Rhizoctonia*. The fungal threads form close coils inside the cortical cells of the roots and tubers and some threads pass out into the surrounding soil. In some of the cells (host cells) the fungus appears to remain in a healthy condition; in other cells the fungus is digested by enzymes. Presumably the products of digestion are utilised by the orchid. The mycorrhizal fungus is important in the germination of the minute orchid seeds. The latter, if sown under sterile conditions and without a supply of organic food fail to germinate. If the fungus is introduced, it enters the cells of the seedling and germination proceeds.

Germination can, however, be obtained in the absence of the fungus by sowing the seeds on a medium containing organic materials, particularly sugars. This method, known as asymbiotic germination, is now largely used in horticultural practice. Under natural conditions it seems that the fungus makes available organic foodstuffs to the minute embryo of the orchid so that germination can proceed.

The bird's-nest orchis (*Neottia nidus-avis*) consists of a short rhizome bearing numerous fleshy roots and a brownish flowering shoot. Brown scale leaves are developed but green leaves are entirely absent, so that the plant can be credited with little or no power of photosynthesis. The cortical cells of the fleshy roots contain coils of fungal threads. As in the green orchids, some of the cells are host cells while in other cells the fungus is digested, presumably thus providing the foodstuffs needed by *Neottia*. Very little is known about the way in which the exchange of foodstuffs between fungus and root takes place in either of the two types of mycorrhiza. It has, indeed, been suggested that the fungus is merely a parasitic which the higher plant keeps in check. Any observed benefit to the host plant is, on this view, thought to be due to the fact that the mycorrhizal fungus, together with other fungi growing in the neighbourhood of the root, break down the organic matter in the humus into soluble forms which the root itself can then absorb directly. The balance of evidence, however suggests that mycorrhiza is in fact a symbiotic relationship.

Root Nodules

For centuries it has been known that one way to enrich the soil is to grow on it plants which belong to the family Leguminosae. Some of the best-known plants belonging to the family Leguminosae are peas, beans, clover, etc. Such plants actually enrich the soil with nitrogen compounds, rather than make them poorer. For this reason, clover is often cultivated on farm land. It is a crop often used in crop rotation. Sometimes, especially if the clover turns out to be a poor crop, it is merely cultivated until the end of its growth and then just ploughed into the soil, where, during the winter, it decays, thus giving a splendid natural manure to the soil. More often, however, the crop of clover is harvested; but much addition of nitrogen compounds to the soil has by then taken place. This beneficial effect of leguminous crops on the soil was recognised by the Greeks and Romans, but it was not explained until towards the end of the nineteenth century.

In experiments then carried out, plants, such as wheat, were grown in soil deficient in nitrates. They were found to become unhealthy, although the atmosphere around them contained, of course, a large proportion of nitrogen. On the other hand, clover and other leguminous plants were found to flourish quite well on such soils. After much investigation, it was concluded that leguminous plants were able to utilise atmospheric nitrogen in their growth, provided that their roots were closely associated with a certain soil bacterium now known as *Rhizobium* (formerly called *Bacillus radacscola*). The Rhizobium organism inhabits the cells of the special structures called *nodules* which are usually to be found on the roots of clover, broad bean, or other leguminous plants. The organism also exists in large numbers in most soils, the exceedingly minute spherical or rod-shaped cells being able to move slowly through the soil by means of flagella. They appear to be attracted to the roots hairs of leguminous plants and proceed to penetrate through the tips of the hairs, forming inside an infection thread which grows down the root hair and enters the cortical cells of the root.

The presence of the bacteria stimulates the cortical cells to commence meristematic division, with the result that a swelling or nodule develops, the central region of which consists of enlarged cells containing large numbers of the bacteria, while in the outer part of the nodule vascular strands linking up with the stele of the root are present. The central infected region in a fresh nodule is usually red in colour, and this has been shown to be due to the presence of haemoglobin, a pigment which, apart from this instance, is confined to the animal kingdom. The *Rhizobium* organism has the property of being able to utilise atmospheric nitrogen and to build it up into organic form, though this activity is only shown when the organism is associated with the leguminous plant in the nodule cells. This utilisation of atmospheric nitrogen is known as *nitrogen fixation*. A large proportion of the products of fixation passes from the nodule into the rest of the plant, with corresponding benefit to nutrition. Indeed, as is suggested by the experiments already mentioned, the leguminous plant will grow vigorously in a rooting medium free of combined nitrogen, the plant deriving, under these conditions, all its nitrogenous materials from the nodules.

Like most other bacteria, *Rhizobium* requires to be supplied with organic carbon compounds such as sugars before it can grow. These are obtained from the leguminous plant. Thus a clear example of symbiosis is provided, both organisms deriving benefit from the association. This fixation of nitrogen explains the beneficial effect of leguminous crops on the soil. Besides the members of the family Leguminosae a few other plants possess nitrogen-fixing root nodules inhabited by a soil micro-organism. Two examples are alder (*Alnus glutinosa*) and bog myrtle or sweet gale.

13

FLOWER

The present and the following two chapters deal with the angiospermous flower and the structures derived from it, the fruit and the seed. The phylogeny and morphologic nature of the flower and its parts are subjects of much discussion in the literature. The old classical theory homologizes the flower with a shoot, that is, it regards the flower as an entity consisting of an axis (*receptacle*) and foliar appendages (floral parts, or organs). The axis is relatively short and has determinate growth. The floral parts are divided into sterile and fertile, or reproductive. Megasporogenesis and microsporogenesis are carried out on separate floral organs, which may occur on the same or different flowers.

The floral parts concerned with megasporogenesis constitute, collectively, the *gynoecium* (from the Greek words meaning woman and house). The basic unit of the gynoecium is the *carpel* (in Greek, fruit), which is commonly regarded as a megasporophyll. One or more carpels may enter into the composition of a gynoecium. *Pistil* (in Latin, pestle) is another term referring to the megasporangial part of the flower. The pistil may consist of one carpel (simple pistil) or of several (compound pistil). If the gynoecium is composed of a single carpel or of several united carpels, the pistil and the gynoecium refer to the same entity. If the gynoecium consists of more then one separate carpel, it also consists of more than one separate pistil. The abandonment of the term pistil has been advocated but it continues to be useful. Some authors substitute *ovary* for pistil, but this word denotes only the lower part of the pistil. The other parts are the *style* and the *stigma*.

The carpels enclose the *ovule* or *ovules* (in Greek, egg) borne on the *placenta* (in Latin, cake, or flat plate). The *nucellus*, which is the central part of the ovule, is usually interpreted as the megasporangium. The functioning megaspore germinates within the megasporangium and gives rise to the female gametophyte, the *embryo sac*. Because of this developmental sequence, the gynoecium is commonly referred to as the female part of the flower.

The floral parts forming the microspores are called, collectively, the *androecium* (from the Greek words meaning man and house). The individual units of the androecium are the *stamens* (in Latin, filament). Classically, the stamen is interpreted as a microsporophyll, and the part of the stamen called *pollen sac*, as the microsporangium. The pollen sacs are contained within the *anther* (based on the Greek word for flowering). A microspore develops into the male gametophyte, the *pollen grain* (pollen from the Latin, fine flour). Since gametogenesis occurs in the anther the androecium is referred to as the male part of the flower.

The sterile parts of the flower are the *petals* (in Greek, flower leaves), collectively called the *corolla* (in Latin, small crown), and the *sepals* (in Greek, a covering) composing the calyx (in Greek, a cup). The calyx and the corolla constitute the *perianth* (from Greek words, about and flower). If the perianth is not differentiated into sepals and petals, the individual members of the perianth are called *tepals* (from the Latin *tepalum*, an anagram of *petalum*). Flowers commonly have nectarines borne on their various parts. Some of these are modified stamens, or staminodes.

The literature dealing with the question of morphologic nature of the flower is extensive and has been more or less comprehensively reviewed. Most of the proponents of the concept interpreting the flower as a modified shoot assume that the floral organs are appendicular structures in the same sense as leaves, both kinds of appendages possibly having undergone parallel evolutionary development. Emphasis is thus placed upon unity of types of structures; that is, foliage leaves and floral organs are both regarded as leaf-like appendages or phyllomes. Discussions on the nature of the flower frequently refer to the concept of leaves as derivatives of branches. The floral organs, though resembling leaves in extant angiosperms, evolved from cauline assemblages similar to those that gave rise to the foliages leaves. From such an aspect, the question to be asked is not how the leaf became a floral organ but how this organ evolved from a branch system.

Apparently plants bore ovules before the leaves–as we know them now–were in existence.

Sepals and petals are basically leaf-like in external form. They may intergrade with one another and with the small bracts (bracteoles) subtending the flower. In some flowers, however, the petals intergrade with the stamens through structures bearing characters of both. Moreover, frequently stamens and petals differ from other floral parts in having a single vascular trace. These features are used to suggest that in some taxa the petals have evolved from the stamens.

The specialized types of stamens, characterized by a distinct differentiation into a filament and an anther, appear rather unlike the leaves, but in many Ranales the stamens are wide, leaf-like structures with no differentiation of a filament. This form is the basis for the view that primitively the stamen may have been leaf-like. The fascicled types of stamens (Malvaceae, Guttiferae, and other dicotyledons), on the other hand, are thought to indicate an origin from primitive dichotomous branch systems–systems of telomes–bearing terminal sporangia. Still another theory proposes that stamens originated from the *gonophyll*, a leaf-like structure bearing fertile branches. Through condensations and deletions the gonophyll gave rise to the modern stamen either through a line with fascicled stamens or through one in which the stamen is laminar.

The classical concept of the carpel interprets it as a leaf-like appendage. By folding and fusion of margins and by unions with one another the carpels are assumed to have evolved into pistils.

The German literature deals extensively with the question regarding the type of a leaf to which the carpel may be compared. The postulate has been advanced that many carpels have the same growth form as a peltate leaf, that is, a leaf in which the stalk is attached to the lower surface of the blade. The degree of peltation is considered to be variable and absent in some forms. Peltation is recognized in stamens and perianth parts as well.

The concept of the carpel as a sporophyll bearing sporangia is frequently criticized because not all gynoecia of angiosperms can be interpreted by reference to it. Some authors consider that the ontogeny of the carpel makes it quite distinct from leaves; others find that the vascular anatomy of many flowers suggests an independence between carpellary and placental vascular traces. The inconsistencies in the sporophyll concept of the carpel are proposed to be resolved by the gonophyll theory. According to this theory, the basic component of the gynoecium is a leaf with an epiphyllous fertile branch, the two together comprising the gonophyll. Evolutionary modifications have resulted in a close association of the fertile branch–the placental axis bearing the ovules–with the laminar part of the gonophyll. If the floral parts are ultimately derived from branch systems, the flower is a condensed and highly modified inflorescence and the term flower covers reproductive structures of angiosperms in various stages of condensation. This interpretation modifies the term flower to one referring to a biological unit rather than to a morphological one and makes it applicable not only to single flowers but also to more or less condensed inflorescences.

STRUCTURE

Arrangement of Flower Parts

The apical meristem of the flower usually ceases its activity after the reproductive structures have been initiated, an expression of the determinate type of growth. In certain groups of angiosperms considered to be primitive the determinate growth is less pronounced than in the more advanced families. In the primitive groups, the activity of the apical meristem is prolonged and therefore the number of floral parts is relatively large and indefinite. Moreover, these parts occur on a rather elongated axis, with sepals, petals, stamens, and carpels succeeding each other acropetally in the order named. The similarity between such a flower and a vegetative shoot is not difficult to visualize, especially if the flower parts are arranged helically.

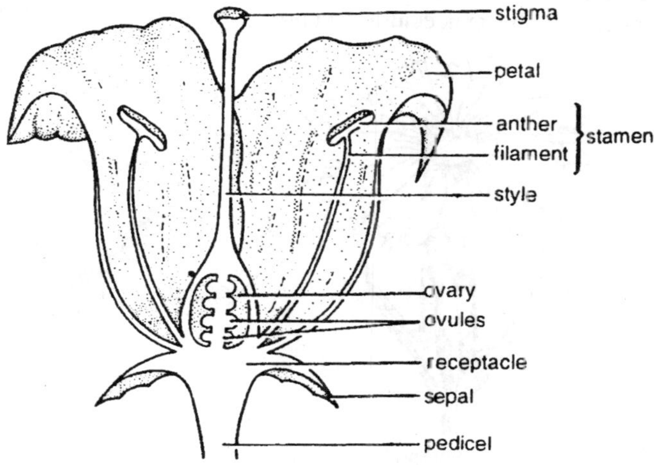

Fig. 13.1. Longitudinal section of a typical flower.

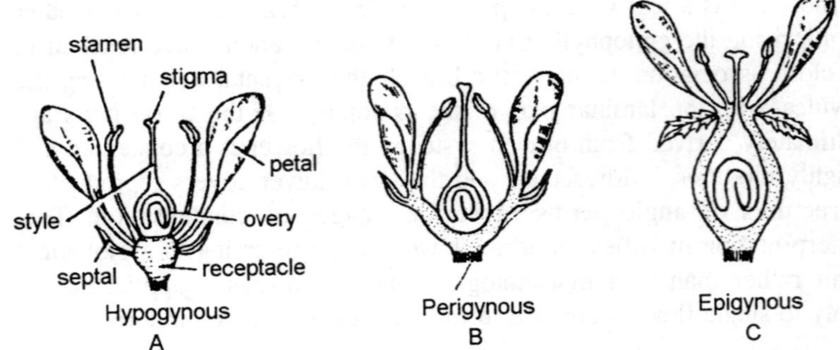

Fig. 13.2. Relative position of floral parts on the thalamus: A, hypogynous flower; B, perigynous flower.

In the more highly specialized flower types the growth period is shorter and the number of floral parts is smaller and more definite. Moreover, the shortening of the period of activity of the apical meristem is associated with a development of distinguishing characteristics that obscure or even efface the evidences of similarity between a flower and a vegetative shoot. Such characteristics are: whorled (or cyclic) instead of helical arrangement of parts; cohesion of parts within one whorl; adnation of parts of two or more different whorls; loss of parts; zygomorphy (bilateral symmetry) instead of actinomorphy (radial symmetry); and epigyny (inferior ovary) instead of hypogyny (superior ovary). The words synsepalous, sympetalous, and syncarpous are used to characterize flowers with united sepals, petals, and carpels, respectively. If the gynoecium occupies a position similar to that in

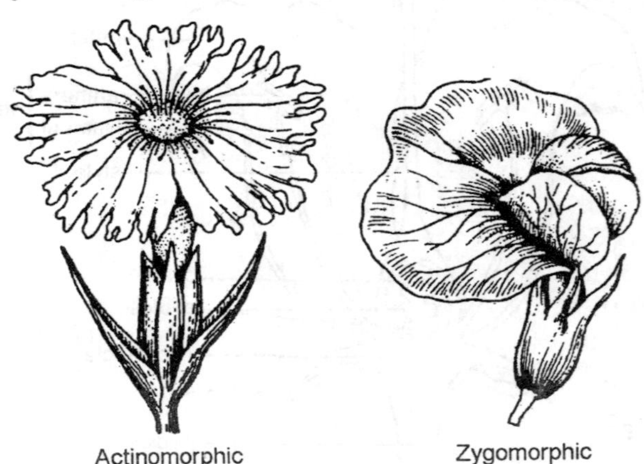

Fig. 13.3. Symmetry of flower.

an epigynous flower but is not adnate to the noncarpellary tissue, the flower is called perigynous and the ovary superior. Epigynous flowers (those with inferior ovaries) are especially difficult to interpret morphologically because the gynoecium is imbedded in noncarpellary tissue and appears to be inserted below the other floral parts.

The flowers of different degrees of specialization form an intergrading series of morphologic types. The degree of fusion of sepals, petals, stamens, and carpels varies widely, and the union is not necessarily equally pronounced in the different whorls of the same flower. The perianth may not be differentiated into calyx and corolla, or the sepals and petals may intergrade with each other. Transitional forms may also occur between the petals and the stamens. The flower may lack certain parts. If it lacks either the gynoecium or the androecium, it is called unisexual.

Vascular System

Investigations on the vascular system of the flower occupy a prominent place in the literature on the anatomy of the flower. A commonly accepted postulate is that the vascular system is conservative and, therefore, might be expected to reveal at least some of the evolutionary changes that have been obliterated in the external form. Thus, the vascular anatomy of the flower has been frequently studied to find explanations of the morphology of flowers, to obtain additional data for establishing taxonomic relations, and to construct evolutionary schemes.

The vascular system of relatively unspecialized flowers with superior ovaries is comparable to that of a vegetative shoot in which strands diverge into the lateral organs from an axial system of bundles. Many authors draw a complete parallel between the patterns of vascularization in the shoot and the flower, and apply the concepts of stele, traces, and gaps with reference to both structures. If the receptacle is elongated, the floral parts may be arranged according to a phyllotactic pattern correlated with an orderly arrangement and interconnection of vascular traces. But the shortness of internodes characteristic of so many flowers, the union of parts, the epigynous condition, and various other modifications in the interrelations of floral parts make the vascular system of flowers less regular than that of vegetative shoots and obscure the relation between the vascular system of the axis and that of the floral organs.

In a hypognous flower with relatively little fusion of parts, the vascular system may be readily depicted in terms of traces to the various floral appendages. The pedicel shows a cylindrical vascular region enclosing a pith and delimited on the outside by the cortex. In the receptacle or torus (the part of the axis bearing the floral parts), at the level of attachment of sepals, traces diverge into these appendages. Each sepal frequently has as many traces as a foliage leaf of the same plant. Above this level, traces diverge into the corolla, one or more to each petal in dicotyledons, one to many to each tepal in the monocotyledons.

Still higher, the traces to the stamens become discernible, predominantly one to each stamen, and finally is found the carpellary supply. The frequent number is three traces to each carpel, one median and two lateral, but more than three traces have been recorded. Small branches of carpellary vascular bundles, often derived from the laterals, connect the carpellary system with the ovules. The placental bundles may also be branches from the dorsal bundles, as in some Ranales, or be independent from the carpellary traces. The vascular system is prolonged into the style.

Some of the common modifications in the arrangement of the vascular system are associated with the fusion of floral parts. In many flowers the lateral bundles of the adjacent carpels are fused with each other. Similar fusions occur in other floral organs. The reduction in the numbers of traces and bundles may also occur if some of them do not develop.

The vascular system of epigynous flowers shows additional complications related to the apparently basal position of the gynoecium. The vascularization of such flowers has been frequently studied with the result that some authors have developed rather definite ideas on the nature of the noncarpellary tissue enclosing the gynoecium. In most epigynous flowers this tissue is interpreted as appendicular in origin, composed of the bases of sepals, petals, and stamens that underwent a concrescence during the evolution of the flower. The vascular system is thought to reflect this structure in that the bundles pertaining to members of different whorls are variously fused but all show the usual orientation of xylem and phloem. In some epigynous flowers (Calycanthaceae, Santalaceae, and probably Juglandaceae), however, the ovary is said to be partially enclosed in receptacular tissue. The vascular bundles are prolonged from the axis to the level below the insertion of floral parts, other than the carpels, where traces to these parts diverge. The main bundles, instead of ending here, continue farther from the periphery in a downward direction–with a corresponding inverse position of the xylem and the phloem–and at lower levels give branches to the carpels. This orientation of the vascular system is interpreted as a result of an invagination of the axis (actually intercalary growth of the tissue enclosing the gynoecium).

In general, the vascular elements in the bundles of the flower are comparable to those in foliage leaves. The tissues are mostly primary, although some secondary growth may occur later, during fruit development, particularly in the pedicel. The vascular system of the sepals, the petals, and the carpels is more or less elaborately ramified. Stamens rarely show a branched vascular system. In general character the venation of the perianth parts of monocotyledons and dicotyledons shows distinctive characteristics similar to those in the foliage leaves of these two groups of plants. Perianth parts of many flowers exhibit an open venation.

Sepal and Petal

The sepal and the petal are essentially leaf-like in form and anatomy but generally simpler in detailed structure than a foliage leaf. They consist of ground parenchyma, often called mesophyll, a vascular system permeating the ground tissue, and epidermal layers on the abaxial and adaxial sides. Crystal-containing cells, idioblasts, and laticifers may occur in the ground tissue or in association with the vascular elements. The sepals of Geraniaceae have a thick-walled hypodermis with a druse in each cell. The sepals are commonly green. The chloroplast distribution in the sepals depends on their position. If the sepals are upright and are closely applied to the petals, most chloroplasts are on the abaxial side; if the sepals are recurved, the chloroplasts are most abundant on the adaxial side. The mesophyll is rarely differentiated into palisade and spongy parenchyma. Commonly it is simple in structure and consists of approximately isodiametric cells loosely arranged into a lacunose tissue. The epidermis of the sepals shows a deposition of cutin and a development of stomata and trichomes similar to those on the foliage leaves. The vascular system resembles that in the leaves but is less elaborate.

Petals show a wider variety of shapes than the sepals and are usually distinguished from the sepals by their color. The vascular system may consist of one or several large veins and a system of small veinlets. The patterns formed by these veinlets vary greatly. Commonly the veinlets are dichotomously branched. The mesophyll is few cells in thickness, except in flowers with fleshy corollas. The tissue is parenchymatic with the cells either closely packed or loosely arranged.

The epidermis of petals shows certain peculiarities in the shape of cells and in the structure of cuticle. The anticlinal cell walls may be straight or wavy or may bear internal ridges. The undulation and ridging vary widely in degree of expression in different plants. In some the anticlinal walls are only slightly wavy; in others the undulations are so deep that the cells are star-like in shape as seen from the surface. The ridges, which arise through a localized centripetal growth of cell walls, may appear as small buttons in sectional views, or as long bars, straight or bent, solid or hollow. The degree of waviness or ridging may vary in the same petal. For instance, the anticlinal walls are usually straight at the base of the petal and along the veins, even if they are wavy elsewhere. Frequently, the undulate walls are restricted to or are more pronounced on the lower side.

Intercellular spaces may develop in the epidermis in connection with the differentiation of ridges. In some species the two wall layers composing a ridge split apart and the space between the two layers becomes filled with air. These spaces are open toward the interior of the petal but appear to be closed with the cuticle on the exterior. Ridged walls occur mainly in the dicotyledons, although they have been found in some members of the Liliaceae also.

The tangential walls of the epidermis may be horizontal or convex to various degrees. The inner tangential wall is commonly slightly convex over the entire extent. The outer wall is often strongly convex, or it may bear one or more capitate or cone-shaped papillae (*Viola*, *Nasturtium*). The papillose structure is more common in the adaxial epidermis than in the abaxial and does not develop at the base of the petals. Various trichomes may occur on the petals, usually similar to those found on the leaves of the same plants. The stomata which occur on the petals either resemble those on the foliage leaves or are incompletely differentiated.

The cuticle of the corolla is rarely smooth. Commonly it is striated, and the lines form various patterns in different plants. The development of these patterns has been suggested as resulting from two phenomena; first, a temporarily excessive production of cutin and the consequent increase in surface and folding of the cuticle; second, a stretching of the cuticle and a reorientation of the initial folds by cell extension. Cuticular patterns formed by folds were seen also at the ultrastructural level.

The color of petals is caused by the presence of chromoplasts or pigments in the cell sap. The pigment color is usually modified by

acids and other components of the cell sap. Starch is often formed in young petals. Volatile oils imparting the characteristic fragrance to the flowers commonly occur in the epidermal cells of the petals, sometimes in parts of flowers differentiated as osmophors.

Stamen

The well-known type of stamen, with a single-veined filament bearing at the upper end a two-lobed, four-loculed anther, is phylogenetically an advanced structure. As was mentioned before, among the Ranales leaf-like stamens are found. In the least modified form, such stamens have three veins and bear the microsporangia on the abaxial surface between the midvein and the lateral veins. The reduction of the three veins to one is apparently a concomitant of the reduction

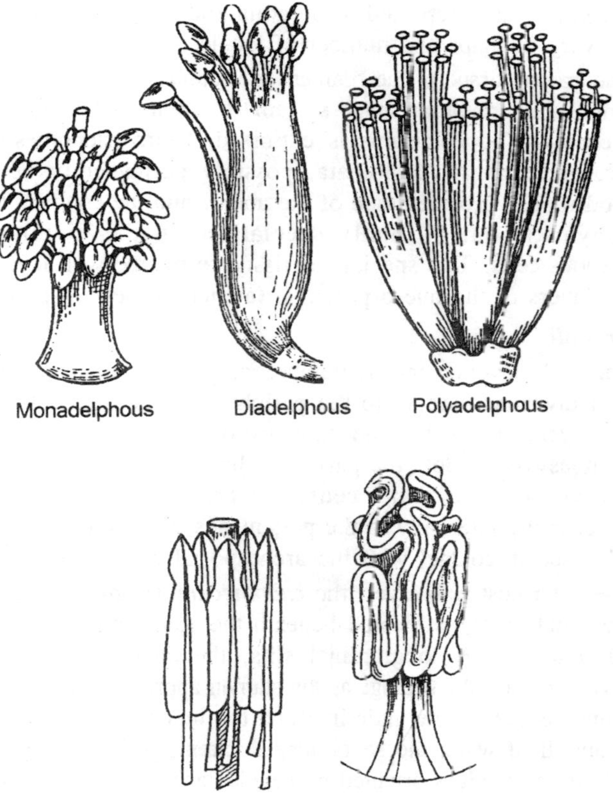

Fig. 13.4. Cohesion of stamens.

in width of the sporophyll, and particularly of the modification of the base of the sporophyll into a filament. The presence of a single vascular bundle is the prevailing condition in angiosperms. An extensive survey has shown that 95 per cent of angiosperms have a single vascular bundle in the stamen. This strand traverses the filament and may end at the base of the anther or may be prolonged into the tissue between the anther lobes, the so-called connective, terminating blindly near the apex. The vascular bundle is not connected by any vascular elements with the sporogenous tissue, but if the ground parenchyma of the anther develops secondary thickenings the cells in the vicinity of the sporogenous tissue remain thin walled and there are also vertical bands of similar thin-walled cells interpolated between the vascular strand and the anther lobes. The vascular bundle of the anther may be amphicribral in dicotyledons, but is reported to be collateral in monocotyledons. The anthers vary in shape and number of locules.

The ground tissue of the filament is vacuolated parenchyma without a prominent intercellular-space system. If often contains pigments in the vacuoles. The epidermis is cutinized, bears trichomes in some species, and may have stomata, possibly permanently open as in hydathodes. The ground tissue of the anther and the connective is also parenchymatic but is highly specialized in the vicinity of the sporogenous cells. This specialized tissue forms the wall layers or the parietal layers of the microsporangia (anther locules, or pollen sacs).

Anther wall

The wall layers vary in number and are established through a series of divisions parallel to the periphery of the anther locule. The parietal layers facing the epidermis are ontogenetically related to the sporogenous tissue. Both the parietal cells and the pollen mother cells arise from the same initial cells, the archesporial cells. The wall layers occurring internally to the pollen sacs, however, arise from the ground tissue in contact with the archesporial cells.

The outermost wall layer, the *endothecium* (from the Greek words for inner and case), is located beneath the epidermis. In anthers that open at maturity by longitudinal slits, the endothecium commonly develops secondary thickenings as the stamen approaches maturity. These thickenings occur on the anticlinal and the inner tangential cell walls. In the anticlinal walls the secondary thickenings frequently have the form of strips or ridges oriented perpendicularly to the epidermal layer. The cell walls facing the sporogenous tissue may have uniform or irregular thickenings. Because of these thickenings the endothecium is often called the fibrous layer. The pattern of the thickening is variable and may be useful in taxonomic studies. The endothecium also may have uniformly thick walls. The protoplasts either disappear as the cell layer completes its development, or they remain alive until the pollen is shed. Wall thickenings similar to those in the endothecium may develop rather generally throughout the ground parenchyma of the anther.

The innermost of the parietal layers is the *tapetum* (from the Greek, carpet). The tapetal cells are characterized by densely staining protoplasts and prominent nuclei. The nuclei show various behavior in different plants. In some, they do not divide after all the tapetal cells have been formed; in others, one or more nuclear divisions occur without being followed by cytokinesis so that the cells become bi- or multinucleate (*Lactuca*, *Taraxacum*). Sometimes the nuclear divisions are not carried to completion: the chromosomes divide but do not form separate nuclei. Such behavior results in polyploidy of the tapetal nuclei. In general, tapetal cells become richer in chromatic material. This occurs through either multiplication of nuclei, restitution of nuclei during various stages of mitosis, or endopolyploidy. The tapetal layer attains its maximal development at the tetrad stage in microspore formation. In some angiosperms the tapetum remains as a discrete layer–apparently functioning as a secretory tissue–until the pollen is mature. In many others, however, the cell walls disintegrate and the cells assume the appearance of plasmodial masses. The latter gradually disintegrate as the pollen develops.

The tapetum is apparently concerned with the nutrition of the pollen mother cells and young microspores. Ultrastructural studies suggest that the material of the outer pollen wall (exine) is synthesized in the tapetum. But autoradiographic methods have failed to show any relation between the DNA of the tapetal nuclei and that of the microspores.

The parietal layers intervening between the endothecium and the tapetum frequently are crushed and destroyed so that, after the maturation of the pollen and the disintegration of the tapetum, the anther locule is bordered on the outside only by the epidermis and the endothecium.

In many plants the release of the pollen occurs through dehiscence (from the Greek, to yawn), that is, spontaneous opening of the anther. The opening, or *stomium*, may be a longitudinal slit located between the two pollen locules of each half of the anther. Before the dehiscence, the partition between the two locules of the same anther lobe may break down. After this event, only one cell layer, the epidermis, separates the locule from the outside in the region of dehiscence. This part of the epidermis consists of particularly small cells and is easily broken when the pollen is mature. Another common type of stomium is oriented transversely near the apex of the anther lobe. When such a stomium is formed, the apex of each anther lobe separates like a cap and leaves a pore (poricidal dehiscence; many Ericaceae, *Solanum*). Pores may also be formed laterally. It has been suggested that the long, slit-like stomium is a more primitive character than one shaped like a pore. In species of *Senna*, the anther is provided with lateral sutures that do not serve as stomia. The epidermal cells along these sutures divide and apparently serve as plugs. Dehiscence occurs at the sterile tip of the anther where short linear stomia are present. The tissue located between these stomia and the pollen sacs breaks down and the pollen emerges through the stomium. In some plants, anthers do not dehisce but open by an irregular breaking and exfoliating of tissue fragments.

Pollen

The development of the sporogenous tissue in the anther involves certain characteristic phenomena of wall formation. The cells that eventually undergo meiosis, the pollen mother cells, are closely packed in their early stages of development. During meiosis these cells usually separate from one another, and the protoplasts round off and become enclosed in a thick gelatinous wall that has been identified as callose. This wall is designated as the pollen mother cell wall or special wall. The megaspore mother cells may assume a distinctive arrangement in the pollen sac. In Gramineae and Cyperaceae, for example, they appear, in transactions of the anthers, like sectors of a circle. As is well known, normal meiosis results in the formation of four nuclei, the microspore nuclei. Each nuclear division may be followed immediately by cytokinesis (successive formation of walls), or the four protoplasts may be walled off simultaneously at the end of meiosis (simultaneous formation of walls). The first type of division is particularly common in monocotyledons, the second in dicotyledons. The simultaneous wall formation may occur by development of cell plates or by furrowing.

The first wall delimiting the microspore protoplasts from each other is of the same material, callose, as the special wall around the entire tetrad of microspores. Later, each microspore forms its own wall, the *sporoderm*.

According to a submicroscopic study of *Tradescantia* anthers, mature pollen grains have abundant mitochondria, dictyosomes, and endoplasmic reticulum; in the earlier stages these entities are not fully differentiated. In younger cells leucoplasts with starch are present; later the plastids become scarce. The number of nuclei in the mature pollen grains is of taxonomic significance and is also associated with certain physiological characteristics of the grains.

The sporoderm is usually described as consisting of two layers, the *exine* (outer wall) and the *intine* (inner wall). The exine is differentiated into a sculptured *ektexine*, or *sexine*, and a nonsculptured *endexine*, or *nexine*. Some workers recognize a third layer, the *medine*, located between the exine and the intine. The exine consists mainly of a lipoidal substance *sporopollenin*, which is less soluble than cutin or suberin. The research on the structure of walls of pollen grains and spores is highly technical and is designated by a special term, *palynology*.

Most pollen grains are *aperturate*, that is, provided with pores or furrows (colpi, sing, colpa). These apertures are not actual openings but places where the exine is very thin and the intine well developed. The pollen tube emerges through the aperture during the germination of the pollen grain apparently by pushing aside the intine. The apertures are also regarded as the flexible parts of the sporoderm that permit the change in shape and size of the pollen grain caused by varying water content. The number of apertures varies from one to many.

As seen from the surface, the exine of many species has spines, depressions, arcolations (division into distinct spaces), and other types of ornamentations. These external markings and the shape of the pollen grains are characteristics that may be utilized in taxonomic studies. Ultrastructurally the ektexine often shows a porous structure.

The intine varies in thickness and in a given species is more or less thickened in the aperture region. It has no ornamentations. The intine consists mainly of polyuronides or a mixture of polyuronides and polysaccharides but its inner part contains also cellulose. In conifers the outer intine is reported to contain callose.

When the pollen tube emerges from the pollen grain it grows by addition of wall material at its apex. The pollen tube wall contains cellulose and is cutinized. It has also been described as having an outer lamella of pectin and an inner lamella of a mixture of callose and cellulose. The cytoplasm accumulates at the tip of the tube and may completely disappear from its basal part. In such instances, the older parts of the elongating pollen tube are successively sealed off by plugs of callose. Accumulation of callose is intensified under conditions of incompatibility, possibly in relation to the reduced growth rate of the pollen tube. In plants forming no plugs of callose (*Fagopyrum esculentum*) the whole tube probably has a thin layer of cytoplasm in addition to the accumulation at the apex. Cytoplasmic streaming has been observed in pollen tubes, even in parts sealed off by the callose plugs.

Carpel

Relation to the gynoecium

Whatever may be the phylogenetic origin of the carpel, in many extant angiosperms with superior ovaries it resembles a leaf. As mentioned before, the carpels may or may not be united with other carpels. If the carpels are free, the gynoecium is apocarpous, if they are united the gynoecium is syncarpous. An apocarpous gynoecium may have a single carpel (*Prunus*, *Leguminosae*).

The carpel of an apocarpous gynoecium appears as a leaf-like folded structure, differentiated, in the specialized condition, into a basal fertile part, the ovary, and an upper sterile part, the style. According to an older concept, the folded carpel has infolded or involuted margins, that is, margins turned toward the interior of the folded carpel, and these margins bear the placentae that give rise to the ovules. A later view, based on studies of the woody Ranales, states that in the primitive form the carpel is a conduplicately (from the Latin, doubling) folded structure, that is, a structure folded lengthwise without involution of margins. Such a carpel shows laminar placentation; the ovules are borne not on the margins but on the inner (ventral) surface, more or less distant from the margins, and may be vascularized by connection with the dorsal bundle rather than the ventral. The apparent involution and marginal placentation are thought to have resulted from phylogenetic change in the ontogeny of the carpel, a decrease in the extension of its folded adaxial part. An argument offered in opposition to the concept of conduplicate carpel states that the surfaces coming in contact in the folded carpel are not ventral but marginal. The evidence on the phylogenetic reduction of the adaxial margins does not support this argument.

The interpretation of the phylogenetic differentiation of the dicotyledonous carpel into ovary, style, and stigma has been consequentially developed with reference to the carpel of the woody Ranales. The unspecialized carpel is a styleless, unsealed, conduplicate leaf-like structure with laminar placentation. The stigmatic tissue occurs on the free margins of the carpel, on its inner surface, and at times also on the outer surface. Successive phylogenetic stages involve closure of the carpel, reduction in the number of ovules and their restriction to the lower part of the carpel (the ovary), and differentiation of the upper part into the style with a stigma localized on its apex. The closure of the carpel occurs through a growing together (concrescence) of the ventral surfaces along the margins that are in contact with each other. The concrescence is ontogenetic and may leave a conspicuous suture; or the union may be so complete that the evidence of a suture is partly or entirely obliterated.

The evolutionary changes in the structure of the gynoecium of the angiospermous flower also involve various manners of union of carpels of the same flower. The carpels may become joined by their margins to the receptacle, or they may grow together laterally in a folded closed condition, or they may become laterally united in a folded open condition. The union of carpels may occur during their ontogeny or they may grow as a unit structure and are then interpreted as congenitally fused, that is, fused from their inception.

The manner of union of carpels is related to differences in internal structure, such as number of locules in the ovary and the arrangement of placentae, the placentation. Each carpel typically has two placentae. If the carpel has a congenitally united lower part, the placentae may fuse in this part and the placental region then has the shape of a U. In syncarpous gynoecia, the junction of carpels in a folded condition may result in an ovary with as many locules as there are carpels and with the placentae arranged around a central column of tissue (*axile placentation*). If the carpels are united with each other in an open condition, the ovary is usually not divided into locules and the ovules are borne on the ovary wall or on extensions from it (*parietal placentation*). The parietal placentation is considered to have evolved from the axile.

Various deviations from the basic structures of the ovary just described are encountered in different angiosperms. Division of the ovary into compartments may occur in other ways than by the folding of carpels. The placentae may be borne upon a central column of tissue not connected by partitions with the ovary wall (*free central placentation*), or may occur at the very base of a unilocular ovary (*basal placentation*). The free central placentation apparently results or has resulted from disappearance of partitions in terms of either ontogeny or phylogeny. Syncarpy and apocarpy may be present in the same pistil if the carpels are joined only at the base. The type of syncarpy may also vary in different parts of the pistil since the individual carpels may have a congenitally united lower part and an open upper part; the type of concrescence of carpels may be different in the two parts.

The controversial views on the nature of the carpel are reflected in the interpretation of the placenta. According to one of the common concepts, the column of tissue bearing the ovules in ovaries with axile or free central placentation may be entirely carpellary or partly axial and partly carpellary. Presence of vascular tissue other than that of the carpels in the central column is one of the evidences used to identify the axial nature of the central tissue. In both situations the placentae would be part of the carpels. When the ovules are borne on the carpels the species is said to be *phyllosporous*. The opposite view, chiefly concerning species with central and basal placentations, considers that the placentae and ovules may be cauline structures. When the ovules are borne on cauline tissue the species is designated as *stachyosporous*.

The structure of inferior ovaries also presents problems of interpretation, especially with regard to the questions whether any carpellary tissue lines the lower part of the ovarian cavity and whether the extracarpellary tissue is axial (receptacular) or appendicular (floral tube). As was mentioned before, the use of vascular anatomy has led to the concept that the cup (hypanthium) of extracarpellary tissue is appendicular in some plants, receptacular in others. Some authors, however, see no distinction between the inferior ovaries and prefer to consider the cup as uniformly receptacular.

The ovary wall is not highly differentiated before and during anthesis (time when fertilization takes place in the flower). It consists largely of parenchyma and vascular tissue and bears a cuticularized epidermis on the outer surface. In the Compositae, calcium oxalate crystals occurring in the cells of the ovary wall were found to differ according to species. The ovary wall undergoes more or less profound changes during the development of the fruit and then may show marked specializations.

Style and stigma

The development of the style occurred as a concomitant of sterilization of the apical part of the carpel. In an apocarpous gynoecium each carpel usually has one simple style. In syncarpous gynoecia the styles of the component carpels may be variously united with each other. The carpels may be united only at their bases, leaving the styles free, or partly so. In highly modified flowers the carpels are united from base to apex and form a gynoecium with a single ovary, style, and stigma. If the styles are free, the stylar portions derived from the individual carpels are often called style branches, a designation giving an erroneous concept of the structure of the compound style; the branches are morphologically entire styles. The term *stylode* has been proposed as a replacement for stylar branch.

The style and the stigma have structural and physiological peculiarities that make possible the germination of the pollen and the growth of the pollen tube from the stigma to the ovules. On the stigma the protoderm differentiates into a glandular epidermis with cells rich in cytoplasm, often papillate in shape, and covered with a cuticle. This epidermis excretes a sugary liquid. Thus, the stigma resembles a nectary in structure and function. The cells beneath the epidermis may be as rich in cytoplasm as the epidermis, and then they constitute a part of the glandular tissue. In many plants, the stigmatic epidermal cells develop into short, densely crowded hairs (cherry, bean) or into long, branched hairs (grasses and other wind-pollinated plants).

An outstanding feature of the organization of the carpel is that the stigma is connected with the inferior of the ovary by a tissue

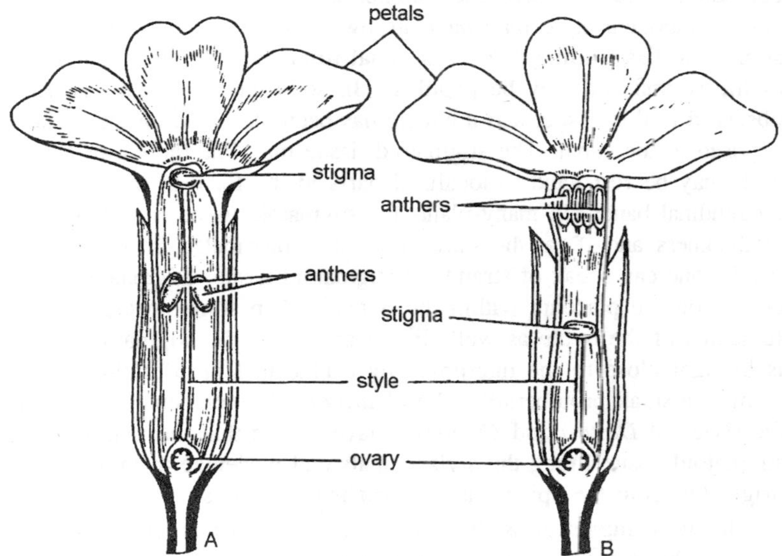

Fig. 13.5. Heterostyly: A, long styled flower; B, short styled flower.

cytologically similar to the glandular stigmatic tissue. This tissue is interpreted as a medium facilitating the progress of the pollen tube through the style and supplying the developing pollen tube with food. It is commonly called conducting tissue, a term easily confused with that referring to the vascular tissue. The terms transmitting tissue and pollen-transmitting tracts serve as substitutes. In the following discussion this tissue is referred to as *stigmatoid tissue* on the basis of its apparent cytologic and physiologic similarity to the tissue of the stigma.

The carpels of the more primitive dicotyledons do not show a differentiation into stigmatic and stigmatoid tissues, for, as was stated previously, the surfaces of the flaring margins and the inner surface of the open carpel are lined with stigmatic glandular hairs. With the increase in specialization of the carpels, characterized by their gradual closure and the development of the style, the stigma proper became restricted to a part of the style, but the continuity of the stigmatic tissue with the placentae was maintained. The internal glandular surfaces became modified into pollen-transmitting, or stigmatoid, tissue.

In relation to the variation in degree of concrescence of carpels and in methods of growth of the styles, the styles may be open or solid in both the apocarpous and the syncarpous gynoecia. The open styles are described as having a canal. In a syncarpous gynoecium the compound style may have one common canal (*Viola*, *Erythronium*), or each component style may have its own canal (*Lilium*, *Citrus*). The stigmatoid tissue lining the stylar canal resembles the glandular tissue of the stigma and may be papillose. In some plants starch has been observed in this tissue, and a cuticle has been identified on the surface exposed to the canal. The stigmatoid tissue may line the entire canal, or it may be restricted to localized parts in the form of one or more longitudinal bands. In many plants the stigmatoid tissue is several cells in thickness and if, at the same time, it is distributed in longitudinal bands, one can speak of strands of stigmatoid tissue. Stigmatoid tissue occurs on the placenta within the ovary and in some species on the funiculus of the ovule as well. In certain plants the stigmatoid tissue is brought close to the micropyle by a placental proliferation in the form of a small protuberance, the *obturator*. Developmental studies on the styles of *Datura* and *Cucurbita* have shown that the multilayered stigmatoid tissue lining the stylar canals and placentae in these plants originates from the epidermis by periclinal divisions.

In most angiosperms the styles are solid; that is, they have no canals. The stigmatoid tissue is present, nevertheless, usually in the form of strands of considerably elongated cells staining deeply with cytoplasmic stains. If the gynoecium with a single solid style is syncarpous, the stigmatoid tissue of the style forms several strands. Commonly the stigmatoid tissue has a course independent from that of the vascular bundles, but it may be associated with the bundles (*Zea*).

Syncarpous gynoecia may have openings that enable a pollen grain germinating on a stigma of one of the styloids or any part of the stigma of a single style to reach any part of the ovary rather than only the one to which a given stigma or part of stigma is related. The opening (*compitum*) may consist of a canal pore, or split in the septum between locules. In unilocular ovaries with parietal placentation the crossing over of the pollen tube may occur in the style itself. Some syncarpous gynoecia have no compital structure and, therefore, function like apocarpous gynoecia with regard to pollination.

With regard to the possible factors that might direct the growth of pollen toward the ovule, some workers stress the evidence that there is a chemotactic attraction between the pollen tube and the tissues of the stigma and of the ovule; others consider that the structure of the stigmatoid tissue and its distribution in the pistil are sufficient to account for the direction of growth of the pollen tube. The presence of pollen tubes in or on the stigmatoid tissue has been repeatedly ascertained in various plants.

The relation of the pollen tube to the stigmatoid tissue is somewhat different in styles with and without open canals. In the former, the pollen tubes may have an entirely superficial course. After the germination of the pollen grain on the stigma the pollen tube grows among the papillae or hairs or on the surface of the nonpapillate cells. The course in the stylar canal is essentially the same as on the stigma. Frequently the cuticle disappears in the stylar canal before pollination, and the walls of the glandular tissue become swollen and soft. The pollen tube may also penetrate the lining of the stylar canal to somewhat deeper layers and proceed there by growing between cells.

If the style is solid, the pollen tube usually passes through the stigmatoid tissues by intercellular growth. Reports that pollen tubes penetrate the cells themselves are not well substantiated. In grasses, the pollen tube may take an intercellular course on the stigma itself. As was mentioned previously, the grass stigma commonly bears long hairs. These may be multicellular columns, both vertically and horizontally. The pollen tube penetrates into the interior of the column of cells and proceeds from there into the stigmatoid tissue of the style. After the pollen tube reaches the ovarian cavity, it follows the stigmatoid tissue lining the ovary wall and the placenta and eventually comes in contact with the ovule.

The intercellular growth of the pollen tube appears to involve a digestion of the intercellular substance. In agreement with this assumption pollen tubes give a positive reaction for an enzyme capable of digesting pectic substances. The stigmatoid tissue, however, appears to undergo a partial weakening in its structure before the pollen tube passes through it. Its walls assume a swollen aspect (the tissue resembles collenchyma in this state), and the connection between cells is loosened, as demonstrated by the ease with which the tissue may be macerated. In fact, the walls appear as though they have been converted into a mucilage. When the pollen tube passes through the stigmatoid tissue, it occupies the space formerly filled with cell wall material. The protoplasts of the stigmatoid tissue may also become exhausted and sometimes even shrivel and die. Because of these relationships the entry of pollen tubes, even if these are very numerous, does not cause the expansion of the stigmatoid tissue. The pollen tubes may be said to replace some of the stigmatoid tissue.

The exhaustion of the protoplasts of the stigmatoid tissue by the pollen tube indicates an effect of chemical nature. In studies on Gramineae the pollen was found to have an effect on the stigmatic tissuè after a short period of contact, that is, even before germination: the cells of the stigma showed increased stainability of nuclei.

The stigmatoid tissue and the vascular bundles constitute the most specialized parts of the style. The ground tissue is parenchymatic, and the outer epidermis shows no peculiar features. It bears a cuticle and may have stomata.

Ovule

The ovule developing from the placenta of the ovary is the seat of formation of the megaspores (or macrospores) and of the development of the embryo sac (female gametophyte) from a megaspore. Sporogenesis, the development of the embryo sac, and the many variations in the details of these phenomena, have been the subject of numerous investigations and are not reviewed here. Concomitant with the development of the embryo from the fertilized egg, and of the endosperm from the product of the triple fusion (two polar nuclei and one sperm nucleus), the ovule develops into a seed. Histologically, the ovule is rather simple as compared with the resulting seed.

Commonly the ovule is differentiated into the following morphologic parts: the *nucellus* (from the Latin, small kernel), a central body of tissue containing some vegetative and some sporogenous cells; one or two *integuments* (from the Latin, covering) enclosing the nucellus; the *funiculus* (from the Latin, rope), the stalk by means of which the ovule is attached to the placenta. The size of the nucellus, the number of integuments, and the shape of the ovule are important distinguishing characteristics of ovules in different groups of angiosperms. If the nucellar apex points away from the funiculus, the ovule is termed *atropous* (synonym of *orthotropous*; a, not; *tropos*, turned, in Greek), that is, not turned. If the ovule is completely inverted so that the nucellar apex is turned toward the funiculus, it is called *anatropous*. Between these two extreme forms of ovules, there are several variously named intermediate ones with various degrees of curvature.

The ovule primordium arises from the placenta as a conical protuberance with a rounded apex. The first sporogeneous cell (archesporial cell) becomes evident, in the still undifferentiated protuberance, by its size and often also by denser appearance of its cytoplasm. This cell occurs beneath the protoderm at the apex of the primordium. Slightly below the apex the inner integument (or the single integument) is initiated by periclinal divisions in the protoderm. It arises as a ring-like welt and grows upward. With the appearance of the integument, the nucellus of the primordium becomes delimited as the part enveloped by the integument. The latter grows faster than the nucellus and encloses it partially or completely. Usually a narrow, canal-like opening remains at the top of the integument. This is the *micropyle*. The outer integument, if such develops at all, arises in the protoderm slightly below the inner integument and develops in a manner similar to that of the inner. It frequently does not reach the apex of the ovule in its upward growth. In the anatropous and other curved ovules the growth of the integuments is asymmetrical, being more pronounced on the side of the ovule which eventually becomes convex.

There is no agreement on the morphologic nature of the ovule and its parts. Some workers consider the ovule a foliar structure, others an axial. The nucellus is commonly regarded as the megasporangium, but the interpretation of the homology of the integuments constitutes a major morphologic problem.

The nucellus, the integuments, and the funiculus cannot be sharply delimited from one another either morphologically or cytologically. The nucellus is usually clearly outlined above the level where the integuments originate. From this level upward the nucellus and the integument (or integuments) have each their distinct epidermal layers. Below this level, that is, at the base of the nucellus, the nucellus and the integuments are confluent with the funiculus. The region of the ovule where all its parts merge with one another is called the *chalaza*.

The ovules of certain plants show considerable deviations from the structure just outlined. Some have no integuments, and others have more than two. The nucellus may be entirely confluent with the integuments, a condition supposedly different from that interpreted as absence of integument. Ovules may have other than the integuments, such as the *aril* (*Euonymus europ* from the funiculus, and the *caruncle* (*Ricinus*), an integument protuberance near the micropyle. In some plants the integument completely overgrows the nucellus that no micropyle remain; in other the integuments do not reach the apex of the nucellus.

The nucellus varies in size in different groups of plants. It may be so small that it comprises little more than an epidermis and the sporogenous tissue enclosed by it. In other plants a more or less massive vegetative tissue envelops the sporogenous tissue. The integuments also show variations in thickness. The thinnest integument is two cells thick; that is, it consists only of two epidermal layers. Sometimes the micropylar end is somewhat thicker in the two-layered integuments. Most angiosperms have two-layered integuments, although some dicotyledonous families have integuments of three and more layers. In relation to the size of nucelli the ovules are classified into *crassinucellate* (*crassus*, thick in Latin) and *tenuinucellate* (*tenuis*, slender in Latin). Crassinucellate ovules with two integuments are considered to be more primitive than the tenuinucellate with a single integument.

The ovules have a vascular system connected with that of the placenta. The presence of integumentary bundles is sometimes considered a primitive characteristic, but such bundles occur in the more specialized as well as in the less specialized angiosperms, and therefore their phylogenetic significance is uncertain. Most commonly there is a single strand ending in the chalaza with no prolongations into the integuments. In some species the bundle extends beyond the chalaza as a single strand or is variously branched. Such an intraovular system occurs in the integument. If two integuments are present, vascular tissue may be found in both integuments or only in the outer. Rarely does vascular tissue occur in the nucellus. The vascular tissue is primary and appears to be in a functioning state during the maturation of the seed.

The distribution of cuticles in the ovules deserves special mention because of their prominence and physiologic importance in the seed that develops from the ovule. The cuticles of the ovules and seeds are called by various names: cuticles, suberized membranes, semipermeable membranes, and fatty membranes. They are here referred to as cuticles in keeping with the most prevalent designation. Cuticles are reported to be present in ovules in relatively early stages of development.

The entire surface of the ovule primordium bears a cuticle. After the development of the integuments three cuticular layers may be distinguished: the outer, on the outside of the outer integument and the funiculus; the median, double in nature, between the two integuments; the inner, also double in nature, between the inner integument and the nucellus. In ovules with a single integument the median cuticle is absent. If the nucellus is small and its vegetative tissue is disorganized during the development of the embryo sac, the cuticle of the micropylar part of the nucellus may also be dissolved.

Parts of the ovule are disorganized during the development of the embryo sac, and the resulting materials are presumably utilized by the growing female gemetophyte. The vegetative tissue of the nucellus is partly or entirely resorbed. In the latter instance the embryo sac comes in contact with the inner epidermis of the integument. Large nucelli may be partially retained, and in some plant groups they form a storage tissue (*perisperm*) in the seed (Centrospermae). The nucellar epidermis is sometimes highly resistant and may proliferate into a nucellar cap with relatively thick walls (*Allium*). The integuments undergo certain histologic changes or are disorganized to varying degrees. Particularly common is the differentiation of the inner epidermis of the integument into the so-called nutritive jacket or *integumentary tapetum* consisting of deeply staining cells elongated perpendicularly with reference to the surface of the embryo sac. Such differentiation is characteristic of families in which the nucellus is early disorganized and the integument comes in contact with the embryo sac (Sympetalae). The physiologic significance of the integumentary tapetum is not agreed upon, and it might be variable. Some connection with the nutrition of the embryo is suggested by the disintegration of the ovule tissue located next to the tapetum and the persistence of the tapetum until the contents of the embryo sac complete their development.

Origin and Development

The change from vegetative to reproductive activity in the apical meristem follows a sequence that is determined by the nature of the plant. Herbaceous annuals pass, during one season, through an uninterrupted sequence of vegetative growth, floral initiation, and floral development. Woody species, at least in the North Temperate zone, commonly initiate the flowers in one season and complete their development during the next. The degree of differentiation that the flowers attain before the end of the first season is highly variable. Floral initiation is affected by external factors, but only the limits of reactivity of the plant to a given environment, for example, characteristic responses to length of day and produce flowers under specific combinations of these two factors.

Flowers arise at the apex of the main shoot, or on lateral branches, or on both. The lateral branches may form further branches of various orders before producing flowers. In different angiosperms the grouping of the flowers, called inflorescences, are highly variable and bear special names. The formation of all types of inflorescences involves, in the activity of a given apical meristem, a cessation of the vegetative stage and the initiation of the reproductive stage. Frequently, the first visible sign of determination of the flowering stage is the enhanced development of axillary buds. In species with cymose inflorescences a change from alternate, five-ranked arrangement of leaves to the single-ranked arrangement of floral primordia occurs at the beginning of the reproductive stage.

Organogenesis

Much can be learned about floral development by comparing flowers in different stages of development in material dissected under magnifications of moderate degrees. Payer (1857) employed this method in his classic comparative study of organ development in flowers, and in modern times it has been applied with particular success to investigations of floral differentiation in the Gramineae. A correlation of the observations on dissected material with those on flowers sectioned with a microtome gives a rather comprehensive picture of the main phenomena in the development of the specific form of flowers and their parts.

Depending on the structure of the flower, the parts may appear in acropetal order at successively higher levels like the leaves on a vegetative shoot (*Ranunculus*), or the parts of a given kind may arise at the same level or nearly so (*Capsella*). In the former instance the floral parts are arranged helically; in the latter they are in whorls (cycles). If the parts arise in a helical sequence, the helices of the various parts are usually not continuous with one another. The calyx

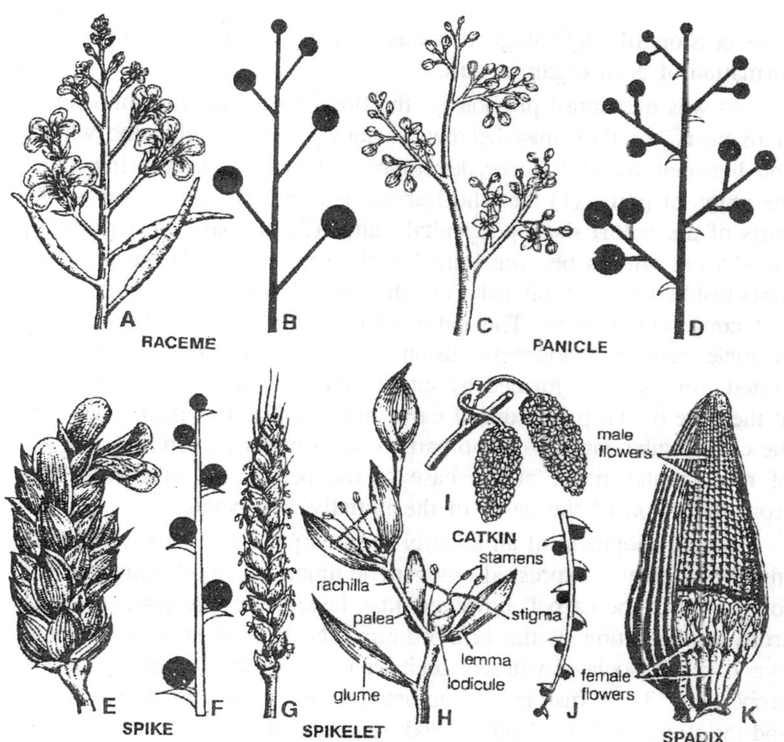

Fig. 13.6. Racemose inflorescence: A-B, raceme of brassica; C-D. panicle of Rhus; E-F. spike of Adhatoda; G-H. spikelet of Triticum; I-J. catkin of Morus; K. spadix of Musa. B, D, F H and J are diagrammatic representations.

members, however, may appear along helices that are continuation of those of the foliage leaves. The flower parts either arise in a continuous acropetal sequence of sepals, petals, stamens, and carpels, or else this sequence is more or less modified. In *Capsella*, for example, the stamen and carpel primordia appear before those of the petals. There may be a difference in the rhythm of development of floral parts. In Papaverceae, for example, the sepals arise considerably in advance of the other parts. Petals, stamens, and carpels appear in rapid succession and overlap one another in timing of their origin.

The successive formation of the different floral parts–as contrasted with that of similar parts during vegetative growth–apparently is governed by complex determination phenomena involving, among other mechanisms, those of hormonal balances. Surgical experimentation with developing flowers of *Primula* indicates that the flower passes through a succession of physiological states which permit and regulate the formation of each organ in turn.

As was mentioned previously, the floral parts may remain discrete at maturity, or they may become variously united within the whorls and between whorls. Three developmental patterns may bring about the union of parts: (1) the whorl arises as a unit structure; that is, the parts of the whorl show congenital unity; (2) the parts of a whorl or of adjacent whorls become joined during ontogeny; (3) the union of parts results from a combination of the two phenomena, the ontogenetic and congenital unions. The calyx and corolla tubes in *Datura*, for example, arise by ontogenetic fusion. Those of *Frasera* are congenitally united, for they are formed by intercalary growth of a ring of tissue at the base of the primordia of calyx and corolla. In *Vinca*, however, the corolla tube consists of two parts, one formed by intercalary growth of receptacular tissue at the base of the petals, the other resulting from the union of the bases of the initially free petals.

The development of an initially open carpel into a closed structure involves a clearly expressed ontogenetic union of carpel margins. The lower part of the carpel, however, may have a sac-like seamless form from the inception of the primordium. The formation of syncarpous gynoecia is associated with congenital and ontogenetic union in varying proportions. There may be also an ontogenetic union between the carpels and the stamens. On the other hand, the perianth parts and the stamens may originate together from unit primorida and become distinct during later growth.

The features discussed above may be elucidated by means of specific examples of floral development. The flower of *Allium cepa* (onion) is relatively unspecialized in having an undifferentiated perianth of free parts and a superior ovary. Its carpels are united, however. The six-parted perianth consists of two whorls of tepals, an outer and an inner. The six stamens occur in the axils of the six perianth members. The three carpels are united into a gynoecium with a three-loculed ovary and an axile placentation. The style is thin and has a slightly three-lobed stigma. An individual flower is a globose protuberance before the flower parts appear. The outer three tepals arise first. The stamens in the axils of these tepals arise simultaneously with the tepals and from the same primordia. The outer tepals and the associated stamens arise in a clockwise direction. The inner tepals and the stamens subtended by them also arise together, but in a counter-clockwise direction. With further growth, the tepals overarch the stamens. When this stage is reached, the carpels are initiated. They occur within the inner staminal whorl in alternation with its members. At first they project over the surface of the receptacle in the form of three horseshoe-shaped welts of meristematic tissue. Then they grow upward and toward the center where their margins meet and fuse. The compound style is formed by apical growth of the three carpels, the three parts uniting completely. The base of the style eventually appears deeply imbedded in the center of the ovary because the carpels bulge upward during the differentiation of the ovules. The ovules are initiated before the carpel margins fuse. They are anatropous and have two integuments.

The flower of *Lactuca sativa* (lettuce) may be used to illustrate the growth of a highly specialized flower, one with an inferior ovary (epigynous flower) and a zygomorphic sympetalous corolla. Lettuce belongs to the Compositae in which the flowers occur in capitate (head-like) inflorescences. The individual flowers arise acropetally on the flattened receptacle, so that the outermost flowers of the head are the oldest, the innermost the youngest. In an individual flower, the petal lobes appear first, as five protuberances on the margin of the floral primordium. However, immediately after their appearance they are thrust upward by intercalary growth of a ring of tissue upon which the corolla is inserted. As a result of this growth the central part of the flower primordium becomes cup-shaped. The stamens, which are initiated after the corolla, seem to be inserted below the corolla, but actually they occur closer to the center or apex of the flower than the other floral parts. The pappus, which is interpreted sometimes as a set of epidermal trichomes, sometimes as the calyx, appears almost at the same time as the stamens. It arises below and opposite the stamens on the outer surface of the rim of the cup-like primordium which higher up bears the corolla and the stamens.

In its further growth the corolla develops as a tubular structure with a unilateral strap-shaped prolongation (zygomorphic ligulate corolla). Two stages may be distinguished in the growth of the corolla tube. First, intercalary growth above the insertion of the stamens forms the upper part of the corolla tube. Second, intercalary growth below the insertion of the stamens forms the lower part of the tube in which the bases of the corolla and of the stamens are congenitally fused (epipetalous stamens). The second stage occurs comparatively late in the development of the flower. In the Compositae with actinomorphic tubular corollas the growth of the upper part of the corolla is uniform throughout. In zygomorphic corollas, as in lettuce, the upper part grows asymmetrically. The free parts of the stamens elongate also, and each becomes differentiated into a filament and an anther.

The carpels develop at the morphologically highest position of the flower, that is, within the cavity of the cup-like primordium. The two carpels become visible as two protuberances located seemingly below the stamens. These two carpel units overarch the ovarian cavity and become prolonged above into a solid compound style with a two-parted stigma. In Compositae the cup enclosing the ovary is commonly interpreted as consisting of adnate bases of the floral whorls joined to the carpel bases; in other words, the ovary is enclosed by the floral tube.

The development of an inflorescence and flower of a representative of the Gramineae may be illustrated by reference to the study on *Triticum* and *Avena*. The wheat inflorescence is a spike and consists of several groups of flowers, each referred to as a spikelet. The spikelets are attached directly to the main axis, the rachis. A spikelet of a grass consists of a short axis, the rachilla, bearing several chaff-like, two-ranked (distichous), overlapping bracts (commonly called glumes). The two lowermost bracts bear no flowers in their axils and are called empty glumes. Above the empty glumes are others that subtend flowers, usually referred to as florets. The wheat spikelet has four to six florets, each subtended by two bracts: the lower or abaxial called the *lemma*, and the upper or adaxial, called the *palea*. The reproductive parts of a grass floret consist of three stamens with thread-like filaments and rather large anthers, and a single, unilocular pistil with a short style and two feathery stigmas. At the base of the ovary and opposite the palea are two lodicules, small scales involved in the opening of the bracts during anthesis.

The reproductive stage of a wheat plant begins while the plant is still in the rosette stage. The initiation of reproductive stage is quickly followed by a sudden and vigorous elongation of the shoot, the subsequent culm. The addition of leaf primordia ceases, and even the further development of the existing leaf buttresses is stopped. Some of the younger buttresses may be obliterated as the apex expands in length and width. Whereas the foliage leaf primordia arise as single ridges gradually encircling the shoot axis, the spikelet development is initiated by the appearance of double ridges. The spikelet proper differentiates from the upper of the paired ridges. A spikelet is interpreted as an axillary bud, and the lower ridge as the subtending leaf. The first spikelets differentiate in the middle of the spike, and differentiation then progresses acropetally and basipetally. Within the individual spikelet differentiation is acropetal, the parts appear in the sequence of empty glumes, first flower, second flower, and so forth. Within an individual floret the parts arise in a close overlapping sequence: lemma, palea, lodicules, stamens, and gynoecium.

The primordia of the lemma, palea, and lodicules are ridge-like; that is, they resemble leaf primordia. The stamen primordia, on the other hand, are rounded like bud primordia, one of the features that is used to interpret the stamen as a cauline structure.

The gynoecium occupies the apex of the floral meristem. A crescent-shaped ridge, which is highest on the side toward the lemma, arises just below the apex. The apical mound itself constitutes the ovule primordium. The ridge grows entirely around the ovule primordium and initiates two styles on two sides of its margin. Continued upward growth of the margins below the styles brings about the closure of the ovarian cavity. The stigmatic hairs are the last parts of the gynoecium to develop. Thus, the grass gynoecium arises as a unit and does not reveal, ontogenetically, the three-carpellate origin usually ascribed to the gynoecium of Gramineae. The same method of origin and growth of the gynoecium has been observed in various other Gramineae and in Cyperaceae, except that in the latter some species have three styles. The apical position of the ovule is used for interpreting it as an axial structure, but the opinions on the number of carpels are divided. The carpellary part, or ovary wall, of the gynoecium is considered to be leaf-like in its method of origin and growth.

The rhythm of development of a flower as a whole has certain distinguishing characteristics that are closely correlated with the important phenomena of mitosis and meiosis occurring during the formation of spores and gametes. The sequence of formation of floral parts is more rapid than that of the foliage leaves so that the ontogeny of the flower may have an explosive character. Morphologic observations and studies on comparative weights of developing flowers and their parts show that these parts may have divergent rates of growth after they are initiated. The petals, for example, may appear before the stamens but may develop more slowly. Sometimes the principal period of growth of the petals occurs only after the stamens complete their growth. Both the petals and the stamens may accelerate their growth rate shortly before anthesis. The remarkable speed with which the stamens may attain their final length is well illustrated by the rate of elongation of 2.5 mm per minute observed in the growing anther filaments of rye. The stamens may lag behind the gynoecium in development at first, then rapidly attain the final length which brings the anther into a most favorable position for release of pollen. The ovary usually enlarges uniformly like a vegetative organ. Sometimes, however, the enlargement slows down before fertilization, and if fertilization fails to take place, the gynoecium dies. Comparative studies on floral parts show that the reproductive parts constitute a relatively large mass of the flower as a whole.

Histogenesis

Research on histogenesis of floral parts is used extensively for the interpretation of the morphologic nature of the flower and in comparative taxonomic studies. The sepals and petals originate, like the foliage leaves, from periclinal divisions in one or more subsurface layers of the apical meristem. Such origin of the perianth parts is apparently common in both the dicotyledons and the monocotyledons. In their upward growth, the perianth parts show apical activity of short duration followed by some intercalary growth. Marginal activity followed by intercalary growth is responsible for the increase in width of the perianth primordia. In *Vinca* the marginal meristem of the petals is more active than that of the sepals and is involved in the formation of the upper part of the floral tube which arises through the ontogenetic fusion of the corolla lobes.

Some workers find that the stamens are initiated just like the members of the perianth. Others report that the stamens have a deeper origin than the perianth parts and are, therefore, axial structures.

After their initiation the stamens undergo apical growth of short duration, followed by intercalary growth. If the stamen filament is flattened, it shows marginal growth; otherwise such growth is suppressed. The anthers have a special form of marginal activity which produces the characteristic two-lobed, four-loculed structure, rather than a flat blade. With regard to the gynoecium, frequently the origin of the placentae and ovules are considered to be distinctive from that of the carpel. Some authors find that the carpel resembles a leaf in the manner it originates from the apical meristem, whereas the placenta or the single, basally attached ovule is initiated like an axial structure. Others postulate that the primary ontogenetic relation of the ovule is with the carpel and that the method of growth, which is obviously correlated with the future form of an entity, is hardly a safe criterion of homology. Still another view has been advanced on the basis of developmental relations in cytochimeras of *Datura*. All parts of the gynoecium, carpels, placentae, and ovules are cauline in nature because they arise in the third layer of the apical meristem, whereas the foliage leaves are initiated in the second. In their future growth, the carpels undergo apical and marginal growth.

Histologic studies have revealed the manner of ontogenetic union of flower parts. As was mentioned previously, the union of perianth parts or of the carpels may be congenital, or it may occur, partly or entirely, during ontogeny. The ontogenetic union is brought about by fusion of the margins of parts that come in contact with each other during growth. In the petals of *Vinca* such union occurs through apposition of two epidermal layers, with the line of union eventually becoming obscured. Evidence of the fusion of perianth parts is thoroughly obliterated if divisions, periclinal and others, occur in the apposed epidermal layers. The degree of union of carpels also varies from a rather loose one to a thorough interlocking of the epidermal cells, accompanied by divisions in these cells and a complete effacement of the suture. In dicotyledons the carpels of syncarpous gynoecia are generally more firmly joined than in monocotyledons.

Vascular Development

Information on vascular development in the flower is meager. Some consideration has been given the question of the direction of differentiation of procambium. The assumption that there is an acropetal differentiation of procambium in the flower and a basipetal differentiation in the vegetative shoot has been used in support of the concept that the flower is a unique structure and not comparable to the shoot. Later research has shown that there is no such simple and straightforward difference between the flower and the shoot. Acropetal differentiation of procambium is common in the vegetative shoot in a wide variety of plants. In the flowers, both acropetal and basipetal differentiation of procambium have been reported.

According to the classic study of Trecul (1881), the xylem in the flower follows a pattern of differentiation similar to that in the shoot; that is, it appears in one or more loci and then progresses bidirectionally toward the distal and the proximal parts of the flower. In *Perilla* vascular differentiation appears to be speeded up when the reproductive stage is induced.

Abscission

The abscission of floral parts has been less intensively investigated than that of the leaves, but the basic phenomena appear to be similar in the separation of all these structures. Abscission of parts or of entire structures occurs at various stages in the reproductive process. The completion of flowering may be followed by the shedding of parts of flowers, of entire flowers, or of inflorescences. Particularly common is the shedding of petals. The petals may fall without previous wilting (*Canna*, *Aquilegia*, *Cydonia*, *Rosa*, *Geranium*, *Linum*). They also abscise in a wilted or dried state, either close to the level of their insertion (*Lilium*, *Tulipa*, most Cruciferae, *Cucurbita*), or a short distance above the insertion, with the basal part remaining attached to the flower (*Althaea*, *Datura*, *Nicotiana*). If the petals are not shed at the end of flowering, they remain temporarily or permanently attached to the fruit in the dry state (*Agapanthus*, *Hypericum*, *Convallaria*). In some monocotyledons the perianth becomes green and persists in the fruit (*Veratrum*, *Eucomis*, *Paris*).

Petals are often constricted in the abscission zone. Usually no cell division precedes abscission, and the separation layer is poorly differentiated. The cells in this layer remain small, little vacuolated, and closely packed. They may contain chloroplasts or chromoplasts, and also raphides. The cells are roundish or polygonal in outline, occasionally tabular, with their long diameters oriented transversely with reference to the long axis of the petal. If the petal is much constricted, collenchyma may be present beneath the epidermis. Apparently the separation results from a softening of the middle lamella. Cell division may occur in the separation layer. The protection of the scar seems to involve an impregnation of the walls with fatty substances without the deposition of a suberin lamella or formation of cork. Sepals, staminal filaments, and styles may abscise after flowering in essentially the same manner as the petals.

The abscission of entire flowers is characteristic of plants with unisexual flowers. The staminal flowers are regularly abscised after the pollen is shed. These flowers may fall singly (Cucurbitaceae) or as entire inflorescences (catkins of the Amentiferae). If fertilization does not take place, carpellate and bisexual flowers may drop also (*Solanum tuberosum*, *Nicotiana tabacum*, *Lycopersicon esculentum*). Floral abscission can be induced artificially by various treatments. The separation layer in pedicels of flowers is, in some species, preformed during development. Surface grooves are sometimes present in pedicels but do not necessarily coincide with the abscission zone.

14

EPIDERMIS

The term *epidermis* designates the outermost layer of cells on the primary plant body. The word is derived from two words of Greek origin, *epi*, upon, and *derma,* skin. Through the history of development of plant morphology the concept of the epidermis has undergone changes, and there is still no complete uniformity in the application of the term. This surface system of cells varies in composition, function, and origin and, therefore, does not lend itself to a precise definition based on any one criterion. In this book the term epidermis is used in a broad morphologic-topographic sense. It refers to the outermost layer of cells of all parts of the primary plant body-stems, roots, leaves, flowers, fruits, and seeds. It is considered to be absent on the rootcap and not differentiated as such on the apical meristems.

The inclusion of the surface layer of the root in the concept of epidermis is contrary to the view that the root epidermis belongs to a separate category of tissue and should have its own name, (*rhizodermis* or *epiblem*). The epidermis of the root differs from that of the shoot in origin, function, and structure, and, therefore, the emphasis of some workers upon the distinctness of the two parts of the epidermis is justified. At the same time, the proper definition of the root epidermis is inseparably connected with the problem of the morphologic relation between root and shoot. As long as there is no commonly accepted concept on this relation, it seems most convenient to use the term epidermis in its broadest sense to mean the primary surface tissue of the entire plant.

The normal functions of the epidermis of the aerial plant parts are considered to be restriction of transpiration, mechanical protection, gaseous exchange through stomata, and storage of water and metabolic products. Some accessory functions, however, may predominate to such an extent that the epidermis assumes characteristics not typical of this tissue. In this category of functions are included photosynthesis, secretion, absorption (other than that of the root epidermis), and possibly also the perception of stimuli and causal association with the movement of plant parts. Some of the functions of the epidermis appear to be related to certain specialized anatomic characteristics.

The meristematic potentialities of the epidermis merit a brief mention. In general, this tissue is relatively passive with regard to meristematic activities. Nevertheless, the epidermis is known to resume such activity during the normal course of development (formation of phellogen) and after injuries to the plant.

Origin and Duration

Briefly, the epidermis of the shoot arises from the outermost cell layer of the apical meristem, either from independent initials or jointly with the subjacent tissue layers. If the shoot apex shows a segregation into zones of surface and volume growth, that is, into a tunica and a corpus, the epidermis originates from the outermost layer of tunica. Such a layer of cells fits the definition of Hanstein's *dermatogen* since its course of development into the epidermis begins in an independent initial region. In plants showing less precise zonations in the apical meristem, as most of the gymnosperms do, the epidermis does not have separate initials. It is a product of the lateral derivatives of the apical initials, which divide both anticlinally and periclinally and are the ultimate sources of the superficial as well as the interior cells of the plant body. In plants with single apical cells the epidermis also have common origin with the deeper lying tissues. In roots the epidermis may be related developmentally to the rootcap or to the cortex.

When the epidermis does not arise from separate initials, it becomes distinct at various distances from the apical meristem, depending on the architecture of the meristem. Haberlandt's terms *protoderm* designates such primordial epidermis, as well as the epidermis arising from separate initials. This term was coined as a morphologic-topographic designation, with no reference to the origin of the tissue. In this book protoderm is used to indicate the undifferentiated epidermis, regardless of its origin.

Organs having little or no secondary growth usually retain the epidermis as long as they exist. An exception is exemplified by some woody monocotyledons that have no secondary addition to the vascular system but develop a special kind of periderm replacing the epidermis. In stems and roots of gymnosperms and dicotyledons and of arborescent monocotyledons having secondary growth, the epidermis varies in longevity, depending on the time of formation of the periderm. Ordinarily, the periderm arises in the first year of growth of woody stems and roots, but numerous tree species produce no periderm until their axes are many times thicker than they were at the completion of primary growth. In such plants the epidermis, as well as the underlying cortex, continues to grow and thus keeps pace with the increasing circumference of the vascular cylinder.

The individual cells enlarge tangentially and divide radially. An example of such prolonged growth is found in stems of a maple (*Acer striatum*) in which trunks about 20 years old may attain a thickness of about 20 cm and still remain clothed with the original epidermis. The cells of such an old epidermis are not more than twice as wide tangentially as the epidermal cells in an axis 5 mm in thickness. This size relation clearly shows that the epidermal cells are dividing continuously while the stem increases in thickness. Another example is *Cercidium toerreganum*, a tree leafless most of the time but having a green bark and a persistent epidermis.

Structure

Composition

In relation to the multiplicity of its functions the epidermis contains a wide variety of cell types. The ground mass of tissue consists of the epidermal cells proper, which may be regarded as the least specialized members of the system. Dispersed among these cells are the guard cells of the stomata and sometimes other specialized cells. The epidermis may produce a variety of appendages, the trichomes, such as hairs and more complex structures. Trichomes with a specific function, the root hairs, develop from the epidermal cells of the roots.

Epidermal Cells

Morphology and arrangement

Mature epidermal cells are commonly described as being tabular in shape because of their relatively small extent in being tabular in shape because of their relatively small extent in depth, that is, in the direction at right angles to the surface of the organ Deviating types, cells which are much deeper than they are wide, also occur, for example, in the palisade-like epidermis of many seeds. In surface view the epidermal cells may be nearly isodiametric or elongated. The three-dimensional shape of the epidermal cells of *Aloe aristata* and *Anacharis densa* approaches that of a tetrakaidecahedron cut in half. The form of epidermal cells is sometimes related to differences in position on the plant organ. Elongated epidermal cells are often found on structures which themselves are elongated, such as stems, petioles, vein ribs of leaves, and leaves of most monocotyledons. Elongated epidermal cells also occur near some hairs and stomata. Frequently epidermal cells are shallow above the strands of subepidermal sclerenchyma. In leaves the epidermal layers on the two surfaces may be dissimilar in shape and size of cells and in thickness of walls and cuticle.

In many leaves and petals the epidermal cells have wavy anticlinal walls, and the undulations may be present in the entire depth of the walls or only in their outermost parts. The cause of this waviness has been the subject of much study and speculation in the literature. One of the explanations of the phenomenon relates the undulations to the development of stresses during the differentiation of the leaf. Another concept is that the waviness is caused by the method of hardening of the differentiating cuticle. The waviness of the walls is variable, depending on the location in the leaf or petal. Often the undulations occur only on the lower side of a leaf or are more pronounced here than on the upper side. The waviness is also affected by environmental conditions prevailing during leaf development. The outer wall of an epidermal cell may be flat or convex, or it may bear one or more localized raised areas.

Some epidermal cells greatly deviate from the main mass of cells. Certain Gramineae, Gymnospermae, Dicotyledoneae, and lower vascular plants (*Adiantum*, *Selaginella*) contain fiber-like epidermal cells (Linsbauer, 1930). The longest epidermal fibers - up to 2 mm - were described in Stylidaceae. In Gramineae such fibers may be over 300 microns in length. Certain Cruciferae contain sac-like secretory cells (myrosin cells) scattered in the epidermis. In Acanthaceae, Cucurbitaceae, Moraceae, and Urticaceae epidermal cells may develop cystoliths. Some of these cystolith-containing cells (the lithocysts) are specialized epidermal cells; others appear to be reduced trichomes.

Sometimes the entire epidermis consists of highly specialized cells. Thus in certain seeds and scales the epidermis is composed of a solid layer of sclereids. The epidermis of the Polypodiaceae is differentiated as a photosynthetic tissue. The epidermal cells project into extensive intercellular spaces and contain chloroplasts.

The epidermal cells are arranged compactly, with rare breaks in their continuity other than those represented by the stomatal pores. Inter-cellular spaces occur in the epidermis of petals, but they appear to be closed on the outside by the cuticle.

Epidermis in gramineae

The morphologic variability of the epidermis of Gramineae is used extensively for taxonomic purposes and for discussions on the evolution of this group of plants. The gramineous epidermis typically contains long cells and two kinds of short cells, silica cells and cork cells. The short cells frequently occur together in pairs. The silica cells are almost filled with SIO_2 which solidifies into bodies of various shapes. The cork cells have suberized walls and often contain solid organic material. They are also silicified. The silica in epidermal cells of oat has been identified as opal. In some parts of the plant the short cells develop protrusions above the surface of the leaf in the form of papillae, bristles, spines, or hairs.

The epidermal cells of Gramineae are arranged in parallel rows, and the composition of these rows varies in different parts of the plant. The inner face of the leaf sheath at its base, for example, has a homogeneous epidermis composed of long cells only. Elsewhere in the leaves combinations of the different types of cells may be found. Rows containing long cells and stomata occur over the assimilatory tissue; only elongated cells or such cells combined with cork cells or bristles or with mixed pairs of short cells follow the veins. In the stem, too, the composition of the epidermis varies, depending on the level on the internode and on the position of the internode in the plant.

The Gramineae and other monocotyledons possess still another peculiar type of epidermal cell, the bulliform cell. The bulliform cells, literally "cells shaped like bubbles," are large, thin-walled, highly vacuolated cells which occur in all monocotyledonous orders except the Helobiae. Bulliform cells either cover the entire upper surface of the blade or are restricted to grooves between the veins. In the latter situation they form bands, usually several cells wide, arranged parallel with the veins. In transections through such a band the cells often form a fan-like pattern, for the median cells are usually the largest and are somewhat wedge-shaped. Bulliform cells may occur on both sides of the leaf. They are not necessarily restricted to the epidermis but are sometimes accompanied by similar cells in the subjacent mesophyll.

Bulliform cells are poor in solid contents. They are mainly water-containing cells, with little or no chlorophyll. Tannins and crystals are rarely found in these cells. Their radial walls are thin, but the outer wall may be as thick or thicker than those of the adjacent ordinary epidermal cells. The walls are of cellulose and pectic substances. The outer walls are cutinized and also bear a cuticle. Bulliform cells may accumulate silica.

According to one view, the bulliform cells are concerned with the unrolling of the developing leaves. Their sudden and rapid expansion during a certain stage of leaf development is assumed to bring about the unfolding of the blade; hence, the term expansion cells, often applied to these cells. Another concept is that, by changes in turgor, these cells play a role in the hygroscopic opening and closing movements of mature leaves; hence, the alternative term motor cells. Still other workers doubt that the cells have any other function than that of water storage. Studies on the unfolding and the hygroscopic movements of leaves of certain grasses have shown that the bulliform cells are not actively or specifically concerned with these phenomena.

Contents

In general the contents of the epidermal cells have been incompletely investigated, but since these cells possess living protoplasts they may be expected to include a variety of substances, depending on the degree of their specialization. The epidermal plastids are usually not definitely differentiated as chloroplasts, but in many plants they appear to contain chlorophyll as determined by tests for fluorescence and reduction of silver nitrate. Starch also may occur in the epidermal plastids. Some ferns, water plants, and a number of higher vascular land plants, particularly those of shady habitats, contain well-developed chloroplasts in the epidermis. The cell sap of the epidermal cells may contain anthocyanin, as in many flowers, leaves of the purple beech and the red cabbage, and stems and petioles of *Ricinus*. Under the electron microscope the epidermal cells of *Allium* bulbs show structures similar to those found in parenchyma cells.

Wall structure

The epidermal walls of different plants and of different plant parts vary markedly in thickness. In the thinner-walled epidermis the outer wall is frequently the thickest. Epidermis with exceedingly thick walls is found in coniferous leaves. The wall thickening is uneven and so massive in some species that it almost obliterates the lumina of the cells. These walls are probably secondary. Thick secondary walls occur in the epidermal cells differentiated as sclereids in seed coats and scales.

The radial and the inner tangential walls frequently show primary pit-fields. The outer wall also may have thin places and markings resembling primary pit-fields. Plasmodesmata have been described not only in the radial and the inner tangential walls but also in the outer walls, where they are called ectodesmata. Although the ectodesmata do not pass through the cuticle they are thought to be the pathways for substances that are discharged through the cuticle.

The epidermal cells of leaves and petals in some plants show internal ridges that resemble folds. The ridges apparently consist of two wall layers cemented together by intercellular material. The two layers may split apart with the formation of a schizogenous intercellular space. In such instances a ridge appears as a loop in transection.

Cuticle

The restriction of transpiration by the epidermis largely results from the presence of the fatty substance *cutin* as an impregnation of the cell walls (cutinization, incrustation with cutin) and as a separate layer, the *cuticle* (cuticularization, adcrustation of cutin), on the outer surface of the cells. The cuticle covers all parts of the shoot. It occurs on the floral parts, on nectaries, and on ordinary and glandular trichomes. Some authors report the presence of a cuticle on the apical meristem and in the absorbing region of the root including the root hairs. A contrasting report states that no cuticle was found on the youngest leaf primordia of certain angiosperms. The cuticle can be removed from plant parts as an unbroken layer, an indication that it is continuous.

Cutin has also been identified on the free surfaces of the leaf-mesophyll cells and on the inner walls of the epidermis where these are exposed to the internal air spaces. The inner layer of cutin is continuous with the external cuticle through the stomatal apertures, whose bounding cells, the guard cells, are covered with a cuticle on their free surfaces.

The cuticle attains variable thickness in different plants. Environmental conditions and other unknown factors affect its development. The surface of the cuticle may be smooth, or it may have various protrusions, ridges, or cracks. The origin of the complicated relief pattern in the cuticle of floral parts has been ascribed to the effects of cell growth. The cuticle of the tomato fruit contains a yellow pigment, probably a flavonoid, whose development depends on the same light conditions that regulate flowering and seed germination in certain plants.

The cutinized part of the outer epidermal wall beneath the cuticle has a complicated structure. In plants with thick outer walls it consists of many lamellae of cutinized cellulose alternating with layers rich in pectin; and a layer of pectin is often recognized between the cuticle and the cuticular layer.

Wax, oil, resin, salts in crystalline form (*Cressa cretica*, *Tamarix*, *Frankenia*), and caoutchoue (*Eucalyptus*) occur as surface deposits on the aerial plant parts. The structure of the wax deposits has been studied by means of the electron microscope. The wax deposits are crystalline and may have the form of granules, rods, often with hooked ends, network of tubes, isolated tufts, and more or less homogeneous layers. An exceptionally thick layer of wax (up to 5 mm) occurs in *Klopstockia cerifera*, the wax palm of the Andes. The carnauba wax is derived from the leaves of the Brasilian wax palm, *Copernicia cerifera*. Wax affects the wettability of the foliage because it prevents the contact of liquid with the leaf surface. Its structure and development are, therefore, of considerable interest in connection with research on agricultural sprays. Wax evidently passes through the cuticle, but the latter shows no pores that could be interpreted as pathways for its discharge. Wax is commonly present also within the cuticle and the cutinized layers beneath. Older leaves have mainly subcuticular wax accumulations which appear to be of greater ecological significance than the surface deposits.

The stratified appearance of the cutinized outer epidermal wall and the increase in the proportion of cutin toward the periphery suggest that the fatty substances do migrate outwards. Some workers think that this movement occurs through ectodesmata. Others have found lipoidal droplets throughout the young wall, which later apparently diffuse to the surface.

The development of the initial cuticle is interpreted as a flooding with a cutin precursor, or procutin (probably unsaturated fatty acids), analogous to a drying oil, and a subsequent hardening of this material through polymerization under the influence of the air oxygen. According to a study of the cuticle of the apple fruit, however, cutin formation is conceived as a process controlled by enzyme action rather than by spontaneous oxidation. The hardening of the cuticle presumably terminates further extrusion of wax and cutin precursor and, therefore, these substances accumulate beneath the cuticle. This subcuticular deposition would occur in the cellulose-containing part of the wall or it could also increase the thickness of the cuticle. Some evidence suggests that the margins of the upper epidermal wall of leaves continue to grow and retain an immature cuticle for some time. The greater susceptibility of young leaves to herbicides is ascribed to the presence of these immature, permeable zones in the cuticle.

Cutin is semihydrophilous because some of its polar groups remain free during polymerization. This feature explains the moderate swelling of the cuticle in water and the occurrence of cuticular transpiration. It is possible to demonstrate excretion of water through the cuticle and its aggregation into droplets; this occurs without any evidence of submicroscopic pores in the cuticle.

Cutin is highly inert and resistant to oxidizing maceration methods. It does not decay since apparently no microorganisms possess cutin-degrading enzymes. Because of its chemical stability the cuticle is preserved as such in fossil material and is very useful in identification of fossil species.

The cuticle occurs not only over the surface of the epidermal cells but also often as rib-like projections in the anticlinal walls. Such ribs develop relatively late in the life of an organ. One of the explanations of these ribs is that, when new cells are produced by anticlinal divisions during the development of the epidermis, each of these cells extends tangentially and produces its own complete wall, while the parent wall becomes stretched and torn. Thus, the outer cutinized layers accumulate as interrupted lamellae of cellulose together with pectic substances and cutin. The interruptions occur over the anticlinal walls and are filled with cutin deposits. The stretching and tearing of the outer cellulose lamellae and their permeation with cutin eventually make it difficult to distinguish the cuticle and the cutinized layers from one another without special treatment.

Most plants produce only epidermal cuticular layers, even if the periderm is formed late in the life of the organ and the epidermis continues to grow. In some exceptional plants like the Viscoideae and *Menispermum* cuticular layers are also formed among cortical cells in successively deeper regions of the cortex.

Other wall substances

Among other common wall substances, lignin is a relatively infrequent component of the epidermal walls in angiosperms. If present, it is sometimes generally distributed, sometimes restricted to a part of the outer wall. Lignification of epidermal cells is comparatively common in the lower vascular plants. It occurs also in the Cycadaceae, Cyperaceae and Juncaceae, and in a few dicotyledons (*Eucalyptus*, *Quercus*, *Laurus nobilis*, *Nerium oleander*), and specifically in the needles of conifers, the rhizomes of Gramineae, and the leaves of Gramineae outside the sclerenchyma strands. Many plants deposit silica in epidermal cells (*Equisetum*, ferns, Gramineae, numerous Cyperaceae, palms, and certain dicotyledons). In some dicotyledonous families (Malvaceae, Rutaceae, Loganiaceae, Gentianaceae, Euphorbiaceae) mucilaginous modifications of walls occur in individual epidermal cells, or in groups of cells; sometimes most epidermal cells are more or less mucilaginous, as, for example, in seeds.

Stomata

The stomata are apertures in the epidermis, each bounded by two guard cells. In Greek, stoma means mouth, and the term is often used with reference to the stomatal pore only. In this book, the term stoma includes the guard cells and the pore between them.

By changes in their shape, the guard cells control the size of the stomatal aperture. The aperture leads into a substomatal intercellular space, the substomatal chamber, which is continuous with the intercellular spaces in the mesophyll. In many plants two or more of the cells adjacent to the guard cells appear to be associated functionally with them and are morphologically distinct from the other epidermal cells. Such cells are called *subsidiary*, or *accessory cells*.

The stomata are most common on green aerial parts of plants, particularly the leaves. The aerial parts of some chlorophyll-free land plants (*Monotropa*, *Neottia*) and roots have no stomata as a rule, but rhizomes have such structures. They occur on some submerged aquatic plants, and not on others. The variously colored petals of the flowers often have stomata, sometimes nonfunctional. Stomata are also found on stamens and gynoecia. In green leaves they occur either on both surfaces (amphistomatic leaf) or on one only, either the upper (epistomatic leaf) or more commonly the lower (hypostomatic leaf). The density of stomata has been established as 100 to 300 per square millimeter for leaves of many species.

In leaves with parallel veins, such as those of monocotyledons and some dicotyledons, and in the needles of conifers the stomata are arranged in parallel rows. The substomatal chambers in each row are coalesced, and the mesophyll cells bounding these chambers form an arch over (or beneath) the intercellular canal. In netted-veined leaves the stomata are scattered.

Guard cells may occur at the same level as the adjacent epidermal cells, or they may protrude above or be sunken below the surface of the epidermis. In some plants stomata are restricted to the epidermis that lines depressions in the leaf, the stomatal crypts. Epidermal hairs may also be prominently developed in such crypts.

The guard cells are generally crescent-shaped with blunt ends (kidney-shaped) in surface view and often have ledges of wall material on the upper and lower sides. In sectional views such ledges appear like horns. Sometimes a ledge occurs only on the upper side, or none is present. If two ledges are present, the upper delimits the front cavity above the stomatal pore, and the lower encloses the back cavity between the pore and the substomatal chamber. The ledges are more or less heavily cutinized. An outstanding characteristic of stomata is the unevenly thickened walls of the guard cells. This feature appears to be related to the changes in shape and volume (and the concomitant changes in the size of stomatal aperture) which are operated by turgor changes in the guard cells. In many species, the posture of the guard cells appears to be determined by turgor difference between the guard cells and the subsidiary cells.

The primary cause of changes in turgor of the guard cells is not definitely established. The immediate cause appears to be the condensation and hydrolysis of starch in their chloroplasts. Photosynthesis alone is not sufficient to explain the rapid changes in osmotic pressure associated with the closing and opening mechanism. Moreover, the chloroplasts may not be well differentiated in guard cells. Among the environmental factors carbon dioxide concentration appears to play a major role in size changes of the stomatal pore.

Judged by the polymorphism of the guard cells, the mechanisms responsible for opening and closing of the stomata are varied. In one common type, the change in the shape of the guard cells occurs because the wall that is turned away from the stomatal aperture, the so-called back wall, is thin and apparently elastic. When the turgor increases, the thin wall bulges away from the aperture, while the front wall (facing the pore) becomes straight or concave. The whole cell appears to bend away from the aperture, and the aperture increases in size. Reversed changes occur under decreased turgor.

Another distinct type of stomatal mechanism is illustrated by the guard cellos of Gramineae and Cyperaceae. These cells are bulbous at two ends and straight in the middle. The middle part has a strongly but unevenly thickened wall; the bulbous ends have thin walls, and the wall between the bulbous ends of two adjacent cells may be incomplete so that the protoplasts of the two guard cells are partially confluent. Increase in turgor causes a swelling of the bulbous ends and the consequent separation of the straight median portions from each other. The nucleus in a gramineous guard cell is extended and simulates the shape of the cell lumen. It has two enlarged ends connected by a thin thread-like middle part.

Coniferous stomata are sunken and appear as though suspended from the subsidiary cells arching over them. In their median parts the guard cells are elliptical in section and have narrow lumina. At their ends they have wider lumina and are triangular in section. The characteristic feature of these guard cells is that their walls, and those of the subsidiary cells, are partly lignified and partly free of lignin. This combination of more and less rigid wall parts, the manner of connection with the subsidiary cells, and the presence of thin wall parts in the subsidiary cells are features that appear to be involved in the working of the coniferous stomata. In *Equisetum* the guard cells occur beneath two subsidiary cells in which the walls that are in contact with the guard cells have conspicuous wall thickenings.

The front cavities of stomata in conifers and some angiosperms are often occluded with finely granular or alveolar material, probably cuticular in nature. The stomata may be occluded on the inner side by parenchyma cells, the so-called obturating cells, that extend into the substomatal chamber.

The structure of the guard-cell walls is comparable to that of the other epidermal cells of the same leaves. They are usually cutinized in the outer layers and covered with a cuticle. As previously mentioned, the cuticle extends through the stomatal aperture into the substomatal chamber where it joins the inner layer of cutin. (The cuticle is reported to be absent on the thin wall facing the pore in *Citrus*) Guard cells show lignification, at least in parts of their walls, in vascular cryptogams, gymnosperms, Gramineae, Cyperaceae, and certain dicotyledons. Ultrastructurally, a longitudinal orientation of microfibrils has been recognized in guard cells of the *Avena* coleoptile. According to some studies, no plasmodesmata occur between the guard cells and the adjacent epidermal cells; according to others, plasmodesmatal connections do occur between the guard cells and the subsidiary cells.

Development

Stomata arise through differential divisions in the protoderm. After several divisions of a given protodermal cell, one of the products of these divisions becomes the immediate precursor of the guard cells. This is the stoma or guard-cell mother cell, that eventually divides into the two guard cells. These enlarge and assume the characteristic crescentic shape. The area which becomes the pore shows a lenticular mass of pectic material just before the walls separate from each other. This appearance probably results from the swelling of the intercellular material preceding its dissolution. The mother cells of the guard cells occur at the same level as the adjacent epidermal cells. If the mature stoma is raised above or sunken below the surface of the epidermis, the change in position occurs during the development of the stoma through mutual cellular readjustments within the epidermis and between the epidermis and the mesophyll. Even in the coniferous leaves, in which the guard cells are so deeply sunken, the stoma mother cells are at one level with the other epidermal cells. More or less conspicuous intercellular spaces occur in the mesophyll during the initiation of stomata. A large intercellular space, the substomatal chamber, develops during the maturation of the stoma.

The sequence of divisions preceding the formation of stomata varies in different species so that the guard cells and the subsidiary cells may be unrelated or more or less closely related. In Gramineae, for example, the immediate guard-cell precursor arises in one row of cells, the subsidiary cells in two adjacent rows. In *Drimys*, a protodermal cell (the primary mother cell of stoma) gives rise, after two divisions, to a guard-cell mother cell and two subsidiary cells. The grass stoma exemplifies the *perigene* (from the Greek for about and offspring) type of subsidiary cells, that is, cells not arising from the primary mother cell. The subsidiary cells in *Drimys* do arise from the primary mother cell and are called *mesogene* (from the Greek for in the middle and offspring). The same stoma may have both mesogene and perigene subsidiary cells, as in *Trochodendron*. The classification into mesogene and perigene requires developmental studies because the mature pattern does not necessarily reveal the ontogenetic relationship of cells; and the distinction between the two kinds of cells may not be significant physiologically.

A discussion of cell relationships brings up the question as to how early in the ontogeny of a stoma cytologic differentiation in the protodermal cell indicates the inception of this ontogeny. The guard-cell precursor has been repeatedly described as being distinguishable by the density of its cytoplasm, and developmental studies indicate that this feature results from cytoplasmic polarization – an accumulation of cytoplasm in one end of the cell – before the primary mother cell of the stoma divides. An asymmetric division occurs across the gradient resulting from the polarization and gives rise to a small guard-cell precursor and a larger, less specialized epidermal cell. A polarization-asymmetry sequence is possibly operative even earlier, during the formation of the primary mother cell of the stoma. In *Populus pyramidalis* the guard-cell precursor appears hypertrophied and its neighbors undergo an accelerated vacuolation.

The nuclei of the guard-cell precursors are denser than those of their sister cells. Apparently this differentiation occurs gradually, through one or more preceding generations of cells. The sequences suggest an establishment of a center with highly specialized cells – the guard cells – surrounded by less specialized cells with gradation in the degree of specialization. The distinctness of guard cells in this scheme is indicated by their inability to respond with division in wounded tissue.

In a given leaf, the stomata arise not all at once but in succession through a considerable period of leaf growth. Two principal patterns may be distinguished in the development of stomata in the leaf as a whole. In leaves with parallel veins, having the stomata in longitudinal rows, the developmental stages of the stomata are observable in sequence in the successively more differentiated portions of the leaf. In the netted-veined leaves the different developmental stages are mixed in mosaic fashion so that mature stomata occur side by side with immature ones. The first pattern is characteristic of most monocotyledons and of a few dicotyledons (*Tragopogon*, *Thesium*); the second, of most dicotyledons and a few monocotyledons (Araceae, Smilacoideae, Taccaceae, Dioscoreaceae). Both developmental patterns are found among the vascular cryptogams.

Classification

The mode of development of stomata and their spatial relation to neighboring cells are considered with reference to problems of classification and phylogeny in the angiosperms and the conifers. The classifications refer to stomatal types but actually they are based on the relation between the stomata and the subsidiary cells.

In the gymnosperms Florin (1931, 1951, 1958) distinguishes two main types of stomatal complexes, the *haplocheilic* (simple lipped), in which the subsidiary cells are perigene, and the *syndetocheilic* (compound lipped), in which the subsidiary cells are mesogene. The haplocheilic type is highly variable in details and is regarded as the more primitive of the two. Although these categories were established by ontogenetic studies, the classification is used with regard to fossils by relying on the mature patterns characterizing the two types.

In dicotyledons the use of the mature pattern formed by the stomata and neighboring cells is suggested for the establishment of the typology. Four main types have been proposed: *anomocytic* (irregular-celled, formerly ranunculaceous), no subsidiary cells are present; *anisocytic* (unequal-celled, formerly cruciferous), three subsidiary cells, one distinctly smaller than the other two, surround the stoma; *paracytic* (parallel-celled, formerly rubiaceous), one or more subsidiary cells occur on either side of the stoma parallel with its long axis; *diacytic* (cross-celled, formerly caryophyllaceous), two subsidiary cells enclose the stoma, their common wall at right angles to the long axis of the stoma. There are variations within these types and some probably merit separate designations as, for example, the *actinocytic* with the subsidiary cells arranged along the radii of a circle.

In monocotyledons, four categories of stomatal complexes have been described; two of these have four or more subsidiary cells surrounding the guard cells (*Rhoeo*, *Commelina*), one has two subsidiary cells (Gramineae), and one has none (*Allium*). The types with many subsidiary cells are regarded as more primitive, the other two as independently derived by reduction in the number of subsidiary cells.

Trichomes on Aerial Plant Parts

Trichomes (a word of Greek origin, meaning a growth of hair) are epidermal appendages of diverse form, structure, and functions. They are represented by protective, supporting, and glandular hairs, by scales, various papillae, and the absorbing hairs of the roots.

Trichomes are usually distinguished from the so-called emergences on the basis that the emergences are formed from both epidermal and subepidermal tissues. The distinction between such emergences and trichomes is not sharp, however, because some plant hairs are raised upon a base originating by division of subepidermal cells. Trichomes also intergrade with nontrichomatous epidermal cells having protrusions in the form of papillae and with cells differentiated as "water vesicles."

Trichomes may occur on all parts of a plant. Either they persist throughout the life of an organ, or they are ephemeral. Some persisting hairs remain alive; others become devoid of protoplasts and are retained in dry state. The epidermal trichomes usually develop early in relation to the growth of the organ.

Trichomes may show wide variations within families and the smaller plant group, and even in the same plant. On the other hand, there is sometimes considerable uniformity in trichomes within a plant group. Plant-hair types have been successfully used in the classification of genera and even of species in certain families and in the recognition of interspecific hybrids.

Trichomes may be classified into different morphological categories. One common type is referred to as *hair*. Structurally, hairs may be subdivided into unicellular and multicellular. The unicellular hairs may be unbranched or branched. Multicellular hairs may consist of a single row of cells or of several layers. Some multicellular hairs are branched in dendroid (tree-like) manner; others have the branches oriented largely in one plane. Commonly a multicellular hair can be divided into a foot, which is imbedded in the epidermis, and a body projecting above the surface. The cells surrounding the foot are sometimes morphologically distinct from other epidermal cells.

Another common type of trichome is the *scale*, also called *peltate hair* (from the Latin *peltatus*, target-shaped or shield-like, and attached by its lower surface). A scale consists of a discoid plate of cells, often borne on a stalk or attached directly to the foot.

Unicellular, multicellular, and peltate hairs may be glandular. Some of the simple multicellular glandular hairs consist of a stalk and a unicellular or multicellular head. The head constitutes the secretory part of the hair. In a peltate glandular trichome the discoid plate is composed of glandular cells. Some glandular trichomes consist of a multicellular core of cells covered with a palisade-like layer of secretory cells.

A trichome is initiated as a protuberance from an epidermal cell. The protuberance elongates, and if it develops into a multicellular structure, various divisions may follow the initial elongation. The cell walls of trichomes are commonly of cellulose and are covered with a cuticle. They may be lignified. Plant hairs often produce thick secondary walls as, for instance, the cotton seed hairs or the "climber hairs" of *Humulus*. The walls of trichomes are sometimes impregnated with silica or calcium carbonate. Cell contents are varied in relation to function; the most complex are probably those of the glandular cells. Chloroplasts are often present, though they may be small and not persisting. Plant hair cells, other than the glandular, are characteristically highly vacuolated. Cystoliths and other crystals may develop in hairs.

Cotton seed hairs, commonly known as cotton fibers, are extremely long epidermal hairs with thick secondary walls of almost pure cellulose. They are formed from the protoderm of the ovule during flowering and continue to arise for about 10 days after anthesis. The elongation lasts for 15 to 20 days, and the hairs become ½ to 2½ in. long, depending on the variety of cotton. A number of other plants produce commercially usable hairs on the seeds or other parts of the fruit.

Root Hairs

Root hairs are tubular structures constituting direct lateral extensions of the cells that originate them. In a study involving 37 species in 20 families, the root hairs were found to vary between 5 and 17 microns in diameter and between 80 and 1,500 microns in length. They are highly vacuolated and contain the nucleus in their parietal cytoplasm. The root hairs are rarely branched. The adventitious roots of *Kalanchoe* growing in air have multicellular root hairs, whereas the same kind of roots growing in soil have unicellular ones. Root hairs are typical of roots, but under certain conditions they may develop on other plant parts.

The ability of root hairs to absorb water has been demonstrated experimentally. The same experiments showed that the hairless epidermal cells also absorb water with a range of velocity comparable to that of cells bearing root hairs.

The principal function of the root hairs is considered to be the extension of the absorbing surface of the root, and from this standpoint the available information pertaining to their numbers and surface area in a rye plant is of interest. In this plant the 13,800,000 roots had a surface area of 2,500 sq ft. Living root hairs numbered 14 billion and had a total surface area of 4,300 sq ft. Thus the combined surface area of the roots and root hairs was 6,800 sq ft, and it was packed in less than 2 cu ft of soil. This total surface was 130 times that exposed to the outside air by the aerial parts of the same plant. If the surface of the mesophyll cells of a foliage leaf facing the intercellular spaces was taken into consideration, the root surface was still 22 times that of the transpiration area of the top. Relating these figures to the absorptive capacity of root hairs, Rosene (1955) calculated that a small number of the total root hairs can obtain all the water necessary for transpiration and growth of the plant.

Wall structure

Although there is general agreement that the chief components of root hair walls are cellulose and pectic substances, the matter of distribution of these substances is still controversial. According to one view, the pectic substances occur as a matrix in the cellulosic microfibrillar system; according to the other, calcium pectate forms a separate layer outside the cellulosic part of the wall. An ultrastructural study of root hairs suggests that the outer layer of the wall consists of randomly oriented microfibrils imbedded in an amorphous matrix, probably composed of pectins and hemicelluloses. The inner layer consists of axially oriented cellulose microfibrils associated with little or no amorphous material. Another study recognizes, from outside in, a layer of mucilage, a cuticle, a layer of pectin, and a layer of cellulose and pectin. Environmental condition may induce the formation of callose within the root hairs.

Development

The root hairs develop acropetally, that is, toward the apex of the root, and apparently never originate among preexisting hairs. Because of the acropetal sequence of initiation, in most seedling tap-roots the root hairs show a uniform gradation in size, beginning with those nearest the apex and going back to those of mature length. The hairs are initiated in the part of the root located behind the zone of most active cell division, but where the longitudinal extension of the epidermal cells may still be in progress. Usually the root hair emerges as a small papilla at or near the apical end of a cell. If the cell continues to elongate, after the appearance of the papilla, the root hair eventually occurs some distance from this end; otherwise its position remains terminal. The root hair grows at the tip where the cellulose microfibrils have random orientation. In the basal part of the hair, where elongation has ceased, apposition of the parallel-oriented microfibrils occurs. The growing tip contains dense cytoplasm. Some workers see the nucleus in constant position near the tip in growing hairs; others report a continuous movement of the nucleus.

The factors affecting root hair development are a subject of controversy. According to Cormack's (1962) theory, a gradual hardening of the wall through calcification of the pectic layer arrests the growth of the hair at its proximal end and confines it to the soft region of the distal end. Ekdahl (1953), on the other hand, ascribes the hardening mainly to the formation of new cellulose microfibrils.

At the ultrastructural level, evidence has been presented that dictyosomes may be concerned with the formation of root hair wall. Vesicles, separating from the dictyosomes and developing dense contents, appear to be transported toward the wall, especially at the tip.

In some plants the root epidermis shows a morphologic differentiation into hair-forming cells (*trichoblasts*) and cells that do not form hairs. This differentiation varies in degree of expression but is sufficiently characteristic of many genera of grasses that it may be used for the study of relationships in this family. In general, the root hair-forming cells are shorter than the others. When strongly expressed, this difference may be visible at the origin of a trichoblast. The precursory protodermal cell divides into a longer and a shorter cell, the shorter being characterized by denser cytoplasm and a lower rate of elongation than the longer cell. Recently formed trichoblasts are also distinguished from their sister cells by more intense enzymatic activity and greater amounts of RNA. Significantly, the physiologic specialization of the trichoblast is observed before its maximum elongation; in fact, it appears to be initiated through polarization phenomena at the time of the asymmetric division producing the trichoblast.

In plants having a homogeneous root epidermis all epidermal cells are potential trichomatous cells, but not all necessarily produce root hairs. The nontrichomatous cells of a root with a heterogenous epidermis may be induced to form root hairs by suitable changes in the environment, and conversely the potentially trichomatous cells may be prevented from developing such structures.

Root hairs are short-lived. Their longevity is commonly measured in days. Old root hairs collapse, and the walls of the epidermal cells, if the cells are not sloughed off, become suberized and lignified. Persisting root hairs, however, have been observed in a number of plant species. Such hairs become thick-walled and are then probably not concerned with absorption.

Multiple Epidermis

One or more layers of cells beneath the epidermis in leaf, stem, and root may be morphologically and physiologically distinct from the deeper-lying ground tissue. The older literature designates all distinctly characterized subepidermal layers as *hypodermis*. The specialized subsurface tissue may be part of the ground tissue, or it may be derived from the protoderm by periclinal divisions. The recognition of the latter possibility has prompted workers to separate the hypodermis originating in the ground tissue from the subsurface layers of protodermal origin by introducing the concept of *multiple* or *multiseriate epidermis*. A study of mature structures rarely permits the identification of the tissue either as multiple epidermis or as a combination of epidermis and a hypodermis. The origin of the subsurface layers can be properly revealed only by developmental studies.

The outermost layer of a multiple epidermis resembles the ordinary uniseriate epidermis in having a cuticle. The inner layers are commonly differentiated as water-storage tissue lacking chlorophyll. The multiple epidermis varies in thickness from 2 to 16 layers of cells. Sometimes only individual cells of the epidermis undergo periclinal divisions. Representatives with multiple epidermis may be found among Moraceae, Pittosporaceae, Piperaceae (*Peperomia*), Begoniaceae, Malvaceae, Monocotyledoneae (palms, orchids), ferns, and others. The *velamen* (from the Latin word for covering) of the aerial and terrestrial roots of orchids is a multiple epidermis (or rhizodermis).

The periclinal divisions initiating the multiple epidermis in leaves occur at different stages of leaf growth but usually when a leaf is several internodes below the apex. In *Ficus*, for example, the leaf has a uniseriate epidermis until the stipules are shed. Then periclinal divisions occur in the epidermis. Similar divisions are repeated in the outer row of daughter cells, sometimes once, sometimes twice. During the expansion of the leaf, anticlinal divisions occur also, and, since these divisions are not synchronized in the different layers, the ontogenetic relation between these layers becomes more or less obscured. The inner cells expand more than the outer. The outer cells remain particularly small because they expand less and undergo more numerous anticlinal divisions than the inner. The cystolith-containing cells characteristic of *Ficus* leaves do not divide but keep pace with the increasing depth of the epidermis and even overtake it by expansion and intrusion into the mesophyll. In some plants (*Peperomia*) the cells of the multiple epidermis remain arranged in radial rows and clearly reveal their common origin.

INDEX

Q

R